Materials in Nuclear Energy Applications

Volume II

Author

C. K. Gupta

Head, Metallurgy Division
Bhabha Atomic Research Centre
Bombay, India

CRC Press, Inc.
Boca Raton, Florida

Library of Congress Cataloging-in-Publication Data

Gupta, C. K.
Materials in nuclear energy applications.

Includes bibliographies and indexes.
1. Nuclear engineering—Materials. I. Title.
TK9185.G86 1989 621.48′33 88-10546
ISBN 0-8493-6772-7 (v. 1)
ISBN 0-8493-6773-5 (v. 2)

Direct all inquiries to CRC Press, Inc., 2000 Corporate Blvd., N.W., Boca Raton, Florida, 33431.

International Standard Book Number 0-8493-6772-7 (v. 1)
International Standard Book Number 0-8493-6773-5 (v. 2)

Library of Congress Number 88-10546
Printed in the United States

FOREWORD

The Indian Atomic Energy Program as it stands now is broadly based with a wide range of activities starting from uranium mine and mill operations to productions particularly relating to manufacture of metallic and ceramic nuclear fuels for research and power reactors, and of reactor structural materials and components, extending to such materials as heavy water. This indigenous status is actually the outgrowth of multidisciplinary research and development encompassing physics, chemistry, metallurgy, engineering, and technology carried out at the Bhabha Atomic Research Centre. I have been closely associated, directly and indirectly, with most of these programs right from their inception. It has, therefore, been a great pleasure for me to write a foreword to this text.

The Materials Group is one of the well-recognized groups of this Research Centre which has significantly contributed and provided valuable inputs to nuclear energy materials, be it for making, shaping, or treating. All these have been crucially important in our materials development program. Dr. Gupta, who is currently heading the Metallurgy Division in this group, has been an important member. I have followed Dr. Gupta's career with interest for a long time. He has, to date, carried out and guided some excellent research in the field of nuclear metals and materials. He has acquired not only through his own research but also through his close interaction with other divisional activities, fuel-cycle facilities, nuclear materials production centers, and research and power reactor stations in the country, and has first hand knowledge of most of the major areas in commercial nuclear power. Dr. Gupta has been on the teaching staff for the post-graduate level of courses that are being given at the Training School Division of our center. On the whole, Dr. Gupta is well equipped to write a book of this scope and orientation.

As may be seen from the content, the text combines an account of scientific and engineering principles with a description of materials and processes of importance in nuclear research and industry. The coverage includes fuel materials for reactors of various types, structural materials including core and pressure vessel materials, control and shielding materials, and so on — in fact, for most of the important parts of a reactor.

I congratulate Dr. Gupta for the initiative he has taken in bringing this present book to this form. It reflects his perception of the subject which is indeed intense because of his long association in this field. He has not only contributed to the development of this area but also planned for its future expansion. It is seldom that research scientists put their experience in the form of a book which could be used by a larger community of scientists and students. It requires a very special effort, especially an unselfish attitude to share the information with others. This book abundantly brings out Dr. Gupta's intentions. I am sure it will be a very useful addition as well as a comprehensive reference book for those who wish to know more about the subject. In fact, the economics, safety, and acceptance of nuclear power in the future may very well depend upon the quality of the materials we invent for applications in nuclear technology.

P. K. Iyengar
Director
Bhabha Atomic Research Centre

PREFACE

Increasing concern over the world's long-term energy problems is laying increasing emphasis on energy sources more than ever before. Half the energy used in the world today comes from oil. The discovery of new reserves is not keeping pace with consumption. Production rates are bound to eventually fall. The same is true for natural gas. Coal is plentiful in some parts of the world but there would be problems in extracting and transporting it to population centers. Thus disposed are the traditional sources of energy around us. Reference must be brought to what renewable sources of energy can offer. Renewable sources of energy such as the wind, waves, sun, and geothermal sources represent a prolific supply of energy. Extensive research is being conducted to explore ways of harnessing this energy economically and reliably. Many problems, however, remain to be solved and overcome. If the technology develops, that will permit us to convert some of this energy to electricity at an acceptable cost and with acceptable environmental consequences. Renewable energy sources could then potentially contribute to our electricity supply, and, in turn, help to conserve valuable reserves of fossil fuels for use for other purposes. The renewable energy sources are intermittent in nature, and thus there will still be a need for a reliable back-up to meet peak demand for electricity. Electricity is a clean, convenient, and easily transported form of power. More than a quarter of the oil, gas, and coal burned in the world is used for generation of electricity. Nuclear energy is already making an important contribution to electricity production.

There is hardly any need to elaborate the role of materials in nuclear energy. The two are, in fact, inextricably linked. To date, a wealth of knowledge has been gained regarding nuclear energy materials, and disseminated exhaustively in the form of published literature, books, and monographs. The present books have been conceived, designed, and developed with the objective of bringing together an account of the diverse materials of vital importance in nuclear research and industry. Materials scientists and engineers, particularly those involved with the nuclear energy field, will find this compiled text basically very useful. To the general readers, the two volumes, of course, will prove to be adequately educative providing a general appraisal and knowledge of the nulear energy materials — the central role that they have played and will continue to do so in the present and future of the nuclear energy developments. The books may, in addition, be used for an overview course with a general audience of junior or senior science and engineering students or for comprehensive coverage with senior or graduate nuclear engineering students. The presentation is also suitable for physicists or engineers who are entering the nuclear power field without previous experience with nuclear reactors.

Volume I is comprised of Chapters 1 through 4. The first chapter serves as an introduction to the basic physics of the atom and the nucleus and to nuclear fission and the nuclear chain reaction. The second chapter presents a general survey of various types of nuclear reactors. The third chapter does this for fast reactor systems. The two chapters together essentially introduce and identify various materials grouped under well-defined categories depending upon their functions in nuclear reactors. Most vital of the reactor core materials are the fuel materials which form the subject matter of the fourth chapter. The principal naturally available fissile nuclear fuel material is uranium which has no other significant commerical use. The other artificially produced fissile material is plutonium. A naturally occurring material which by itself is not fissile, but from which a fissile material is produced, is thorium. The fourth chapter mainly deals with the trio, uranium, plutonium, and thorium and their ceramic derivatives that are principally being used or considered for the solid fueled nuclear reactors. Among the other topics that are intimately connected with the nuclear fuel materials, special mention may be made of fuel cycles, uranium enrichment technologies, reprocessing technologies, and radiation effects which have been covered in the chapter. Among the various fuel cycles, the main emphasis has been given to the uranium-plutonium fuel cycle. The current generation of nuclear power is dominantly based on this cycle. Volume II contains Chapters 1 through 6. Chapter 1 concerns itself with structural materials. Aluminum, magnesium, zirconium-based alloys, stainless steels serving as cladding for various nuclear fuel materials, and pressure vessel materials have been dealt with. The moderator and reflector materials are important thermal reactor core materials and they have formed the subject matter for the second chapter. Heavy water and graphite are the principal materials that have been considered. Chapter 3 that follows is on coolant materials. They perform the important function of transporting the nuclear fission heat outside the reactor core system for raising steam in established reactor systems. The list includes light and heavy water, carbon dioxide, helium, and liquid metal. Chapter 4 deals with control and shielding materials. The two materials functionally differ greatly. Materials used in these application areas must have high absorption cross sections for thermal neutrons. This has provided the justification to deal with them together in one chapter. The chapter generally describes the principles and methods of reactor controls, and then refers to the various control materials. It similarly continues with the shielding materials. Boron carbide and the various products incorporated with it, cadmium and its alloys, hafnium, and rare earths are among the important control and shielding materials considered in the chapter. The radioactive waste management, though not strictly, comes directly under nuclear energy generation, but one cannot overlook the importance of the topic. The physical bulk of waste from nuclear-power generation is vanishingly small when compared to the wastes arising from conventional generation or from other industrial processes, but its high activity makes its storage and ultimate disposal a subject of considerable importance. Wastes, depending upon their activity levels, have been classified under different categories, and disposals are implemented accordingly. Development of advanced waste forms — SYNROC is drawing considerable attention in the context of disposal of highly active wastes. Chapter 5 treats all these aspects of nuclear wastes. Chapter 6 reviews the current state of knowledge of fusion reactors. The coverage given includes the fundamental principles of energy generation out of nuclear fusion, practical methods and principles of obtaining terrestrially nuclear fusion energy, nuclear characteristics of fusion fuels, and a conceptual description of a nuclear fusion power plant with emphasis given on various materials.

These books are the culmination of over 3 years of effort. The basic premise of the book evolved from my close association with the nuclear metallurgy course offerings to post-graduate level teaching cum training program being conducted at the Training School Division of our center. It has however, been an ambitious task, and a bold one too, to cover the diverse aspects of diverse materials of importance in nuclear reactors in a book of reasonable length. With this submission, I am leaving the present work to the hands of my readers to judge to what extent the intended objectives and scope have been accomplished. I look forward to receiving insightful comments and suggestions. I am sure they would add measurably towards improving these books in the future.

In the preparation of these books, I have drawn heavily upon the published popular literature from many nuclear industries. I would like to place on record my grateful acknowledgment particularly to the United Kingdom Atomic Energy Authority (UKAEA), British Nuclear Fuels Limited (BNFL), Central Electricity Generating Board (CEGB), South of Scotland Electricity Board (SSEB), Nuclear Industry Radioactive Waste Executive (NIREX), Rheinnish-Westfalishces Electrizitatswerk AG(RWE), Electricite De France (EDF), British Nuclear Forum (BNF), Electricity Council (EC), National Nuclear Corporation (NNC), and Department of Energy (DOE) for readily supplying me with the excellent published literature covering almost all aspects of nuclear energy. I have had the advantage of interacting with many professionally mature scientists and engineers, both from this research center and elsewhere in nuclear materials production centers and atomic power stations in the country, who have generously helped, and given me excellent advice which compensates for any lack of personal familiarity in many areas. Since it would be a long list, I am not identifying anyone by name, but I wish to assure all of those people that I am indeed most thoroughly grateful.

So much for the help which I obtained with the subject. Equally important is the help which I had with the practical business of preparing the manuscript. Very significant acknowledgment belongs to Poonam Khattar who handled all my correspondence and skillfully deciphered the hand-written text and produced the typed version of the manuscript with care and patience. I am most thankful to V. V. Kulkarni for smilingly putting up with me and helping me in many ways until the last day of submission of the manuscript to the publisher. Special thanks are also due to I. D. Walile and Y. M. Thakur for their art work. I would be failing in my duty if I did not include N. Krishnamurthy and A. K. Suri in the given list of names of my colleagues in this work. They have taken extreme care to go through the typed manuscript and provided many helpful suggestions. I also sincerely appreciate and warmly acknowledge (B. J. Starkoff, Chairman and Director of Publishing, Marsha Baker, Associate Managing Editor, Sandy Pearlman, Director of Editing, Barbara Brownlee, Administrative Manager, and Sharon Morey, Coordinating Editor) of CRC Press from whom I have received personal attention and exemplary cooperation. They have gone to the minutest of details and have done a superb job in the final production of this book.

A preface for a book customarily appears. Authors heartily welcome this tradition. It gives one a chance to say some words which cannot be put in the body of the text. The preface as given for the present book would remain hardly complete without referring to a few other names. First, I would name C. V. Sundaram, Director, Indira Gandhi Centre for Atomic Research (IGCAR) with whom I have had a very long association. He has constantly encouraged and influenced me in my contributions to scientific and technical literature and to get inducted into the authorship ranks. Second in my list are two names, my wife Chandrima, and my son Chiradeep. The inspiration which both gave meant a lot to me. They instilled confidence in me and helped me in many intangible ways to make life bearable during the long period of preparing and seeing this work through to completion.

C. K. Gupta
Bombay
April 1987

THE AUTHOR

C. K. Gupta, Ph.D., is presently Head, Metallurgy Division, Bhabha Atomic Research Centre (BARC), Trombay, Bombay, India. He received his B.Sc. in 1962 and his Ph.D. in 1969 — both in metallurgical engineering from Banaras Hindu University, Varanasi, India.

Dr. Gupta has primarily concerned himself with research, development, and production programs of special metals and materials of relevance to nuclear energy applications. He has also involved himself with research in collaboration with laboratories overseas, in which special mention may be made of the IAEA coordinated research programs on Thermodynamics and Transport Properties of Nuclear Materials, and INDO-US Collaboration programs in the fields of Ceramics and Corrosion. He has, in addition, been teaching subjects connected with chemical metallurgy of nuclear metals and materials at the Training School Division of BARC, and providing guidance for M.Sc. (Tech.) and Ph.D. degrees in metallurgical engineering of the Bombay University.

Dr. Gupta has been serving as member-secretary of the Materials Science Committee of the Board of Research in Nuclear Sciences of the Indian Department of Atomic Energy. He is a member of the American Association for Advancement of Science (AAAS), Indian Institute of Metals (IIM), and Indian Nuclear Society (INS). He is the editor of the Transactions of the IIM, and is also serving as a member of the Editorial Advisory Boards of the journals, *High Temperature Materials and Processes, Mineral Processing and Extractive Metallurgy Review,* and *Minerals Engineering*.

Dr. Gupta's publication list totaling 150 to date is comprised of research papers, reviews, monographs, popular scientific writings, articles in book, books, and proceedings publications. He has received several national awards for his contributions and accomplishments in his professional career to date.

TABLE OF CONTENTS

Volume I

Chapter 1
Elements of Nuclear Energy ... 1

I. Introduction ... 1
II. Elements of Atomic Structure ... 3
III. Nuclear Energy Release ... 3
IV. Nuclear Reactions ... 6
A. Radioactivity ... 6
1. Rate of Radioactive Decay ... 9
2. Half-Life ... 10
B. Neutrons and Their Interactions with Matter ... 10
1. Neutrons and Neutron Sources ... 11
2. Neutrons-Nucleus Interactions ... 12
C. Neutron Reaction Cross Sections ... 12
1. General ... 12
2. Types of Interaction ... 13
3. Energy Dependence ... 16
V. Fission ... 17
A. General ... 17
B. Fission Fuels ... 19
C. Burners, Converters, and Breeders ... 22
D. Products of Fission ... 22
1. Fission Energy ... 22
2. Fission Fragments ... 24
3. Neutrons ... 25
E. The Nuclear Chain Reaction ... 27
1. Four-Factor Formula ... 28
2. Homogeneous and Heterogeneous Reactor System ... 30
3. Criticality ... 31
4. Reactor Kinetics with Delayed Neutrons ... 32
VI. Radiation ... 34
A. General ... 34
B. Forms of Radiation, Radiation Units ... 34
C. Radiation Detectors ... 35
D. Biological Effects of Radiation ... 36
VII. Summary ... 36

References ... 42

Chapter 2
Nuclear Fission Reactor ... 45

I. Simple Concept of Reactor ... 45
A. Outline of Reactor ... 45
B. Reactor Description ... 47
II. Reactor Classification ... 50

III. Research Reactor Systems 52
A. Graphite-Moderated Research Reactor 53
B. Tank-Type Reactor 53
C. Swimming Pool Reactor 56
D. Aqueous Homogeneous Reactor 58
E. Research Reactor Example 59
F. Concluding Remarks 61
IV. Power Reactors 61
A. Light Water Reactor 61
1. Pressurized Water Reactor 64
2. Boiling Water Reactor 71
3. Comparisons 79
B. Heavy-Water Reactors 81
1. Pressurized Heavy-Water Reactor 81
2. Steam-Generating Heavy-Water Reactor 86
C. Graphite-Moderated Reactors 87
1. Magnox Reactor 87
2. Advanced Gas-Cooled Reactor 90
3. High-Temperature Gas-Cooled Reactor 95
4. Soviet Union's RBMK 100
5. General 100
V. Overview 103
VI. Nomenclature 110
VII. Summary 111
References 113

Chapter 3
Fast Reactors 117

I. Introduction 117
II. General Aspects 118
III. Reactor Feature 119
IV. Leading Characteristics 122
A. Fuel Utilization 123
B. Breeding Ratio, Breeding Gain, And Doubling Time 124
V. Fast Reactor Types and Parts 125
A. Reactor Types 125
B. Liquid Metal Fast-Breeder Reactor 125
1. Core 125
2. Fuel 125
3. Coolant 126
4. Heat Transport 128
5. Blanket 130
6. Structurals 131
VI. Fast Breeder Reactor Plant 131
VII. Fast Breeder Reactor Program Development 137
VIII. World Status of Fast Reactor Development 138
References 140

Chapter 4

Nuclear Fuel Materials 143

I. General 143
II. Nuclear Fuel Cycle 143
  A. Types of Fuel Cycles 144
III. Reprocessing of Nuclear Fuel 147
  A. Reasons, Needs, and Requirements 149
  B. Reprocessing Technology 151
  C. Reprocessing Processes 152
    1. Pyro Processes 152
    2. Aqueous Processes 152
      a. Precipitation 155
      b. Solvent-Extraction Processes 155
      c. LMFBR Spent Fuel 159
  D. Summary, Status, and Examples 160
IV. Uranium Enrichment 166
  A. General 166
  B. Electromagnetic Separation 168
  C. Thermal Diffusion 168
  D. Gaseous Diffusion 169
  E. Gas Centrifuge 173
  F. Aerodynamic Processes 176
    1. Nozzle Process 176
    2. Helikon Process 177
  G. Advanced Enrichment Technologies 178
    1. Laser Isotope Separation Process 178
      a. Atomic Vapor Laser Isotope Separation 178
      b. Molecular Laser Isotope Separation 179
    2. Plasma Separation Process 180
  H. Chemical Exchange 180
  I. Summary 181
V. Uranium 183
  A. Minerals and Resources 183
  B. Uranium Mining and Milling 187
    1. Acid Leaching — Solvent Extraction 189
    2. Acid Leaching — Ion Exchange 193
    3. Alkali Leaching 194
  C. Chemical Conversion 195
    1. Traditional Processing Route 195
    2. Newer Processing Route 199
    3. Present Status 200
  D. Uranium Metal 200
    1. Production 200
    2. Properties 201
    3. Irradiation Effects 202
  E. Uranium Oxide Fuel 203
    1. General 203
    2. Properties and Advantages of UO_2 204
    3. Irradiation Processes of Oxide Fuel 204
      a. Development of Temperature Gradients 205

b. Restructuring ... 208
c. Fuel Densification and Swelling ... 209
4. Uo_2 Fuel Production Process ... 213
VI. Plutonium ... 214
A. Plutonium Generation ... 214
B. Nuclear Properties and Implications ... 220
C. Recovery and Quality of Plutonium ... 222
D. Plutonium Fuel Materials ... 224
1. Metallic Fuel ... 224
2. Oxide Fuel ... 228
a. General ... 228
b. Mixed Oxide Fuel Production ... 228
c. Co-Precipitation Process ... 229
d. Sol-Gel Process ... 230
3. General Comments ... 232
E. Carbide and Nitride Fuels ... 233
1. General ... 233
2. Preparative Aspects of Mixed Carbide and Nitride Fuels ... 236
F. Plutonium Utilization ... 237
1. Thermal Reactor Fuel ... 237
2. Fast Reactor Fuel ... 238
VII. Special Nuclear Materials ... 238
VIII. Thorium ... 243
A. General ... 243
B. Neutronic Features ... 243
C. Thorium Utilization ... 248
D. Occurrence and Extraction ... 250
E. Thorium Fuel Material ... 253
1. Metallic Thorium ... 253
2. Thorium Oxide ... 253
3. High-Temperature Reactor Fuel Elements ... 254
F. Thorium Fuel Reprocessing ... 254
IX. Concluding Remarks ... 257
References ... 259

Index ... 265

TABLE OF CONTENTS

Volume II

Chapter 1
Structural Materials 1
I. Introduction 1
II. Aluminum 2
A. Extraction 2
B. Properties 3
III. Magnesium 3
A. Properties 4
B. Extraction 4
C. Magnox Fuel Element Production 5
IV. Zirconium 10
A. Occurrence and Extraction 10
B. Properties 17
C. Applications 21
D. Fabrication 23
E. Performance 29
1. Creep Strength 29
2. Pellet-Clad Interaction 29
3. Irradiation Growth 32
F. Summary 33
V. Niobium 34
A. Extractive Metallurgy 34
B. Properties 35
C. Applications 36
D. Physical Metallurgy 38
E. Fabrication 38
F. Structure-Property Correlation 39
VI. Stainless Steel 42
A. General 42
B. Types and Characteristics 43
C. Applications 45
VII. Reactor Pressure Vessels 51
A. Pressure Vessel Types and Materials 51
B. Irradiation Effects 56
References 57

Chapter 2
Moderator and Reflector Materials 61
I. Introduction 61
II. Ordinary Water 65
A. Activation 67
B. Radiolysis 68
III. Heavy Water 69
A. Irradiation 70
B. Production of Heavy Water 70
1. Distillation 72

a. Hydrogen Distillation 72
b. Water Distillation 73
2. Water Electrolysis 74
3. Chemical Exchange Processes 76
a. Hydrogen Sulfide and Water 76
b. Hydrogen and Water 78
c. Hydrogen and Ammonia 78
4. Summary 83
IV. Graphite 83
A. Preparation of Nuclear Graphite 84
B. Properties of Graphite 85
C. Compatibility 88
D. Irradiation Effects 89
V. Beryllium 90
A. Production 91
B. Properties 93
C. Irradiation Effects 93
D. Conclusion 93
References 94

Chapter 3
Coolant Materials 97
I. Introduction 97
II. Coolant Systems 98
A. Pressurized Water Reactor Coolant System 98
B. Boiling Water Reactor Coolant System 99
C. Pressurized Heavy Water Reactor Coolant System 100
D. Liquid Metal Fast Breeder Reactor Coolant System 101
III. General Characteristics of Reactor Coolants 101
A. Coolant Types and their Characteristics 103
B. Coolant Materials 104
IV. Water Coolant 106
A. Specific Features 106
B. Water-Cooled Reactor Systems 106
C. Coolant Water Chemistry 107
1. Primary Water Coolant Chemistry 107
2. Secondary Water Coolant Chemistry 109
V. Gaseous Coolant 110
A. General 110
B. Gaseous Coolant Reactor Systems 111
C. Carbon Dioxide 112
1. Corrosion under Irradiation 112
2. Induced Radioactivity 115
3. Reactor Material Corrosion 115
a. Steels 116
b. Magnesium 116
c. Zirconium 116
D. Helium 116
VI. Liquid Metal Coolant 118
A. General 118
B. Liquid Metal Coolant Materials 119

C. Sodium 120
1. Corrosion Aspects 121
2. Summary 123
References 123

Chapter 4
Control and Shielding Materials 125
I. Introduction 125
II. Methods of Control 125
A. Movable Control Rods 125
1. Control Element Functional Classification 126
2. Control Rod Geometries 126
3. Control Rod Design and Operating Criteria 127
a. Thermal Worth 127
b. Epithermal Worth 128
c. Depletion 128
4. Other Factors 128
B. Burnable Neutron Poisons 129
1. Modes of Burnable Poison Utilization 130
2. Examples of Reactor Systems 130
C. Chemical Shim 131
III. Control Rod, Burnable Absorbers, and Chemical Shim 133
A. Boron-Based Control Materials 133
1. Chemical Metallurgy of Boron 133
2. Applications 136
B. Cadmium-Based Alloys 137
C. Hafnium and its Alloys 138
D. Rare Earth-Based Control Materials 140
E. Summary and Comparisons of Control Materials 142
IV. Shielding Material 143
A. General 143
B. Interaction of Radiation with Matter 143
1. Alpha Particles 143
2. Beta Particles 143
3. Gamma Rays 144
4. Neutrons 146
C. Reactor Shielding 146
V. Shielding Material Categorization 149
A. Heavy and Moderately Heavy Elements 149
B. Hydrogenous Materials 149
C. Boron-Containing Substances 151
VI. Summary 151
VII. Nuclear Power Plant Operation 154
A. Plant Start-Up 154
1. Initial Start-Up 154
2. Start-Up after Hot Standby Conditions 155
3. Start-Up after Severe Power Reduction 156
B. Shutdown 156
C. System Chemistry 156
D. System Transient Control 156
E. Power Distribution Transients 157
References 158

Chapter 5

Radioactive Waste Management

Radioactive Waste Management 161
I. Introduction 161
II. Waste Management Objectives and Principles 162
III. Waste Sources and Classification 163
A. Sources 163
B. Waste Classification 164
C. General 164

IV. Treatment of Radioactive Wastes 166
A. High-Level Waste 167
B. Highly Active Solid Waste 167
C. Highly Active Liquid Waste 167
D. Highly Active Liquid Storage 168
E. Highly Active Liquid Solidification 170
1. Drying and Calcination Processes 170
2. Vitrification Processes 170
a. Laboratory Development Process 172
b. Industrial-Scale Conversion 173
f. Intermediate- and Low-Level Wastes 174
1. Volume Reduction 174
a. Noncombustible Solid and Liquid Wastes 174
b. Combustible Wastes 176
2. Immobilization 176
G. Gaseous Wastes 177
H. Miscellaneous Wastes 179
V. Radioactive Waste Disposal 179
A. Intermediate-Level Waste Disposal 180
1. Shallow Burial 180
2. Deep Underground Burial 180
B. Highly Active Waste Disposal 181
1. Geological Disposal 182
2. Ocean Bed Disposal 183
3. Wastes in SYNROC 183
VI. Summary 184
References 187

Chapter 6

Fusion Reactor Materials

Fusion Reactor Materials 189
I. Introduction 189
II. Fusion Energy Sources 189
A. Fusion Reactions 190
B. Fusion Fuels 192
III. Confinement 193
A. Magnetic Confinement 193
1. Plasma Heating 194
a. Ohmic Heating 194
b. Neutral Beam Injection Heating 194
c. Radio-Frequency Waves 194
d. Adiabatic Compression 195
2. TOKAMAK 195

3. Magnetic Mirrors 196
4. Pinch-Type Systems 196
a. Z-Pinch System 198
b. Theta-Pinch System 198
B. Inertial Confinement 198
C. Ion Beam Focus 199
IV. Thermonuclear Fusion Reactor 201
V. Fuel Materials For Fusion 201
VI. Moderator, Reflector, and Blanket Materials 203
a. Lithium 203
b. Tritium Breeding and Neutron Multiplication 205
VII. Coolant and Shielding Materials 205
VIII. Structural Materials 206
A. Plasma First-Wall Materials 207
1. Refractory Metals and their Alloys 207
2. Steels 208
3. Nickel-Based Alloys 209
4. Aluminum-Based Materials 209
5. Carbon-Based Materials 209
IX. Materials For Magnetic Devices 210
A. Superconducting Magnet System 210
X. Irradiation Effects 210
A. First-Wall Material 210
B. Superconducting Material 211

XI. Summary 211
References 215

Index 217

Chapter 1

STRUCTURAL MATERIALS

I. INTRODUCTION[1-5]

Structural materials can be shown as a group listed in seven main categories into which the primary components and materials of nuclear fission reactors are classified. The structural materials of nuclear reactors provide the physical containment (for fuel protection), mechanical strength, and structural support for reactor components. The main basic materials consist of the fuel cladding, pressure vessel, fuel coolant channels, core support plates, coolant piping system, and control element mechanism. Selection of structural materials depends upon the functional requirements and varies with different types of nucelar fission reactors.

For example, reference can be drawn to cladding materials. The term cladding refers simply to the outer jacket of the nuclear fuel and serves as a barrier to the release of fission products. The cladding materials should fulfill the following: (1) mechanical and chemical stability with the fuel on one side and with the coolant on the other side, (2) high-temperature mechanical strength, (3) good heat transfer properties, (4) ability to accommodate radiation damage, (5) ability to withstand high heat loads without melting and losing mechanical integrity, and (6) favorable nuclear properties. The most important among these for a material to be used as fuel cladding in a thermal reactor is that it should have reasonably low thermal neutron absorption cross section. If the absorption cross section is too large, the material cannot be chosen no matter what mechanical and other advantages it may possess. This consideration alone limits the list of materials to a great extent. From consideration of the neutron absorption cross section of materials decreasing with increasing neutron energies, one of the advantages of the fast reactor system is the fact that the neutron economy is not seriously affected by absorption in structural materials. As a result, the materials list need not be as limited as in the case of thermal reactor systems. In fact, volumes up to 20% of a wide variety of structural materials can be incorporated without appreciable sacrifice of neutron economy.

The fissile nucelar fuels with the nonfissile cladding materials on them, as one unit, are called the fuel elements. They are recognized as the fundamental reactor core material. It is in this site where the most important event takes place. Within the core of an operating reactor, the fission heat is generated in the fissile fuel material and conducted into the coolant that flows past the fuel elements. The fuel elements have long lives as compared to the conventional types of fuel. They are the replaceable components of the reactor. The residence of fuel elements in reactors varies widely. It varies, for example, for about 1 year in fast reactors to 7 years in Magnox reactors to between 3 to 5 years in light water reactors. The criteria of an ideal fuel element require (1) thermal, irradiation and mechanical stability; (2) good corrosion resistance; (3) ease of fuel fabrication and fuel reprocessing; (4) good neutron economy and high fuel burnup; and (5) long lifetime service and low cost. To date, there is a large variety of nuclear fuel elements. These may be classified according to such diverse criteria as fuel material composition, fuel element shape, type of fuel-cladding contact, reactor type, etc. Table 1 gives the classification of fuel elements on the basis of the various reference points as stated. From the standpoint of heat transfer, heat removal, and ease of fabrication, the basic considerations of geometry and economics of the fuel elements are (1) large surface to volume ratio, (2) coolant removing the heat with maximum efficiency from the fuel element surface, (3) high coolant pressure-drop configuration permissible only if it is accompanied by a proportional gain in the heat transfer rate, and (5) simple geometry with ease of fabrication and low cost.

Table 1
CLASSIFICATION OF FUEL ELEMENTS

Basis of reference points	Classification	Remarks/example
Fuel composition and cladding materials	Metallic fuel/metallic clad elements	Metallic uranium is used in a magnesium alloy cladding
	Ceramic fuel/metallic clad elements	UO_2, $(U,Pu)O_2$, or $(U,Th)O_2$ is clad in zirconium alloy or in stainless steel; UC or (U,Pu)C is clad in stainless steel
	All-ceramic fuel elements	UO_2, $(U,Pu)O_2$, or $(U,Th)O_2$ grains are coated with SiC and pyrocarbon layers embedded in a graphite matrix
Fuel nature	Metallic fuel elements	The fuel is a metal or an alloy
	Ceramic fuel elements	The fuel is a nondiluted ceramic
	Dispersed fuel elements	The fuel is a highly diluted alloy or a ceramic material with a low density of fissile material
Shape	Cylindrical fuel elements	
	Solid fuel elements	Metallic fuel elements or ceramic fuel pellets are used
	Tubular or annular fuel elements	A metallic fuel rod or tube is used
	Plate fuel elements	
	Flat	The element is generally made of highly enriched ^{235}U fuel clad with aluminum
	Curved	
Type of contact	Mechanic-contact fuel elements	The contact is achieved by either pressing or rolling
	Metallurgical-bonding fuel elements	The materials in contact are bonded by diffusion
	Intermediate-layer fuel elements	An intermediate liquid, solid, or gaseous medium is used for improvement of fuel cladding heat transfer

Structural materials comprise (1) metals and alloys, (2) ceramics, and (3) cermets. Metals and their alloys, because of their ability to withstand the primary requirements, dominate as structural materials in different types of nuclear reactors. Most of the reactor structural materials are made of metals and their alloys, such as aluminum, magnesium, zirconium alloys, stainless steels, carbon steels, and nickel alloys. This chapter presents a description of these major categories of nuclear reactor structural materials.

II. ALUMINUM

Aluminum has been and still continues to be used as a cladding and structural material for teaching and research reactors. The chief requirement of a teaching or research reactor is high neutron flux for neutron economy, but not for power generation. The teaching and research reactors are therefore operated at low temperatures. The choice of aluminuum for application as cladding or structural material in thermal research and teaching reactors is based on advantages such as relatively low thermal neutron absorption, high thermal conductivity, high stability under irradiation, good corrosion resistance to air and water, fabricability and weldability, low cost, and abundant availability over disadvantages such as low melting point and low mechanical strength at elevated temperature.

A. Extraction

Aluminum oxide, of alumina ($Al_2 0_3$), is the main source from which extraction of aluminum is accomplished by electrolysis. For the electrolysis, purified anhydrous Al_2O_3 is dissolved in cryolite (Na_3AlF_6) in an iron tank with carbon which functions as the electrolytic cell cathode. Inside the tank there are large blocks of carbon/graphite to serve as anodes. When electric current is passed through the cell, molten aluminum forms at the wall and bottom of the tank while the oxygen is liberated at the carbon anode to form carbon dioxide. The reactions at the cathodes and anodes are represented as

$$\text{At the cathode:}\quad AlF_3 \rightleftharpoons Al^{3+} + 3F^-$$

$$Al^{3+} + 3e^- \rightarrow Al$$

$$\text{At the anode:}\quad 3F^- \rightarrow 3F + 3e^-$$

$$2Al_2O_3 + 12F = 4AlF_3 + 3O_2$$

$$C + O_2 \rightarrow CO_2$$

B. Properties

Aluminum used as cladding or structural material in thermal research and teaching reactors has excellent corrosion resistance to air and water coolant. Aluminum-clad metallic uranium fuel elements operated in such reactors for a number of years are only slightly corroded. The corrosion resistance of aluminum lies in the fact that aluminum and oxygen have a high affinity to form alumina which forms an adherent and impervious layer on the metal surface. The oxide film protects the metal from further attack unless removed or penetrated by chemical and mechanical means. Aluminum corrodes uniformly in the coolant water of thermal research reactors up to abut 220°C. At higher temperatures, hydrogen atoms produced by the radiation decomposition and corrosion reaction ($H_2O \rightarrow H + OH$, $2Al + 3OH \rightarrow Al_2O_3 + 3H$, $H + 3H \rightarrow 2H_2$) diffuse into the metal and combine as molecular hydrogen. This can enhance the corrosion rate during the development of corrosion products and gas blisters spread on the metal surface. At relatively low temperatures, 200 to 250°C, the corrosion rate is small, and at relatively high temperatures, above 400°C, the corrosion rate increases with temperature and exposure and tends to break away.

Compatibility between aluminum and uranium is generally good. During the operation of the uranium-fueled, aluminum-clad fuel elements, formation of intermetallic compounds, UAl_2, UAl_3 and UAl_4, can take place at the right formation time and temperature. These compounds have different densities, melting points, and crystal structures.

Irradiation effects on aluminum are relatively slight. No drastic changes in mechanical properties and dimensional stability have been observed. Tensile and yield strengths increase somewhat, while ductility decreases. There is negligible effect on density, thermal expansion, and dimensions. Structural ^{27}Al placed near the reactor core often gets activated as per the nuclear transmutation reaction, ${}_{13}Al^{27} + {}_0n^1 \rightarrow {}_{13}Al^{28} + \gamma$. The activity decays with a half-life of 2.3 min, ${}_{13}Al^{28} \xrightarrow{2.3\ \text{min}} {}_{14}Si^{28} + \beta$. This certainly lessens the health hazard involved in working with the aluminum structural components. However, immediately after removal from the neutron field, the activity associated with aluminum structural components is very high, as is expected.

III. MAGNESIUM[6]

In the nuclear field, magnesium occupies an important place as a reducing agent in the production of pure uranium. Besides this, it has found application as a cladding material. Its low thermal neutron absorption cross section is one of the principal reasons for its selection. As to the thermal neutron absorption cross section, magnesium is inferior only to

Table 2
SOME PHYSICAL, THERMAL, AND MECHANICAL PROPERTIES OF MAGNESIUM

Density at 20°C (g/cm^3)	1.7388
Crystal structure	hcp
Melting point (°C)	650
Boiling point (°C)	1110
Specific heat at 100°C (cal/g °C)	0.2518
Thermal conductivity at 20°C (cal/cm s °C)	0.410
Ultimate tensile strength at 20°C (kg/cm^2)	1890
Yield strength at 20°C (kg/cm^2)	980

beryllium. It scores higher than aluminum in some respects. The thermal neutron absorption corrosion of magnesium is about one third that of aluminum. As far as compatibility to uranium is concerned, magnesium is superior to aluminum. Magnesium, however, has poor corrosion resistance in hot water, unlike aluminum. This shortcoming makes it unfit for use in water- or heavy-water-cooled reactors. Magnesium alloys, particularly Magnox A-12 and ZA alloys, have been used as cladding materials on a large scale since the development of the British Calder Hall type of gas-cooled reactor. Magnox series have been used in British gas-cooled reactors. The French gas-cooled reactors, on the other hand, use alloy of ZA composition.

A. Properties

Some physical, thermal, and mechanical properties of magnesium are shown in Table 2. The melting point of magnesium is 650°C. Its corrosion resistance to oxidation from air or CO_2 is up to 400°C. Therefore, most of the gas-cooled reactors are normally operated at low temperatures and low power density.

Magnesium is the most electronegative of the fabricated industrial metals and thus suffers anodic attack whenever it is in contact with other metals in the presence of an electrolyte. This property is utilized in cathodic protection of ship's hulls and buried pipelines, and cables where magnesium slabs are fastened to make electrical contact to the part being protected. When a conducting medium such as seawater or damp earth has completed the circuit, current will flow and the magnesium anodes will preferentially be attacked. During an underwater storage period, Magnox fuel should not corrode. It has been established that in the absence of galvanic couples, corrosion of magnesium can be suppressed altogether by adjusting the pond water to a pH $<$11 by the addition of caustic soda and by restricting the concentration of chlorides and carbonates which are the principal common accelerating ions.

Magnesium and its alloys used as cladding and structural parts in gas-colled reactors have relatively good irradiation stability. Up to the neutron fluence 10^{20} n/cm^2, no significant irradiation effects, such as irradiation swelling, irradiation creep rupture, and changes in mechanical strength and ductility have been observed during the reactor operation.

B. Extraction

Magnesium, though, does not occur naturally in the metallic form; it is abundant in the form of 150 combinations, with its average content in the earth's crust being 2%. It ranks sixth among constituent elements thereof and third among metals, after iron and aluminum.

The main magnesium sources used in industrial production are dolomite ($MgCO_3CaCo_3$), magnesite ($MgCO_3$), carnallite ($KMgC1_3 \cdot 6H_2O$), and brucite [$Mg(OH)_2$]. Sources are not limited to solid minerals; for example, all brines, including seawater and certain lakewater, contain a sufficient percentage of magnesium. Seawater constitutes and inexhaustible source;

it was once said that working on only 1% of its reserve would allow the production of 20 million t of magnesium per year for 10 million years — a unique and fabulous situation which does not exist for other metals.

Production processes for magnesium can broadly be divided into two categories: electrolytic reduction and thermic reduction. The electrolytic processes include, (1) the Dow process, (2) the Norsk hydro process, and (3) the Amax process. In the Dow process, the raw materials consist of seawater and dolomite. The main stages of the process are (1) dolomite calcination, (2) magnesium hydrate precipitation, (3) hydrate chlorination by hydrochloric acid, (4) refining and concentration of magnesium chloride, and (5) electrolysis of hydrous magnesium chloride. Chlorine obtained druing the electrolysis is partly used for the production of hydrochloric acid, while the remnant is liquified. The Norsk hydro process utilizes seawater and dolomite, or brine with high content of magnesium chloride. The Norsk process differs from the Dow process in that hydroxides are calcined. The oxide is pelletized with coke and chlorinated with elemental chlorine to magnesium chloride. In this process it is to be noted that chloride cell feed is fully dehydrated before introduction into the electrolytic cells, unlike the Dow process, which employs a hydrous cell feed. In the Amax process, the raw materials are brine from the Great Salt Lake concentrated in solar ponds. They are further concentrated, refined, and dehydrated prior to electrolysis. The essential steps of the three magnesium extraction processes as outlined are shown in Figure 1.

Thermic processes include (1) the Pidgeon process, (2) the Balzano process, and (3) the Magnetherm process. The Pidgeon process uses dolomite as the source for magnesium and ferrosilicon as the reductant. The two constituents are briquetted and heated externally to a temperature of about 1100°C in a retort kept under high vacuum. The process is being used in Canada and Japan. The Balzano process was derived from the Pidgeon process; the dimensions of the retorts are larger, and their electrical heating is internal. The by-product is a magnesia and lime silicate. The raw materials used in the Balzano process are the same as those of the Pidgeon process. However, the feeding system, the furnace shape, and the recovering operation of products and by-products are entirely different. Briquetts of mixed dolomite and ferrosilicon are introduced into an electrically heated furnace under vacuum. The magnesium is reduced and condensed in the solid form every day in a block of 2000 kg or more. This process has been developed and is operated by SAIM in Italy. The Magnetherm process consists of the reduction of calcined dolomite by ferrosilicon without any prior briquetting. The reduction is achieved in a light electrical furnace under reduced pressure at 1600°C, a temperature definitely higher than in the other two thermic processes. Through the addition of a fluidizing agent, alumina, a liquid slag is obtained. By-products are this slag, which is used in cement industries, and some low-grade ferrosilicon, used as thermic load in steel production. The process is used by SOFREM in France and its licensees in the U.S. and Yugoslavia. Figure 2 shows schematically the Magnetherm process flowchart.

C. Magnox Fuel Element Production[7]

The British-designed gas-cooled reactors popularly known as Magnox reactors use CO_2 as reactor coolant with metallic uranium cladded with Magnox as fuel elements. The fuel elements occupy vertical cylindrical channels in the graphite moderator block. The heat transfer coefficients attainable with gases are quite low when compared with those attainable with water or liquid metals. Because of the poor heat transfer performance of CO_2 and the need to keep maximum cladding temperatures below about 450°C, it is necessary to take steps to promote more effective heat transfer from the fuel element to the coolant by the use of extended heat transfer surfaces. This is accomplished by employing finned cladding. The function of the fins is to increase the surface area for heat transfer and also promote turbulance and better mixing of the gas stream. All these effects increase the heat transfer rate. However, for the fins to be really effective they should be of a material of high thermal

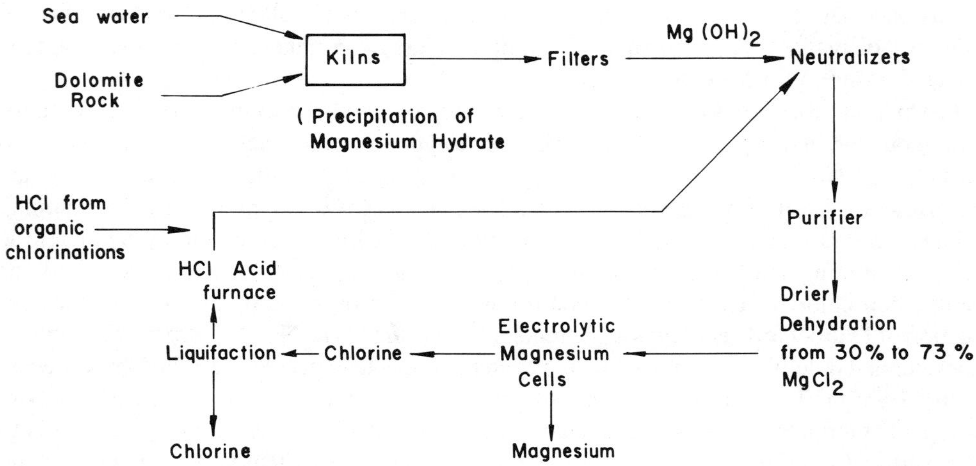

a

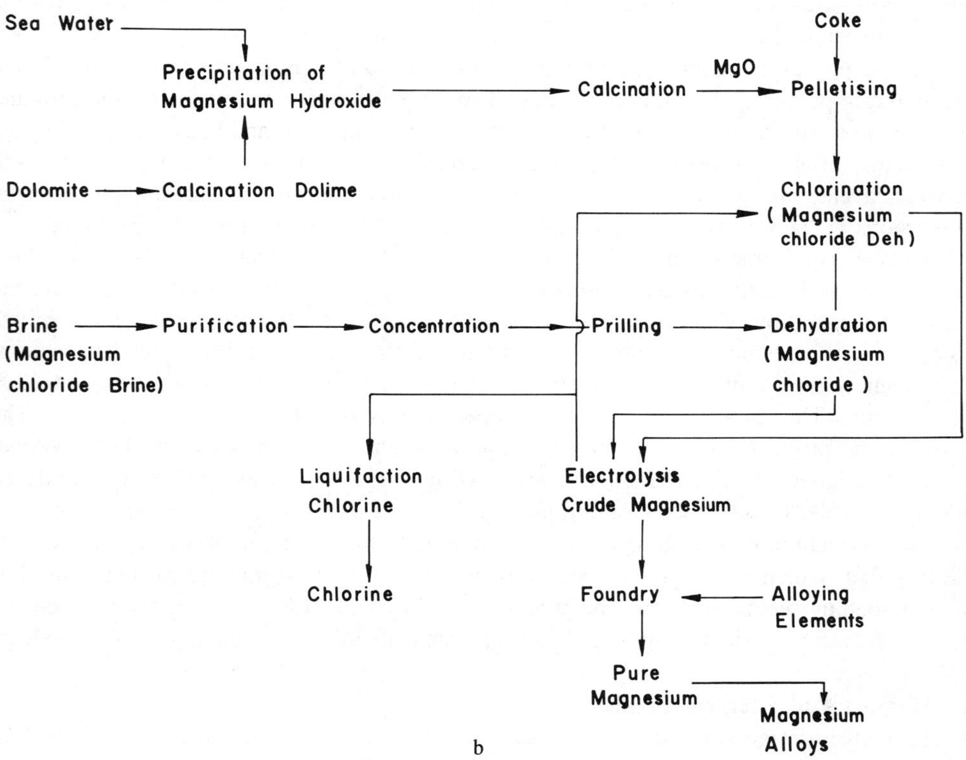

b

FIGURE 1. Flowcharts for magnesium extraction processes. (a) Dow extraction process; (b) Norsk hydro process; (c) Amax process.

conductivity, and the design of the fins should aim to promote circulation of gas between the heated surface and the main gas stream. The introduction of fins results in certain disadvantages. First, the power required to circulate the coolant through the core is increased as a result of the increased turbulence; and second, unless the cladding material has a very low neutron capture cross section, the additional neutron absorption in the fins may be unacceptable from the physics point of view. Thus, fins are frequently used for Magnox-

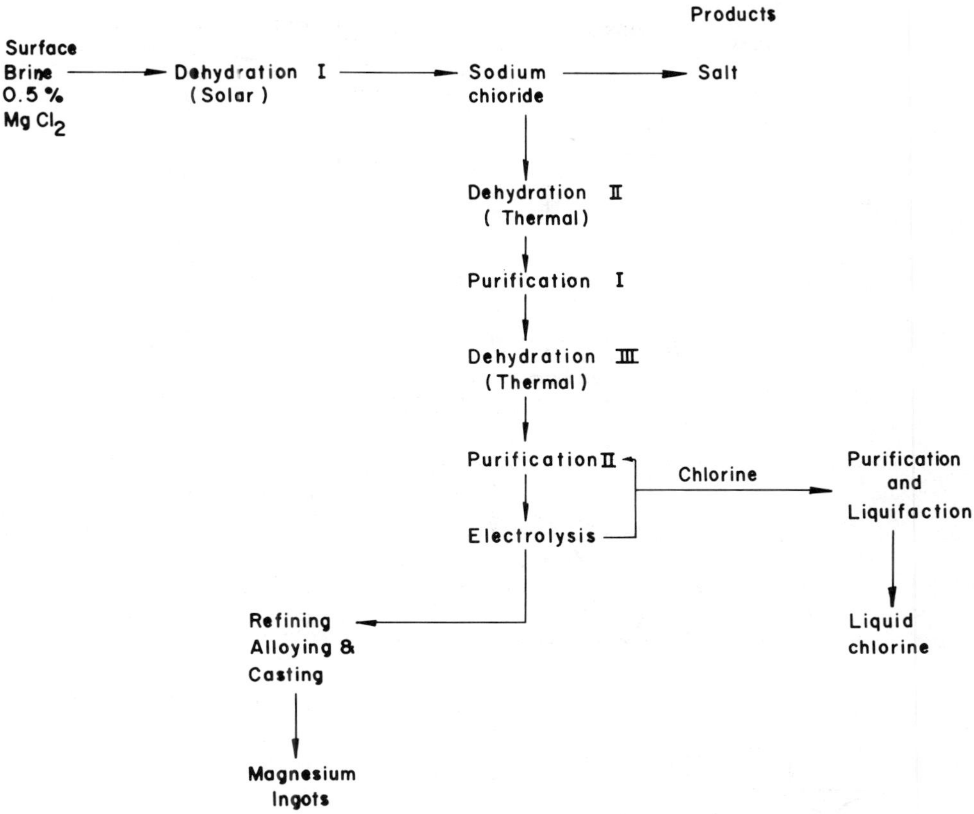

FIGURE 1c

clad fuel (Σ_c for Magnox = 0.0027/cm), but are not used to any extent for stainless-steel-clad fuel (Σ_c for stainless steel = 0.26/cm). Heat transfer can also be improved without any increase in the amount of cladding used by roughening the surface of the cladding, and the heat transfer surface area can be increased by using a larger number of smaller-diameter fuel rods. The use of small-diameter fuel rods is preferred if the thermal conductivity of the fuel is low. The fuel cladding in Magnox reactors thus has a large extended heat transfer surface and demands a more complex manufacturing operation than any other.

As far as the metallurgical structure of the cladding is concerned, special mention may be made of the fact that a fine grain size is needed to avoid creep-induced grain boundary cavitation. The concern is that the incidence of these cavities can lead to a reduction in ductility or, if they linked up, to the formation of permeable paths along grain boundaries leading to coolant ingress into the cladding. The incidence of cavitation appears to be associated with those temperatures at which severe grain boundary deformation occurs. With an increase in temperature, crystallites outlining the grain boundaries become resolvable and progressively larger. The number of cavities increases with the extent of this crystallite formation. At high temperatures, where this is no longer visible, there is no cavitation. Generally speaking, the following facts can be cited: (1) cavitation occurs principally, if not entirely, at the grain boundaries; (2) cavitation is temperature dependent; (3) cavitation is dependent on strain rate, increasing in general with decreasing strain rate; and (4) cavitation is less severe in fine-grained materials. As already mentioned, fine grain size is needed to avoid creep-induced grain boundary cavitation. Such cavitation may lead to leakage in elements to be used near the bottom of the reactor channels, where cladding temperature is

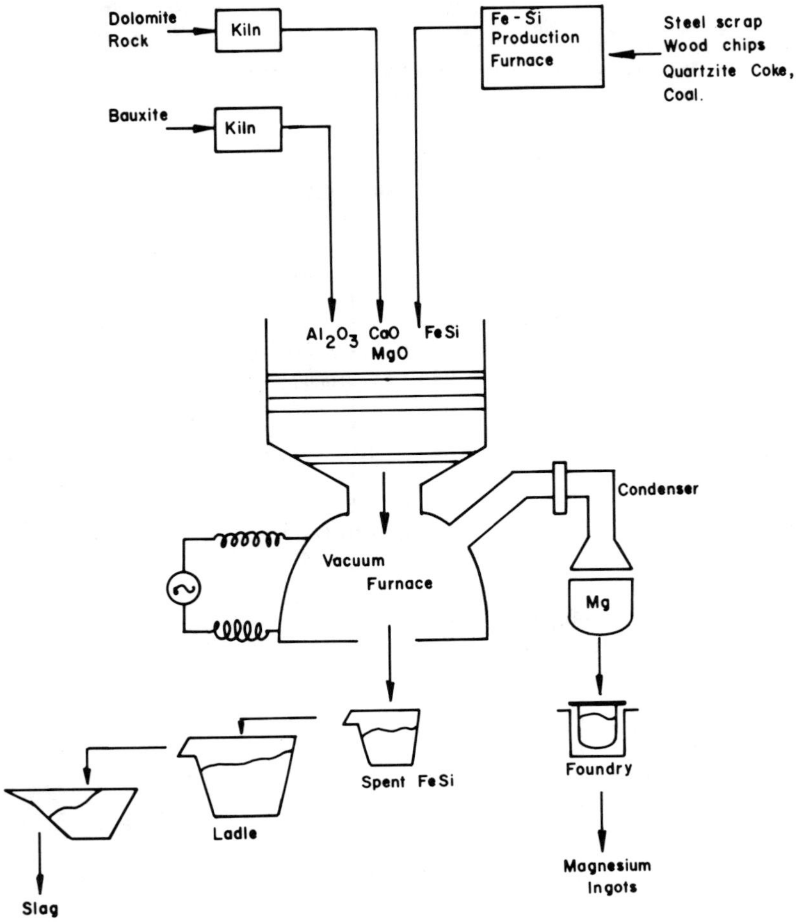

FIGURE 2. Magnesium extraction by the Magnetherm process.

relatively low (200 to 350°C). It should also be remembered that cans produced with a fine grain size tend to undergo rapid grain growth to very large grain size in channel positions where they operate at higher temperatures (350 to 420°C); thus the very fine grain size has to be sacrificed in the upper elements. It is, however, less important in these positions because the irradiation growth responsible for the creep-induced cavitation is less at higher temperatures. These special requirements have led to the use of two different types of Magnox fuel elements depending on their service temperature, the high-temperature elements having a stable, coarse grain size with several grains across the wall. The two principal types as used in reactors in the U.K. are known as helical and herringbone. These designs have been produced with the aim of promoting turbulance. Heat transfer coefficients for such surfaces cannot be predicted from empirical equations for smooth surfaces, and experimental data are necessary for each design. It is sufficient to reiterate that the use of extended surfaces has the effect of increasing the heat transfer rate at the expense of increased power to circulate the coolant through the core and increased neutron absorption in the cladding.

The helical design is manufactured from an extruded hollow billet by an extrusion process. This is followed by a warm twisting operation, in which the straight finned can is twisted on a mandrel to preserve the correct bore diameter. The twisting conditions are controlled to leave the right amount of cold work in the can to recrystallize to a fine grain size when those destined for low-temperature service are subsequently pressurized onto the uranium rods. The cans are finished by four axial slots machined at 90° intervals for insertion of gas flow splitters. The herringbone design is produced entirely by machining from a thick-walled, hollow extrusion with four thick ribs at 90° apart. The ribs are partially machined away to generate the channel support lugs and the fins machined by fly cutter. The fuel assembly process involves insertion of the cleared machined rods of uranium into Magnox cans after which a ceramic disc and magnesium alloy end cap are fitted to each end of the can. A small quantity of helium is then charged into the can, and the end caps are sealed by argon arc welding on automatic rotary welding machines. The integrity of wall welds is checked by X-ray and a mass spectrometer leak test.

The elements are next pressurized, either hydraulically at 250°C, 105 MP, to impart a fine grain size for low channel positions, or in a gas pressure of CO_2 at 515°C, 7 MPa, for high channel positions. This operation deforms the can and locks it into antiracheting groves machined on the uranium rod at intervals along its length and is close-packed end arrays. In general, it can be said that selection of the pressurizing technique is determined by the can grain structure required, which is appropriate to the position of the fuel element in the reactor. X-ray radiography is used to check that effective locking of the can into the rod grooves has been achieved. Susequently, the elements are cleaned by a citric acid pickle, which also decontaminates them with respect to uranic surface contamination. After drying, the canned rods are ready for inspection and final assembly, consisting of making the necessary attachments to the canned rod to provide its support features.

A complete flow diagram from uranium ore concentrate to Magnox fuel production sequences is given in Figure 3. The U.K. practice is shown. The uranium-metal-making process involves a blend of uranium tetrafluoride and magnesium turnings compressed into pellets which are charged to a reaction vessel. A thermit-type reaction takes place, and the resultant billet is cleaned of slag, remelted in a vacuum induction furnace, and cast in graphite molds into rods about 25 mm in diameter and 1 mm long. Small amounts of iron and aluminum are added for the purpose of reducing the tendency to swelling in the reactor. Any other alloying elements can be added at this remelting stage. The case structure at this stage is textured, which shows strong irradiation growth effects in the reactor. This structure needs modification by heat treatment. This is done by accomplishing grain refining by rapid induction heating to the beta phase, about 660°C, and quenching. In some cases, particularly for the longer elements, bars are subsequently annealed at 550°C. This step minimizes the bowing tendency. The beta quenching is a continuous process, the bar traversing through the induction coil an quenching spray. The product is not texture free, but has a balance of positive and negative growth textures in the outer and inner regions, which prevents any large net growth due to irradiation. The outer region of the heat-treated rod should be fine grained in order to avoid the development of a very rough surface by growth of individual grains at low irradiation temperatures. Finally, the urainium rods are machined and grooved; and from then on the sequence to Magnox element production proceeds as described. The metal has relatively high coefficients of thermal conductivity and thermal expansion, both being very similar to those of aluminum. While the high coefficient of thermal conductivity is an advantage in the transfer of heat from the fuel to the coolant, the high coefficient of thermal expansion (28×10^{-6}) relative to uranium (15×10^{-6}) can lead to the development of stresses due to differential length changes during reactor temperature fluctuations. It is necessary to ensure that these stresses are distributed as uniformly as possible over the body of the cladding and are not allowed, for example, to concentrate on the welded joints to the end fittings. It is for this reason that antiratcheting grooves are machined at intervals in the uranium bar and, after sealing, the claddings are, as mentioned, pressurized into the grooves.

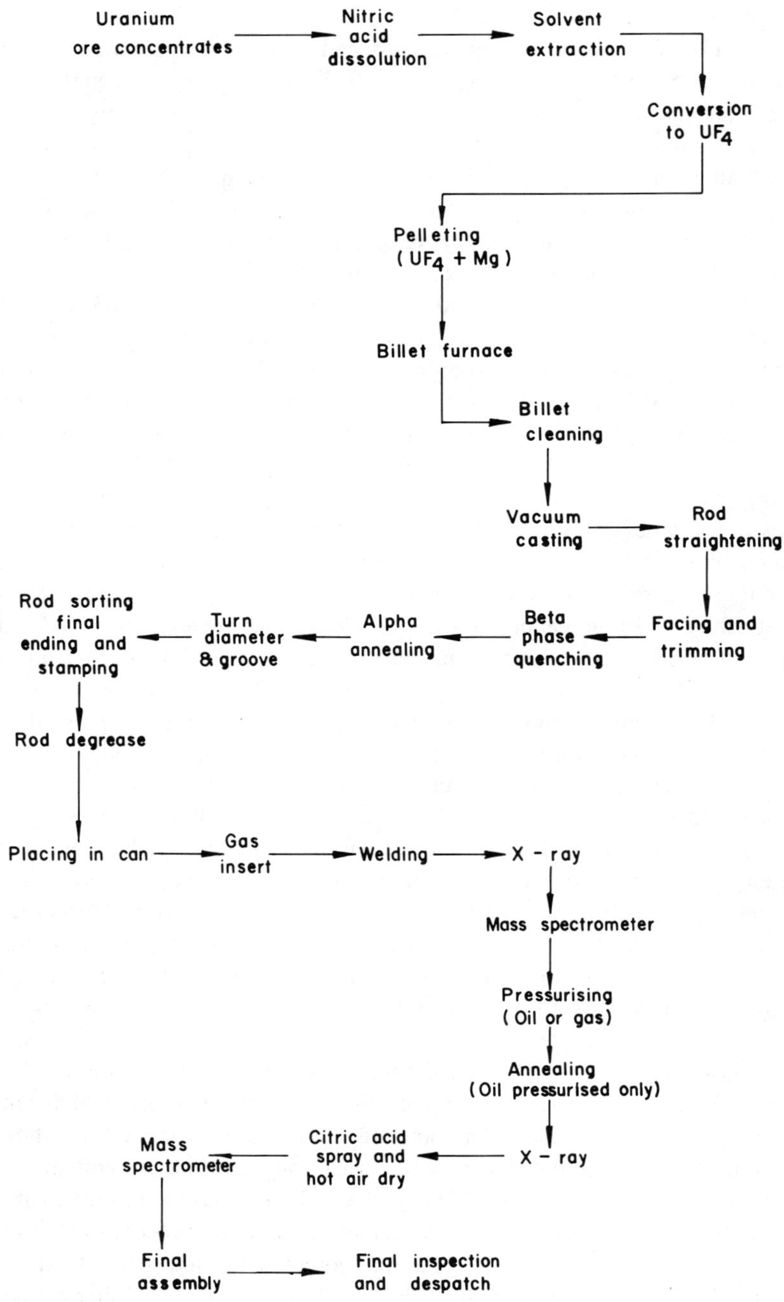

FIGURE 3. Magnox fuel element production.

IV. ZIRCONIUM[8-14]

A. Occurrence and Extraction

Zirconium is the eleventh most abundant element in the earth's crust, which contains 0.028% of this element. Although zirconium is considered a rare metal, it is more abundant than copper, lead, zinc, or nickel.

The principal commercial sources of zirconium and hafnium are the minerals zircon, $(Zr,Hf)SiO_4$, and baddeleyite, $(Zr,Hf)O_2$. Zircon is produced commercially principally from

Table 3
ANALYSIS OF ZIRCON (%)

$Zr0_2 + Hf0_2$	65.5—66.5
$Si0_2$	31—32
Al_20_3	0.75
Fe_20_3	0.1
$Ti0_2$	0.3—2.5
$Th0_2$	0.02
U_30_8	0.035
P_20_5	0.1
$[Hf0_2/(Zr0_2 + Hf0_2)] \times 100$	2.5

beach sands and stream beds, where it accumulates along with other heavy minerals produced in the gradual weathering of granite. Large amounts of these beach sands are found in Florida. The sands are worked similarly for their titanium content, the zircon being obtained as a by-product. Other known sources of zirconium are beach and gold-placer gravels in California and along the Oregon coast, the titanium-bearing beach sands in India and the east coast of Australia, and placer-tin deposits of Malaya.

Zirconium possesses one very important property that has made it most attractive from the point of view of reactor applications. Like aluminium, it has a low neutron absorption cross section: 0.18 b per atom. Only magnesium is lower, at 0.06 b, while iron, nickel, and copper absorb thermal neutrons from 10 to 20 times more strongly. This attribute, coupled with its greater corrosion resistance, fits zirconium admirably for use in cladding fissile material. One serious drawback to the use of zirconium in nuclear reactors lies in the company it keeps. The zirconium resource minerals always contain from 0.5 to 2% hafnium, which behaves chemically almost exactly like zirconium. Unfortunately, its neutron absorption cross section is almost 700 times that of zirconium; therefore, unless it is removed during processing, the advantage for thermal neutrons is lost. The reactor-grade zirconium contains no more than 0.01% hafnium, since the cross section is raised by about 0.01b for every 0.01% of hafnium present. The mechanical properties of zirconium are strongly influenced by small concentrations of oxygen, nitrogen, or carbon as impurities. The metal is quite ductile when free of these elements, but hardens and embrittles with only minor amounts being present. Since oxygen is soluble in the metal and since the oxide is a very stable compound, it is impossible to remove the oxygen easily by any known means once the metal has been contaminated by it. Consequently, rigorous exclusion of oxygen is taken as a cardinal principle in every stage of a metal-producing producing processing operation to produce zirconium possessing good mechanical properties. Zirconium and hafnium oxides are refractory compounds. They are unattacked by any acid except hydrofluoric acid. They react with alkalis to form zirconates, such as (Na_2 ZrO_3), or corresponding hafnium compounds. The hydroxides may be precipitated from aqueous solutions of the sulfate by adding alkalis. Aqueous solutions of zirconium and hafnium salts may be obtained in the form of oxychlorides or sulfates. Zirconium sulfate, $Zr(SO_4)_2{\cdot}4H_2O$, is soluble in water containing sulfuric acid to the extent of 1106 g/l at 18°C. Addition of potassium or ammonium flourides to aqueous solutions of these elements precipitates salts of the type K_2ZrF_6. The tetrachlorides of these metals sublime without melting around 330°C. The difference in vapor pressure between the tetrachlorides of zirconium and hafnium is too small to serve as a practical means for separating one from the other. This general outline of chemical properties is of considerable relevance in the field of zirconium extraction.

A combination of specific gravity, flotation, and electrostatic methods at first produces a zircon concentrate. A chemical analysis of a typical Indian zircon concentrate is given in Table 3. The three most important processes for breaking down zircon chemically and

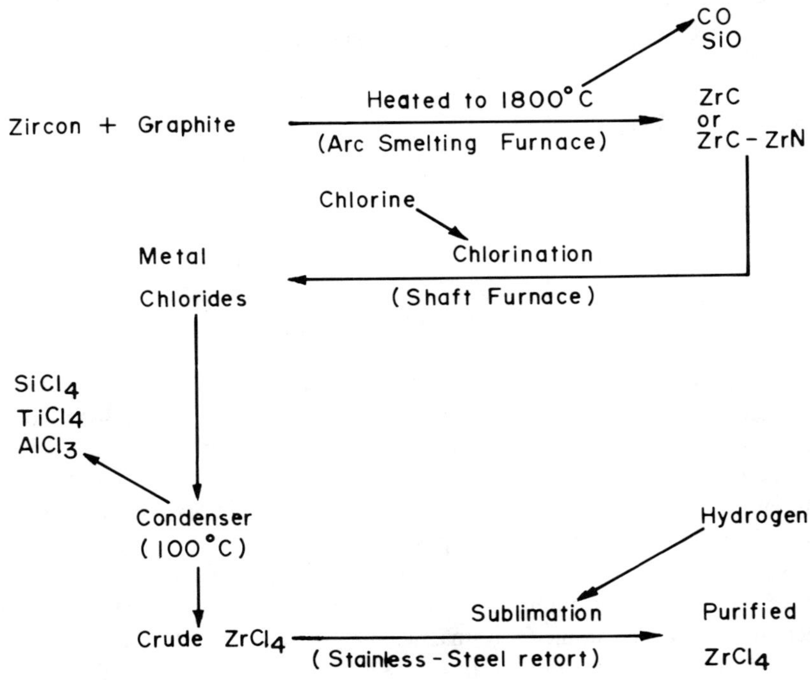

FIGURE 4. Production of $ZrCl_4$ from zircon.

separating zirconium and hafnium from silicon are (1) carbide formation, followed by chlorination; (2) alkali fusion, followed by acid dissolution; and (3) fusion with potassium silicofluoride. The most suitable process will depend, in part, on the chemical form in which zirconium is desired and on the subsequent steps to be used in separating hafnium and preparing metal. The chloride formation process flowchart is shown in Figure 4. The alkali fusion process flowchart is shown in Figure 5. The chemical reaction $4NaOH + ZrSiOI_2 \rightarrow Na_2ZrO_3 + Na_2SiO_3 + 2H_2O$ yields a porous and friable solid which is ground and water-leached to extract the Na_2SiO_3. The residue is then leached with acid, which dissolves the $Na_2\ ZrO_3$. The unreacted zircon which remains as the final residue may be recycled. Any desired zirconyl salt can be made by using the appropriate acid in the final leaching step. This process is especially suitable for preparation of feed for a hafnium-zirconium separation step using solvent extraction from an aqueous solution. Fusion with potassium silico fluoride involves reaction with ground zircon in the presence of potassium chloride, which acts as promoter. Sintering is accomplished in a rotary furnace at 650 to 700°C, and the following reaction takes place: $ZrSiO_4 + K_2SiF_6 \rightarrow K_2ZrF_6 + 2SiO_2$. The sinter is cooled, crushed, leached in hot hydrochloric acid, filtered, cooled to crystallize K_2ZrF_6 $(+K_2HfF_6)$, filtered off, and finally given a wash with water.

The next step in the extraction scheme is the separation of zirconium from hafnium. Among the large number of known processes, three — fractional crystallization of double fluorides, solvent extraction of thiocyanates by hexone, and solvent extraction of nitrates by tributyl phosphate — are important industrially.

Fractional crystallization of the double potassium fluorides is actually the method originally used to separate hafnium from zirconium. Since these salts form solid solutions and the ratio of solubilities is close to unity (1.54 at 20°C), multiple recrystallizations are necessary to attain the required degree of separation. The solvent extraction process is by far the best accepted one. Two processes are in vogue. The first one treats aqueous feed as zirconyl chloride containing hydrochloric acid and ammonium thiocyanate with the organic extractant

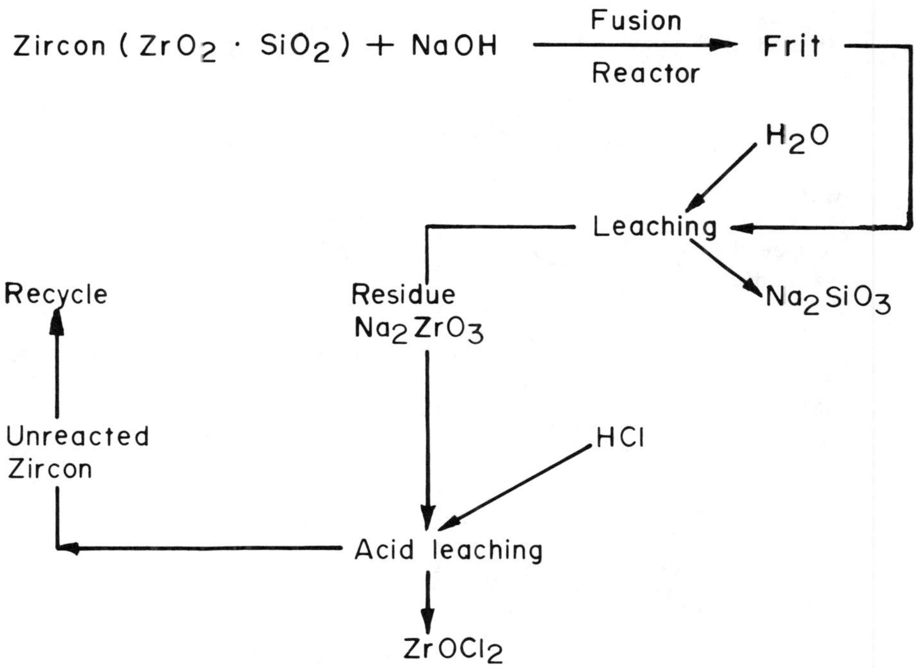

FIGURE 5. Caustic fusion process.

methylisobutyl ketone (MIBK). The second one treats aqueous feed as zirconyl nitrate containing nitric acid with the organic extractant tributyl phosphate (TBP) diluted with kerosene. The reactions that take place can be shown as (1) $Zr0_2^{2+} + 4\ NO_3^- \rightleftharpoons Zr(NO_3)_4$, $K_1 = [Zr(NO_3)_4 / [(Zr0_2^{2+}) \times (NO_3^-)]^4$; and (2) $Zr(N0_3)_4 + 2TBP \rightleftharpoons Zr(N0_3)_4 \cdot 2TBP$, $K_2 = [Zr(N0_3)_4 \cdot 2TBP]/[Zr(N0_3)_4][TBP]^2$. The ratio $[Zr(NO_3)_4 \cdot 2\ TBP]/[Zr0_2^{2+}]$ is called the distributing coefficient of zirconium; representing the ratio by D_{Zr}, it can be shown to be equal to $K_1K_2(NO_3{}^-)^4(TBP)^2$. In all stages of nitric acid concentration D_{Zr} is greater than D_{Hf}, and this is so because $Zr(NO_3)_4$ dissociation to zirconyl $(Zr0)_2^{2+})$ and nitrate) $(N0_3^-)$ ions takes place to a lower degree than hafnium in aqueous medium. The degree of dissociation of $Hf(NO_3)_4$ into constituent ions, however, decreases sharply as the nitric acid concentration increases from 6 *N* to 9 *N* and the separation factor represented as B, the ratio of D_{Zr} to D_{Hf}, falls off very rapidly. To achieve the best results, control of nitric acid concentration is necessary. Figure 6 shows the extraction, scrubbing, and stripping stages for separation of zirconium from hafnium. The solvent extraction process is carried out in mixer-settler type of contactors as shown generally in Figure 7. The aqueous solution containing zirconium nitrate is treated with ammonia and zirconium recovered as zirconium hydroxide, which is calcined to zirconium dioxide.

To the list of the two solvent extraction processes, one must now add a new one developed recently.[15] In the new process, zircon is similarly fused with a caustic to convert it to sodium zirconate, which is digested into zirconium oxysulfate, $H_2ZrO(SO_4)_2$, by sulfuric acid. The feed thus prepared contains zirconium plus hafnium oxysulfate, free sulfuric acid, and some sodium sulfate. Separation of zirconium and hafnium is carried out by solvent extraction with tri-n-octyl amine (TNOA). A flowchart of the process is shown in Figure 8. At first, all zirconium and a part of the accompanying hafnium are extracted into the organic solvent consisting of amine, tridecyl alcohol, and diluent (extraction process). The organic phase loaded with zirconium, $(R_3NH)_2Zr0(SO_4)_2$, is then scrubbed with an aqueous solution containing hafnium-free zirconium oxysulfate (scrubbing process). Finally, zirconium in the

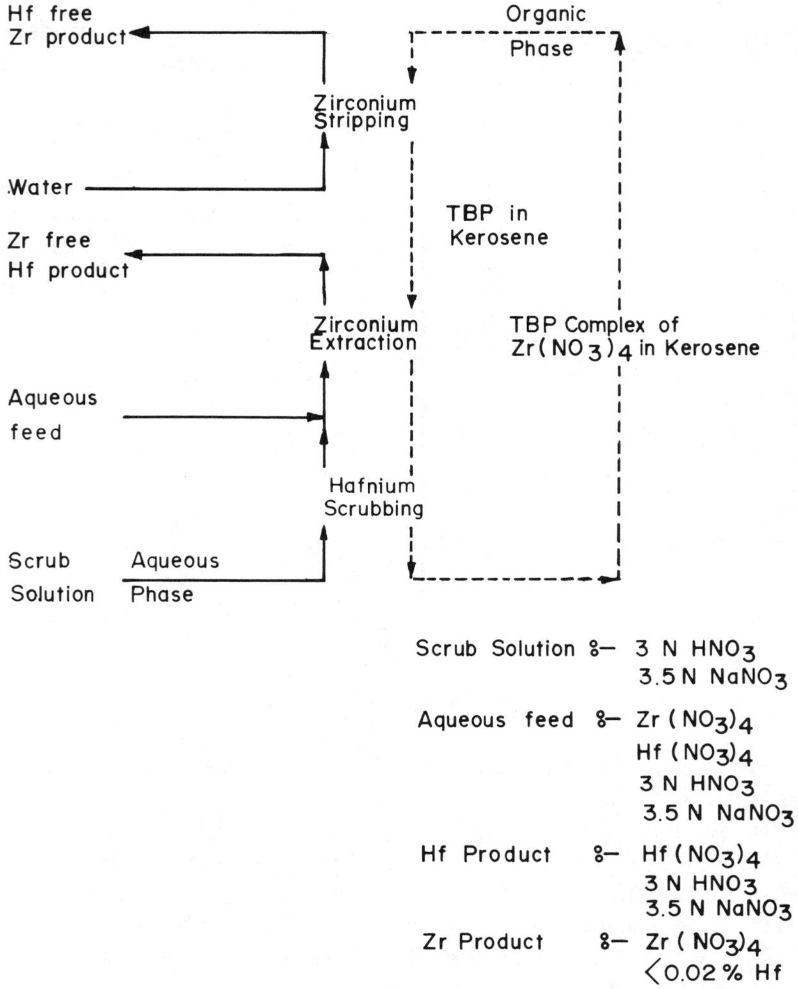

FIGURE 6. Process for separation of zirconium from hafnium by solvent extraction with tributyl phosphate.

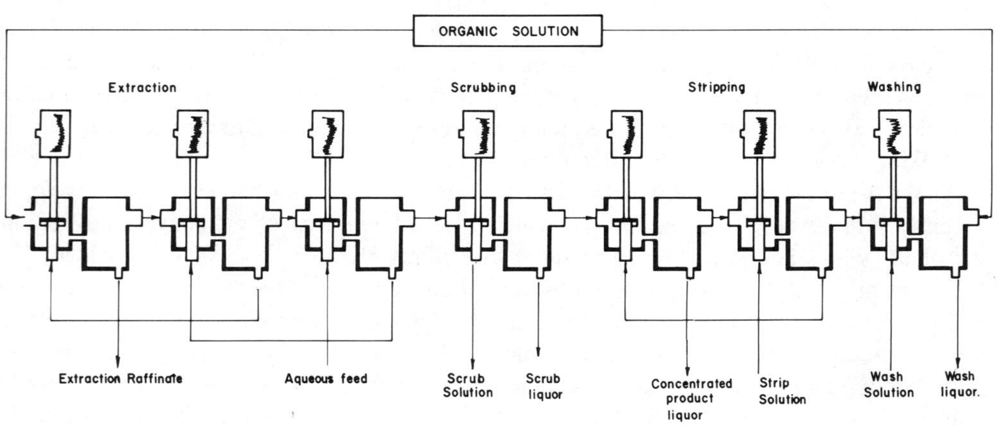

FIGURE 7. Battery of mixer-settler equipment.

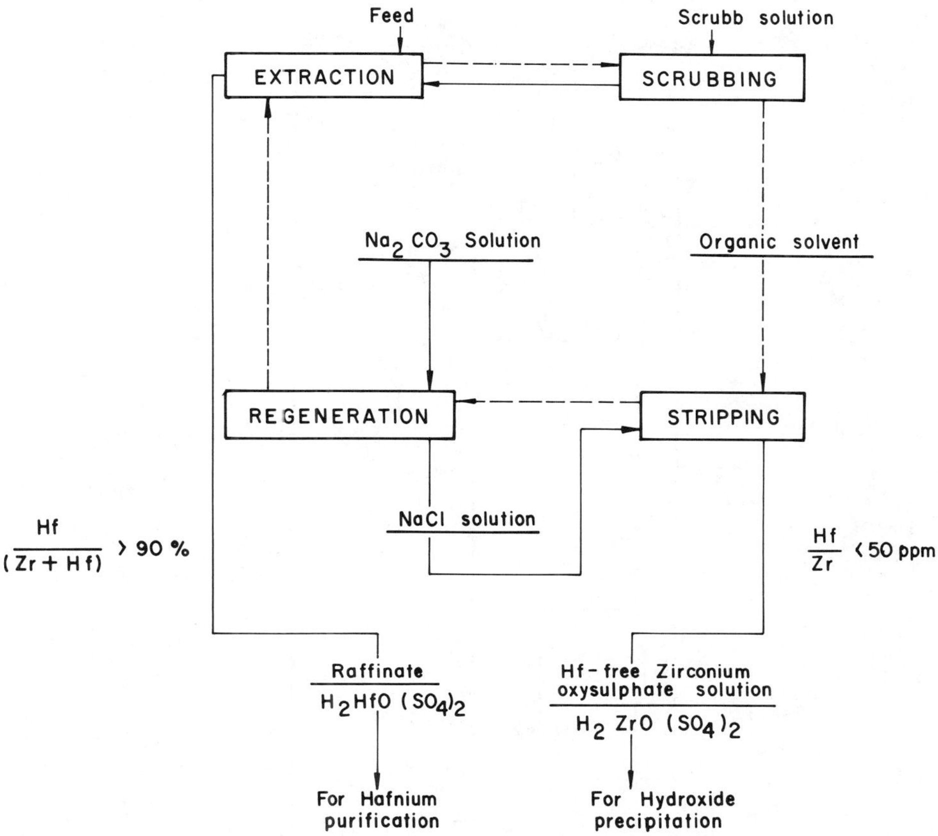

Feed :- Zirconium, Hafnium, oxysulphate $H_2Zr(HfO)(SO_4)_2$;
Sulphuric acid, H_2SO_4 ; same sodium sulphate Na_2SO_4

Extractant :- Tri - n - octyl amine

Scrubbing solution :- Hafnium free Zirconium oxysulphate $H_2ZrO(SO_4)_2$, Sulphuric Acid, H_2SO_4

Strip solution :- Alkali chloride

Regeneration solution :- Alkali carbonate

FIGURE 8. Separation of zirconium from hafnium by solvent extraction with tri-n-octyl amine (TNOA).

organic solvent is stripped with a solution of alkali chloride and concentrated in liquid phase as hafnium-free zirconium oxysulfate (stripping process). Stripped organic is regenerated by contacting with sodium carbonate; the regenerated organic goes to the extraction circuit, and the sodium chloride goes to the stripping circuit (regeneration process). The zirconium solution recovered in the stripping process is neutralized directly with alkali, precipitated, filtrated, and calcined by conventional methods to reactor-grade zirconium oxide, with a hafnium content of less than 50 ppm. The hafnium obtained from the raffinate in the extraction process can be fed to various applications, as hafnium oxide and hafnium metal, by removing contained impurities such as iron and titanium. This flowchart using the new extractant ha been developed by the Nippon Mining Company Ltd. (NMC), and a plant based on the flowchart for zirconium production is currently in operation in Japan.

Yet another new addition to zirconium-hafnium separation technology that has been developed and implemented on plant-scale production in France is based on extractive distillation.[16,17] In the process, the vapor of $ZrCl_4$ and $HfCl_4$ is fed to the upper middle of a column (Figure 9). A solvent made of a molten KCl-AlCl$_3$ mixture is circulated from the

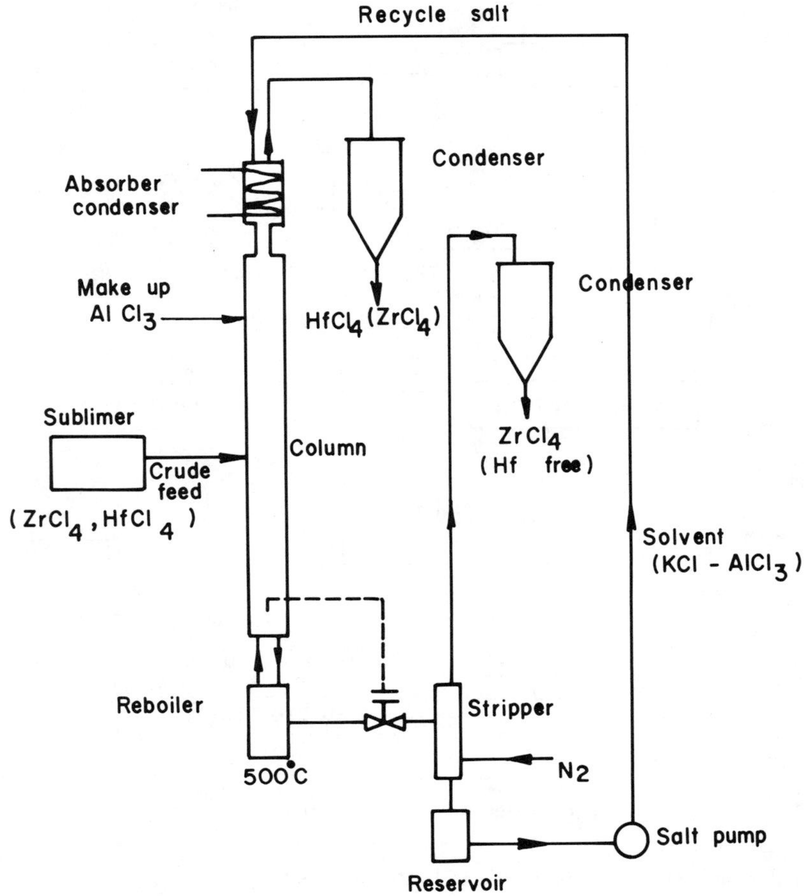

FIGURE 9. Extractive distillation process for separation of zirconium and hafnium.

top to the bottom of the column (chloride vapor is generated in a reboiler at about 500°C). The upgoing flow of vapor is progressively enriched in $HfC1_4$. This vapor goes through an absorber-condenser so that a recovery of a highly enriched $ZrC1_4$-$HfC1_4$ mixture in the 30 to 50% range is obtained. At the bottom, the stream of hafnium-free liquid is fed to a stripping column, about 400°C, where a flow of nitrogen recovers $ZrC1_4$ from the solvent. The solvent is then recirculated to the top of the distillation column. The resulting $ZrC1_4$ is condensed and fed to a Kroll process, as described later, for metal production. This route avoids the second chlorination step that is followed for all the previously described processes. This pyrochemical route avoids the burden of a perfect recovery of any atmospheric pollutant and the neutralization and treatment of the chemical reagent of the standard process. Introduction of these two new processes has brought an added dimension to the zirconium extractive metallurgy.

The zirconium oxide intermediate from all other than, obviously, the pyrochemical process is next chlorinated to zirconium chloride. In the process the oxide powder is mixed with carbon and kneaded with a starch solution. The wet dough is extruded into briquettes which are subsequently dried and cooked at about 800°C in an inert atmosphere. The porous product is next chlorinated in externally heated silica tubular furnace. Production-scale chlorination is done in silica-brick-lined furnaces. The loaded charge in the chlorinator is self-resistively heated.

The reactor-grade zirconium metal is produced by the well-known Kroll process, which is essentially a batch process involving reduction of anhydrous zirconium chloride with magnesium. The reduction reaction, $ZrCl_4(g) + MG\ (l) \rightarrow Zr\ (s) + MgCl_2(l)$, is carried out in a specially designed stainless stell vessel heated externally by a three-zone resistance furnace. Zirconium chloride and clean magnesium pigs are loaded in separate containers into the reduction retort. Zirconium chloride, which has a sublimation point of 331°C, is distilled off and allowed to react with liquid magnesium maintained at a reduction temperature of about 850°C. During distillation of zirconium chloride, its *in situ* purification takes place with respect to oxygen (in the form of $ZrOCl_2$) picked up during the handling of the highly hygroscopic chloride powers. At the distilling temperature, zirconium oxychloride breaks up to yield volatile zirconium tetrachloride and zirconium oxide, which being nonvolatile, is retained in the chloride container. In the course of reduction, zirconium settles on the bottom and side walls of the crucible because of its being heavier than both magnesium chloride and reductant magnesium. Between them, the latter is heavier, which means the reductant magnesium always remains floating at the top, thereby making itself physically available for reaction with the incoming vapors of zirconium chloride. The reduction results in sintered zirconium sponge covered with magnesium chloride and excess magnesium. Zirconium sponge is separated from the reacted charge by pyrovacuum distillation in another heavy-walled retort capable of evacuation down to less than 10^{-3} mmHg at 900°C.

The foregoing processes for extracting zirconium from zircon, separating zirconium from hafnium, and reducing zirconium compounds to metal can be combined in a number of ways as shown in Figure 10.

B. Properties[18-22]

The nuclear, physical, thermal, and mechanical properties of zirconium as a structural material are summarized in Table 4.

Corrosion behavior of zirconium presents special features. The course of corrosion of relatively pure zirconium is indicated schematically in Figure 11, which is a logarithmic plot of weight gain vs. exposure time. It is seen that thee is an initial period in which the rate of corrosion decreases with time followed by a period in which the corrosion rate is constant. The change in kinetics has been referred to as ''breakaway''. In the initial period the corrosion product is tightly adherent and black-colored in appearance. In the second stage the corrosion product is white in appearance and flakes off, resulting in weight losses. The corrosion product in both stages is monoclinic zirconium dioxide. The kinetics of the corrosion reaction may be described empirically by $\Delta W = kt^n$, where ΔW is the weight gain per unit area, t is the time, and k and n are constants. The logarithmic form of the equation is convenient in that the corrosion data are usually presented in logarithmic plots: $\log \Delta W = \log k + n \log t$. Thus, a plot of $\log \Delta W$ against $\log t$ yields a straight line having slope n and intercept log K, as the given figure on corrosion shows. The value of the exponent of the equation $\Delta W = kt^n$ for zirconium is about 1/3; thus, the corrosion of zirconium obeys approximately a cubic rate law during the first stage.

It has been found that, even for the purest zirconium (iodide zirconium), 'breakaway' or failure occurs quite early. Such behavior has been ascribed to the presence of impurities in zirconium. In general, the initial mode of corrosion is interrupted when an oxide thickness corresponds to a weight gain of approximately 40 mg/cm². This critical oxide thickness for the onset of 'breakaway' is indicated schematically in Figure 11, which also depicts the behavior of impure zirconium. The critical oxide thickness is reached much sooner in the case of impure zirconium than in the case of pure zirconium. Among the impurities, the most striking effect on corrosion in zirconium is by nitrogen. Material containing more than 0.004 wt% suffers 'breakaway' within a short time, the time required for 'breakaway' decreasing with increasing nitrogen content. Consequently, nitrogen must be controlled to a maximum of about 0.005 wt% in all stages of metal production and fabrication. Similarly,

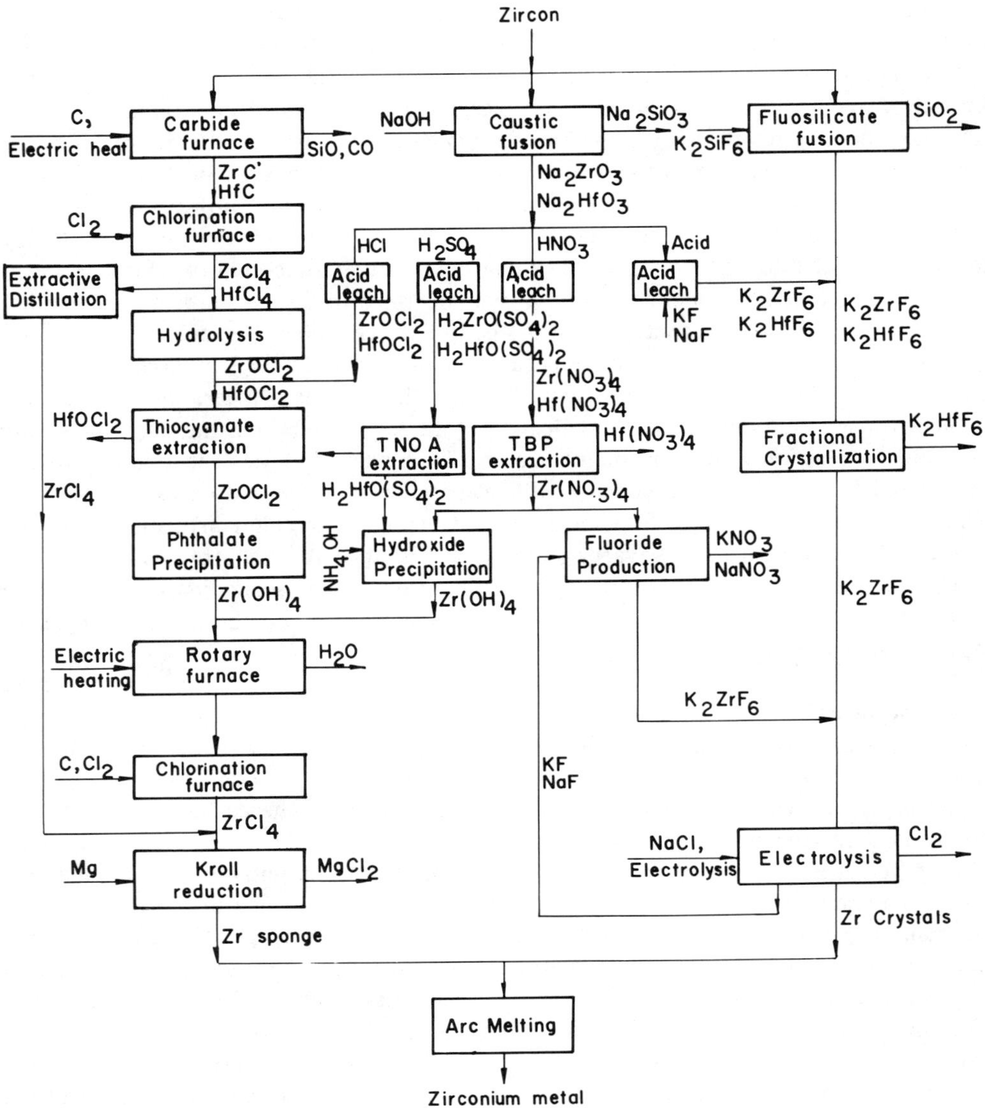

FIGURE 10. Routes from zircon to zirconium metal.

the corrosion rate increases with increasing content of carbon, which must be kept to a minimum if corrosion resistance is desired, a fact which precludes induction melting in graphite crucibles.

The corrosion rate of zirconium in aqueous medium is limited by the diffusion of oxygen ions via oxygen in vacancies from the oxide-water interface to the metal-oxide interface. This process is interrupted at the breakaway point, which occurs when a critical film thickness has been attained. Thus, the rate of diffusion of oxygen ions through the oxide governs the time necessary to reach 'breakaway'. If nitrogen is present in metal and it is incorporated into oxygen ion sites in the growing oxide film as N^{3+}, additional vacancies are created which increase the rate of diffusion of oxygen ions and thus hasten 'breakaway'. Similarly, if tin is present and is incorporated into the oxide lattice on zirconium ion sites as Sn^{3+}, additional oxygen vacancies are created with a similar effect upon the diffusion of oxygen ions. If both tin and nitrogen are present, Sn^{3+} ion in the oxide lattice should tend to associate

Table 4
PROPERTIES OF ZIRCONIUM

Thermal neutron absorption cross section (b)	0.180
Thermal neutron scattering cross section (b)	8.0
Density, theoretical (g/cm³)	6.50
Structure	
Alpha phase (20 to 862°C)	hcp
Beta phase (862 to 1845°C)	bcc
Melting point (°C)	1845 ± 25
Boiling point (°C)	3580
Specific heat, C_p, at 20°C (cal/g °C)	0.067
Thermal conductivity (W/m·k) at ambient temp	20.9
Yield strength (10^2 MPa)	
Annealed	1.42—2.14
Cold-worked	9.28—10.71
Tensile strength (10^2 MPa)	
Annealed	4.28—5
Cold-worked	11.42—12.85
Elongation (%)	
Annealed	30—40
Cold-worked	11.42—12.85
Rockwell hardness	
Annealed	35—45
Cold-worked	87—98

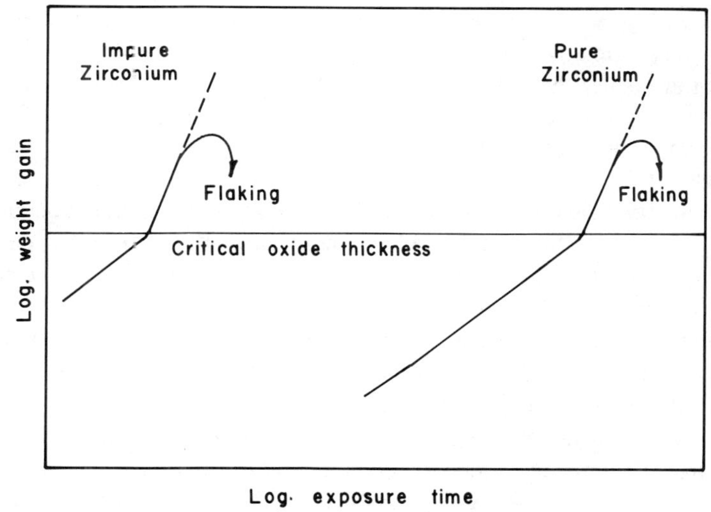

FIGURE 11. Schematic representation of the corrosion of unalloyed zirconium in high-temperature water.

with an N^{+3} ion and an oxygen ion vacancy, thus tending to nullify the contribution of that vacancy to the diffusion current.

It is seen in the foregoing that zirconium exhibits the undesirable phenomenon of 'breakaway', which may occur at widely ranging exposure times. Furthermore, the corrosion rate after 'breakaway' is too high for practical use. Thus, a major purpose in alloying is to provide more reproducible corrosion behavior, to eliminate or delay 'breakaway', and to reduce the post-'breakaway' corrosion rate. A further aim in alloying is to make it possible to employ the industrial zirconium produced by the Kroll process rather than the costlier pure zirconium produced by the iodide process. Zirconium containing tin, iron, chromium and nickel is known under the generic name zircaloy. Zircaloys have a much better corrosion resistance, a superior mechanical strength, sufficient ductility, and nuclear characteristics close to that of pure zirconium.

From what has been said in the previous paragraph, it is apparent that tin forms the essential principal constituent to alloy with zirconium. With tin as the principal constituent along with small amounts of iron, chromium, and nickel in the zirconium metal, it has been possible to overcome the adverse effects of impurities with improvement of corrosion resistance properties. In order to increase the strength, a low but definite amount of oxygen has also been specified; and, taken altogether, the first phase of zirconium alloy development efforts has been found to have culminated with the establishment of the alloy zircaloy-2. The corrosion resistance of zirconium has also been found to be seriously impaired by hydrogen, one of the sources being the metal-water reaction at elevated temperatures ($Zr + 2H_2O) \rightarrow ZrO_2 + 2H_2$). It has been observed that nickel in zircaloy enhances hydrogen uptake during corrosion in water both in and out of the reactor. This finding has led to reduction of the nickel content and an equivalent increase of iron and chromium to give rise to the development of zircaloy-4. This alloy shows a smaller hydrogen pickup fraction during corrosion than zircaloy-2, the other properties remaining mainly unchanged.

The corrosion in zircaloy cladding in the reactor core can occur either at the water side or at the fuel side. The fuel-side corrosion occurs at the inner surface of the cladding tube, as described later, and given full play it can be performance limiting. The water-side of external corrosion, as it may be called, of zircaloy cladding has never restricted the operating strategies of pressurized water reactors (PWRs). However, proposed design changes leading to higher burnup, higher coolant temperatures, and longer in-reactor residence times may increase corrosion of zircaloy. The observed in-reactor PWR water-side corrosion of zircaloy-4 fuel cladding is greater than that estimated using models developed based on laboratory measurments of ex-reactor corrosion. Some of the possible reasons for the in-reactor enhancement of water-side corrosion include (1) radiation effects in the oxide layer, (2) radiolysis of water, (3) coolant chemistry and local boiling effects, and (4) modification of oxide layer chemistry. Also, under heat transfer conditions, the thermal conductivity of the oxide layer will influence the rate-controlling temperature at the metal-oxide interface. If irradiation reduces the thermal conductivity of the oxide layer, this may cause a further increase in corrosion.

Pure zirconium, as shown in Table 4, has two allotropic modifications, alpha zirconium with a hexagonal close-packed (hcp) crystal structure (stable up to 862°C), and beta zirconium with a body-centered cubic (bcc) crystal structure (stable from 862°C to melting point). When the alloying elements are added to zirconium, the $\alpha \rightleftharpoons \beta$ transition occurs over a temperature range. For zircaloy-2 and zircaloy-4 the transformation occurs in the temperature range 825 to 985°C on heating and 945 to 780°C on cooling. On cooling from the beta down into the alpha region, a fairly fine precipitate is formed; the particles measure about 1 μm in diameter. The alloying elements in solution and as precipitated secondary phase increase in deformation resistance of the polycrystalline metal.

The hcp metal zirconium has a crystal lattice with a unique direction, the basal plane normal or basal pole [0001], implying a pronounced anisotropy. This results in plastic deformation that occurs either by slip or twinning. The slip systems are of the type $\{10\bar{1}0\}$ $\langle 1\bar{2}10 \rangle$ with all slip directions perpendicular to the [0001]. Thus, tension or compression along [0001] cannot be accomplished by slip. Instead, twinning occurs, and this requires a higher resolved shear stress than slip, making [0001] a strong direction of the crystal. The preferred orientation of the crystallites, i.e., the texture, carries the anistropy of single crystals over to the polycrystalline zirconium alloys. In zircaloy, the hcp alpha phase is dominant, giving zircaloy the same characteristics with regard to texture and anisotropy as

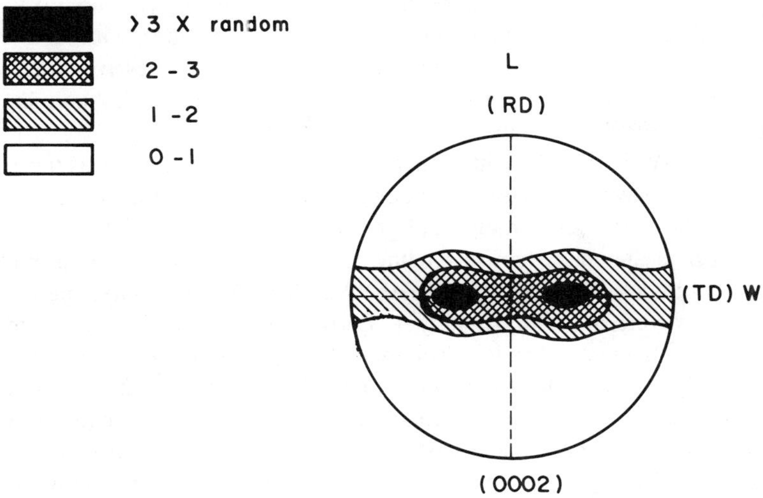

FIGURE 12. Texture of hot-rolled and annealed zircaloy plate (L, longitudinal direction = rolling direction [RD]; W, width direction = transverse direction [TD]).

pure zirconium. Figure 12 shows a typical texture of hot rolled and annealed zircaloy plate. The rolling direction is L and transverse direction is W. The basal poles of the hexagonal lattice are about 30° away from the normal direction, and there is some spread toward the transverse direction. As a result of the thermomechanical treatment during production and the development of pronounced crystallographic textures, zircaloy fabricated products always show anisotropic mechanical properties.

C. Applications[23-25]

Zirconium alloys have emerged as the principal materials for nuclear fuel cladding, fuel channels, and pressure tubes for aqueous power reactors.

The fuel element in a PWR is essentially a long straight zircaloy-4 tube containing close-fitting UO_2 pellets and having a short unfuelled length (gas plenum) at one or both ends, usually at the top. The gas plenum in most elements contains a helical stainless steel spring, whose sole purpose is to hold the UO_2 pellet stack together during transport and handling operations. The fuel tube is sealed at both ends by zircaloy-4 end plugs. All PWRs have cladding which is elastically stable under the coolant pressure of about 150 bar. This, combined with a desire to minimize the creep-down rate, leads to thickness-diameter ratios of about 0.05 or greater. A typical design has a clad outer diameter of 9.5 mm with a wall thickness of 0.57 mm. The gap between the fuel and cladding is generally kept within the range 0.05 to 0.25 mm. It is generally desirable to avoid immediate contact at power between the UO_2 and the cladding tube, and in the PWR this is usually delayed to sometime within the second cycle. A typical design clearance is about 0.15 mm diametral. Helium is used universally as a fuel pin filling gas because of its good thermal conductivity. In the original practice, filling was at atmospheric pressure. There were two major advantages with this practice. First, the clad rapidly crept down onto the UO_2 under the high coolant pressure and would then creep into any gaps in the UO_2 pellet stack, sometimes resulting in clad rupture. The mode of creep-down ovalizes the cladding, which leads to high local bending stresses on a power increase, again with the risk of clad rupture. Second, the gap thermal conductance reduced rapidly as fission gases, xenon and krypton, were released from the UO_2, causing a feedback effect on UO_2 temperature and fission product release. For these reasons it has become the practice to prepressurize the PWR fuel element with helium to pressures in the range 20 to 30 bar (cold). The fuel element filling gas helium, with its hot pressure in the range of 60 to 90 bar, straightly halves the initial clad compressive stress. The prepressurization reduces the dilution by the released fission product gases, thus minimizing the adverse influence they have on gas gap conductivity.

The boiling water reactor (BWR) fuel element bears essentially the same features as those of the PWR fuel element. However, because of somewhat lower ratings, the two-phase (steam-water) coolant and the lower temperature and pressure result in some differences. The fuel elements are larger in diameter, typically being 12.5 mm and clad in a 0.86-mm wall of zircaloy-2 instead of zircaloy-4, as in the case of PWR. The channel containing the BWR fuel assembly is, however, made of zircaloy-4. Early BWR channels made from zircaloy-2 suffered from embrittlement at low temperature after long exposure times of about 10 years. On the basis of the fact that the zircaloy-2 structural component picks up more hydrogen than zircaloy-4, it is understandable that components designed for long exposures in boiling water environments were later made from zircaloy-4 that was deleloped for a lower hydrogen uptake. The BWRs usually operate with purified water. Decomposition of water takes place in a radiation field to H_2 and O_2 (or D_2 and O_2 in the case of D_2O reactors). As a result, the liquid phase will have an oxygen content of 0.05 to 0.2 ppm, and the vapor phase about 5 to 20 ppm with about stoichiometric amounts of hydrogen. In the case of PWRs, the radiolytic decomposition of the water is suppressed by adding about 1 ppm of hydrogen to the water, which leads to a very low oxygen content. The water in a PWR is further kept at a pH of about 10 by LiOH additions. As hydrogen is chemically recognized as a reducing agent, these water conditions are referred to as reducing, in contrast to the BWR, which is called oxidizing due to the presence of oxygen. Since zircaloy-4 hydrides less than zircaloy-2 under typical PWR conditions, only zircaloy-4 has become the cladding material of choice for PWRs. In BWRs, because of the reduced coolant pressure, the fuel elements are not commonly prepressurized, as is the case with the PWR fuel element. The recent trend however, is toward filling of helium at pressures of around 5 bar. This additional pressure results in lower fission gas release and avoids extreme situations in which a gas release can run away by a feedback mechanism resulting from deteriorating gap conductance.

Apart from PWRs and BWRs, the two variants of liquid water reactors (LWRs), there are other reactor systems, such as the ones of Canadian design, of natural-uranium fueled, heavy-water-moderated power reactors using pressurized heavy water (CANDU-PHW) or boiling water (CANDU-BLW) as the heat transfer medium, which makes extensive use of zircaloys. The CANDU reactors use zircaloy-4 fuel sheaths (thickness-to-radius ratio equals 0.06) which are collapsed onto the UO_2 fuel by the pressurized coolant. The fuel element in the pressure tube reactors has slight variations. The steam-generating heavy water reactor (SGHWR) fuel element, for instance, incorporates, as do PWR and BWR fuel elements, a fission gas plenum at the top end. The CANDU-PHW has no end plenum as such, but a void volume for fission gas storage is provided by the use of hollow end caps. The pressure tubes are pressure vessels of the CANDU reactors. They are about 10 cm in diameter and 0.3 to 0.5 cm thick by 6 m, and contain the fuel element bundle and coolant operating at about 9.6 MPa and 300°C. The requirements for pressure tube alloys differ from those for fuel cladding, largely because the pressure tube is part of the reactor's permanent structure, whereas the fuel element is a consumable component with a design life of only 1 or 2 years. A pressure tube alloy must be able to be fabricated into tubes, have adequate neutron economy, have adequate in-reactor creep strength, resist corrosion by the reactor coolant, have adequate tensile strength, and retain adequate ductility in the reactor. The criterion that operating stresses should not exceed one third of the ultimate tensile strength has determined the wall thickness for pressure tubes in CANDU reactors. The initial choice for CANDU pressure tube material has been zircaloy-2. The SGHWR system is based on the use of a

Table 5
TYPICAL COMPOSITIONS OF ZIRCALOYS AND THEIR APPLICATIONS AS STRUCTURAL COMPONENTS IN VARIOUS REACTORS

Alloy	Typical compositions (wt%)	Melting range (K)	Thermal neutron cross section (b)	Comments	Applications
Zircaloy-2	Sn, 1.5 Cr, 0.1 Fe, 0.14 Ni, 0.05 0—1400 ppm	2090	0.193	Thermal conductivity 30% less than pure zirconium	Fuel cladding (BWRs); fuel cladding, pressure tubes, and calandria tubes (SGHWRs); calandria tubes (CANDUs)
Zircaloy-4	Sn, 1.5 Cr, 0.1 Fe, 0.2 0—1400 ppm	2090	0.194	Possesses better resistance to hydrogen pickup	Fuel cladding (PWRs, CANDUs); fuel channels (BWRs)

series of zircaloy-2 pressure tubes. Table 5 shows the typical compositions of commercial zircaloys and their applications as structural materials in various reactor components in various reactors, as indicated.

D. Fabrication[26]

The details of the production of flow of tubular and flat products of zircaloy are given in Figures 13 and 14. Zircaloy cladding tubes inherit their mechanical properties from the production schedule which principally consists of a melting stage, a hot-working stage, and a cold-working and annealing stage, as summarized in Table 6.

Pressing and welding are done with care to assure uniform distribution of alloying elements in the ingot. The homogeneity is further improved during remelting. High vacuum has to be maintained in the melting as well as in the annealing furnaces because of the strong affinity to oxygen and nitrogen shown by zirconium metal.

Heating to the beta phase range and quenching are undertaken to dissolve coarse secondary-phase particles and to finely reprecipitate them. In the subsequent heat treatments in the alpha range, the particles grow only slightly, and in the final tubing the particle size is about 1 μm. An even distrubution of such particles is considered important for high corrosion resistance. Extrusion is made with the material clad in copper to prevent contact with the air. The extrusion must be made with perfect control of the dimensions to make sure that the final tubing fulfills the dimensional tolerances.

Cold rolling is done in pilger mills. Two grooved roles deform the tube, which is supported by an internal mandrel. The tube elongates, which is the natural way of tubemaking. This inherent elongation determines the dominant texture feature of all zircaloy tubing: that there are no basal poles in the axial direction. Annealing in the upper alpha range makes the material recrystallize. This does not change the basal pole texture markedly, but causes only a slight shift of the positions and intensity of the maximum density orientations. Also, there is a rotation of crystals 30° around the basal pole, but this does not affect the basal pole texture.

Of prime importance for the texture, i.e., how the basal poles are distributed between the radial and tangential directions, is the relation between wall thickness reduction and diameter reduction during cold rolling. Most important in this respect is the final cold rolling, but the preceding cold-rolling steps may also affect the final texture. A parameter called Q, defined as the ratio of the percent reduction in wall thickness divided by the percent reduction in diameter, has been widely used to characterize a reduction step in the manufacture of zircaloy tubing. This parameter, which characterizes a reduction sequence and its relationship

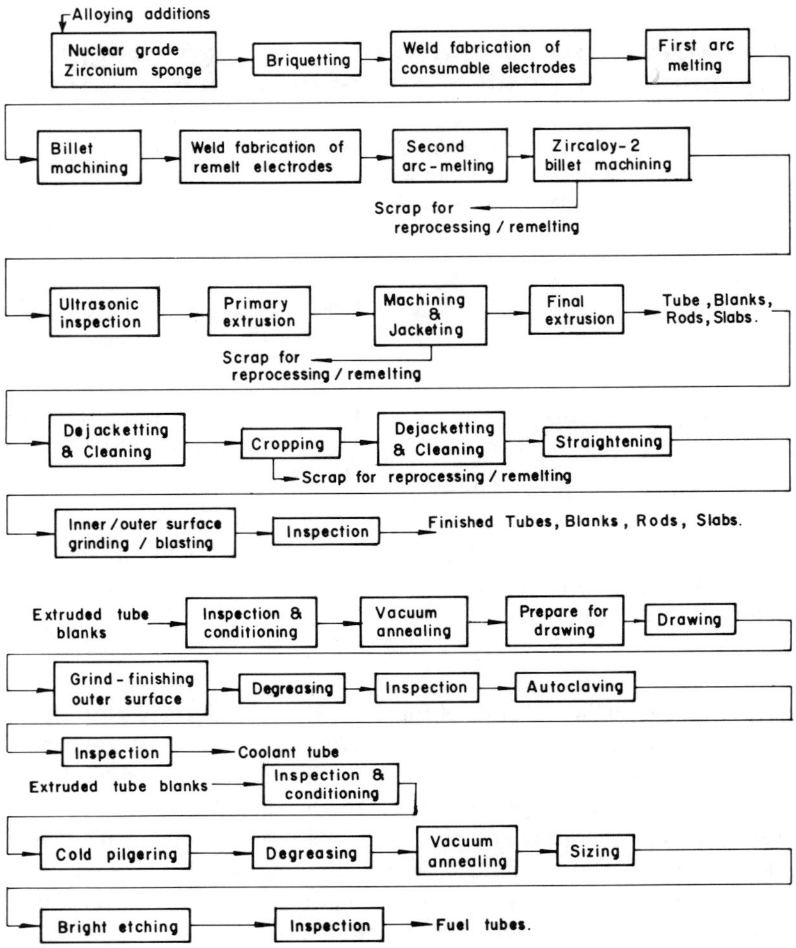

FIGURE 13. Production of extruded zircaloy tube blanks, rods, and slabs from nuclear-grade zirconium sponge.

to final tube crystallographic texture, has been studied extensively because of its great importance in determining the morphology of the hydride precipitation. The relation between Q and the texture appears simple. High values of Q yield a texture with the basal poles closer to the tube normal than do low values of Q. Tubes produced with a high Q value possess a combination of strength and ductility superior to that of tubes produced with a low Q value. Zircaloy cladding tubes in nuclear reactors must have ductility balanced against high strength. The purpose of high strength is to keep the tube from deforming, thus increasing the ductility margin. The strength can be adjusted with a small loss of ductility by controlling oxygen concentration. In the context of ductility in zirconium-based alloys, however, a more important consideration is given to hydride orientation.

Zirconium hydrides[27] precipitated in the zircaloy tubing have a deleterious effect on the ductility. Three potential sources of hydrogen exist in reactor systems. (1) corrosion product hydrogen, e.g., $Zr + 2H_2O \rightarrow ZrO_2 + 2H_2$; and similar reactions for other metals in the reactor primary system, including (2) radiolytic hydrogen from decomposition of water, ammonia, or hydrazine; and (3) hydrogen overpressures imposed to maintain a reducing atmosphere. The BWR coolants typically are oxygen rich (<5 cm^3 H_2 per kilogram, 0.1 to 1 ppm O_2), while the PWR coolants generally are hydrogen rich (15 to 60 cm^3 H_2 per

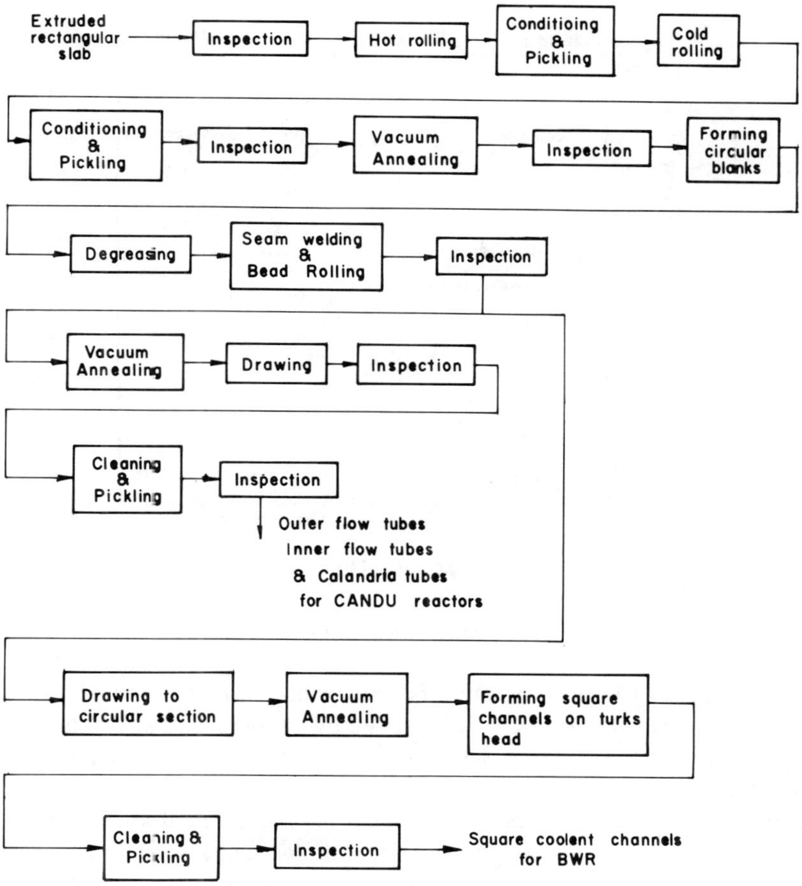

FIGURE 14. Production of flow and calandria tubes for CANDU and coolant channels for BWR.

Table 6
MAIN STEPS IN PRODUCTION OF ZIRCALOY CANNING TUBES FROM ZIRCONIUM SPONGE

Stage	Operation
Melting	Pressing of Zr sponge and alloy additions into briquettes; welding of briquettes into electrodes
	Melting } High-vacuum electric arc
	Remelting } consumable electrode
Hot-working	Hot-forging in upper alpha range; heating to beta range and quenching; extrusion in upper alpha range
Cold-working and annealing	Cold-rolling; annealing in upper alpha range
	Cold-rolling Annealing } Optional, depending on fuel tube size
	Final cold-rolling, 50—80%
	Final annealing, 475—575°C
	Straightening

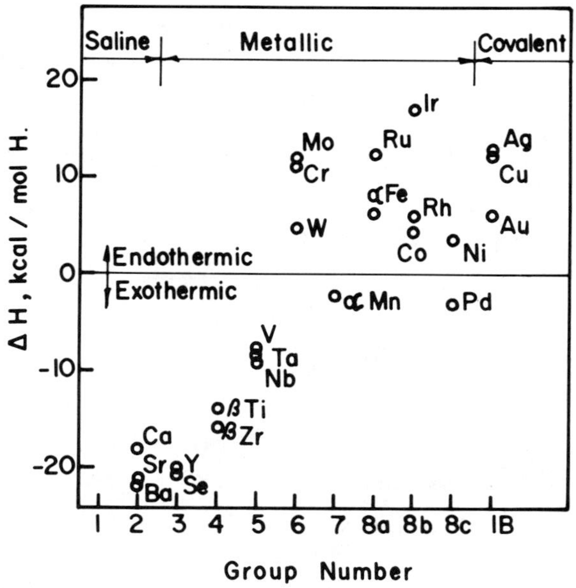

FIGURE 15. Heat of solution of hydrogen in metals.

kilogram, 0.01 ppm O_2). In a system where the only hydrogen source is corrosion, a percentage of the corrosion hydrogen is absorbed by zirconoium and its alloys. In reactor systems, the source of hydrogen is less certain, but the percentage of corrosion product hydrogen still is a useful reference.

It will be of interest to briefly refer to some of the fundamental aspects of hydrogen in metals.[28] The hydrogen atom consists of a single proton and a single electron. Being the only element in the first row of the periodic table before the inert gas helium, hydrogen acts as both the metal and nonmetal of this row. With one valence electron, ionization to H^+ is possible; alternatively, acceptance of an electron to complete the 1 s shell can also occur. These unique properties enable hydrogen to combine with atoms from all parts of the periodic table to form hydrides with characteristics ranging from saline (i.e., Na^+H^-) through metallic (i.e., PdH) to covalent (i.e., BeH_2). The simplest hydrogen compound is molecular hydrogen, H_2. Gaseous hydrogen at 1 atm pressure condenses at 20.2 K into a liquid, which in turn freezes at 13.9 K into a molecular solid. Hydrogen dissolves in metals in the atomic form according to the reaction $1/2H_2 \rightleftharpoons (H)_M$. With an equilibrium constant K related to the free energy of solution ΔG by $\Delta G = -RT\ln K$, we have, according to standard thermodynamic relations, $K = a_{(H)}/a_{(H2)}{}^{1/2}$, where a is the activity. Under conditions of dilute solution we may write $a_{(H)} = \gamma_H C_H$, where C_H is the concentration of hydrogen atoms in solution and γ_H is an activity coefficient. For a gas, the activity is identical to the fugacity, and so for conditions of near ideal gas behavior we have $a_{H2} = P_{H2}$. Thus, $C_H = \frac{K}{\gamma_H}(P_{H2})^{1/2}$, which is well known as Sievert's law. Experimental values of the heat of solution $\Delta H = \Delta G + T\Delta S$, where ΔS is the entropy of solution, are displayed in Figure 15. Two features of this figure are worth a reference. First, metals may be classified as either endothermic or exothermic occluders according to the sign of ΔH. The latter react with hydrogen at atmospheric pressure to form stable hydrides, having important effects on their mechanical properties. Second, there is a general trend of ΔH for the transition metal series to become more endothermic toward the middle of the series. The sign of ΔH is important to the temperature dependence of solubility. At a given hydrogen pressure, from the relations above we can deduce $d\ln C_H/dT = \Delta H/RT^2$. Hence, C_H either increases or decreases with increasing

Table 7
FACTORS DETERMINING HYDRIDE ORIENTATION

Factor	When effective	Hydride orientation
Grain shape	Dominant	Parallel to flattened grains
Texture	When grains are equiaxed	Perpendicular to basal poles
Stress	When grains are equiaxed and texture is indecisive	Perpendicular to tensile stresses

temperature according to whether ΔH is positive (endothermic) or negative (exothermic). However, for exothermic systems, hybride precipitation can occur with decreasing temperatures, and so the solubility in equilibrium with the hydride decreases. Because of its small size and mass, hydrogen can move more readily within metal lattices than other solutes.

Turning the focus to zirconium, it will be seen from Figure 15 that zirconium is typical among the exothermic occluders of hydrogen. Unalloyed zirconium and zirconium-rich alloys take up to 450 ppm hydrogen in solid solution at temperatures of 500°C; the solubility decreases rapidly with decreasing temperature, becoming 65 ppm hydrogen at 300°C and perhaps 0.05 ppm hydrogen at 20°C. In the fabricated condition, zirconium alloy structural components in the nuclear reactor core contain about 10 ppm hydrogen. During reactor service, additional hydrogen from the corrosion process can diffuse into the tubes, and when present in quantities above the solubility limit, it precipitates as a brittle, solid zirconium hydride. The hydride forms either as single platelets or as elongated groups of platelets. These hydride platelets will have an embrittling effect. The severity of hydride embrittlement depends not only on the volume fraction of hydride present, but also on its distribution and morphology, particularly the orientation of platelets with respect to any residual or applied stress. Oriented hydrides in zircaloy fuel cladding tubes may have a deleterious effect on the in-reactor service of the tube. They may be performance limiting. When the hydride platelets are parallel to the radial direction of the tube, a drastic decrease of ductility may occur. As a result, studies of hydride orientation are of major importance.

Texture is considered important to hydride orientation, since there are habit planes for hydride precipitation and preferential precipitation may occur on some grain boundaries. However, taking total stock, it may be said that the main shape, texture, and stresses are factors that determine the hydride orientation, as shown in Table 7. Grain shape is the dominant factor. When the grains are equiaxed, the texture is the determining factor. When the grains are equiaxed and the texture is of an indecisive type, possible stresses define the hydride orientation. In longitudinal specimens of tubing, zirconium hydrides oriented parallel to the tensile axis have little effect on the room temperature properties, but in transverse specimens with hydrides normal to the ultimate tensile strength (UTS) and ductility are considerably reduced compared with hydrogen-free specimens. The two types of hydride orientation that are found in nuclear reactor pressure tubing are circumferential and radial hydrides, as shown in Figure 16. The radial orientation of hydrides causes a marked decrease in the circumferential ductility and can provide a ready crack path under a loop stress. Zirconium alloy tubular products used in power reactors today are fabricated under conditions controlled so that hydrides in the tube are circumferentially oriented. In evaluating effects of the orientation of the hydride platelets, it is usually necessary to have some quantitative measure of the degree of orientation. One of the most commonly used parameters is f_x, which is defined as a percentage of hydrides whose traces in the plant of observation make an angle with a reference direction in the range $x \rightarrow 90°$. The general principle of defining f_x is illustrated schematically in Figure 17. Defined in this way, a high value of f_x gives a very low tensile ductility.

There are three main factors which alone or in combination affect hydride orientation: (1)

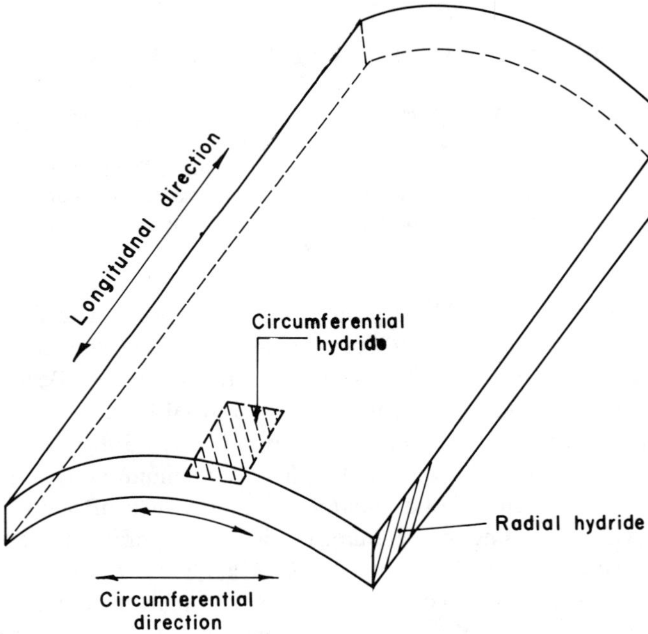

FIGURE 16. Segment of pressure tube showing orientation of circumferential and radial hydride platelets relative to principal directions in the tube.

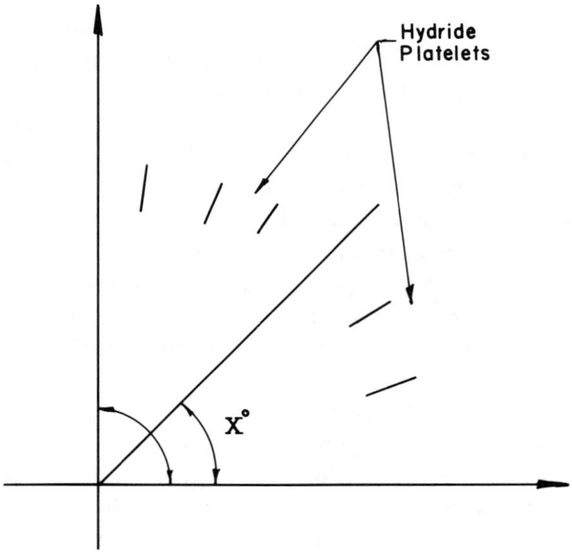

FIGURE 17. Quantitative description of hydride orientation.

crystallographic factors (it has been generally observed that within given grains a few habit planes are preferred notably $\{10\bar{1}0\}$ and $\{10\bar{1}7\}$); (2) prior strain (hydride platelets tend to form parallel to the direction of drawing of the wire); and (3) stress (hydride platelets tend to become aligned perpendicular to the tensile stress axis). For most reactor applications interest is in the hydride orientation in tubular products, mainly fuel cladding and pressure tubing. Since the principal stress direction in these tubes is circumferential, radial hydrides are undesirable and the designer aims to have any hydride present in the circumferential planes. To do this, it is first necessary to ensure that the hydride platelets in the absence of an applied stress are circumferentially oriented (strain orientation) and then to minimize any reorientation under stress (stress orientation). In practice it has been found easier to achieve the desired strain orientation than the desired stress orientation. It is now well established that the desired strain orientation is achieved by having a large amount of compressive strain in the radial direction in the final stages of the fabrication process.

E. Performance

1. Creep Strength

A good creep strength, in the real conditions of use, is needed for zircaloy cladding of PWR and BWR fuel elements. In PWRs the cladding tubes are subjected to creep stress, demanding high creep strength. During power cycling and during the fuel lifetime, either the outer reactor pressure or the inner fission gas or fuel swelling pressure dominates. This makes the transverse creep strength in the tangential direction most important. Since creep is highly temperature sensitive, possible temperature peaks contribute significantly to the creep strain. This makes the creep strength at a temperature as high as 400°C an interesting matter.

Most available zircaloy creep data originate from the Canadian development of pressure tubes. The service temperature of these tubes is up to 300°C. The creep strength at this temperature has been found to depend on metallurgical structure (cold work and annealing) and texture in roughly the same manner as the short-time strength. Neutron irradiation greatly influences the 300°C creep strength by enhancing creep rate at low creep stress and retarding creep rate at high creep stress.

In contrast to this, the creep strength at 400°C does not depend on structure and texture in the same manner as the short-time strength from 20 to 400°C and the creep strength at 300°C. In addition, the neutron irradiation effect is considered to be small at 400°C because the irradiation defects anneal out rapidly.

2. Pellet-Clad Interaction[29-40]

A special form of interaction between the fuel and the cladding under certain circumstances occurs in fuel rods with zircaloy cladding. In a power reactor, the fuel pellets are inserted into the cladding tube with some filling-in clearance. The fuel pellets of an unirradiated fuel rod that is gradually brought to full power start off as perfect cylinders, but expand thermally and distort to hourglass-shaped bodies with convex ends, and a typical crack pattern forms. The cracking may shift outward, a phenomenon called relocation. A differential thermal expansion between the UO_2 fuel pellets and the metal cladding arises because of the steep temperature gradient through the pellet and the temperature difference between pellet and cladding. With increasing burnup due to all these effects, the gap between the fuel and cladding will narrow and will ultimately disappear, giving rise to mechanical interaction between fuel and clad. The fuel tube takes a bamboo-like appearance, as shown in Figure 18.

The stresses on the cladding are highest in the vicinity of the pellet-to-pellet interfaces (where the pellet diameter is greatest). The effect can be mitigated, for example, by using shorter or barrel-shaped pellets, chamfered edges, dished ends, different clearances, etc., but all these devices have corresponding performance penalties from some other standpoint, such as increased manufacturing costs, reduction in uranium inventory per unit run of the fuel tube, or loss of heat output due to impairment of heat transfer. During a power ramp after a significant fuel burnup, the fission products are released (after substantial burnup at low linear power there is a significant inventory of fission products held within the ceramic fuel pellets), particularly at pellet cracks and pellet-to-pellet interfaces. With the simultaneous

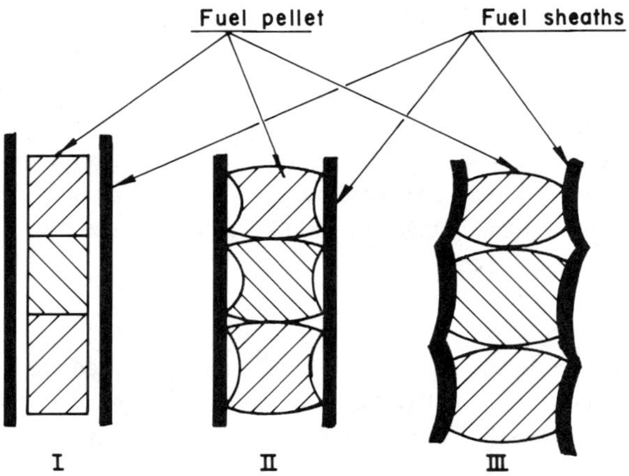

FIGURE 18. Interaction between fuel and cladding: the bamboo effect (manifested through an increase in sheath length and diameter and in the formation of ridges at pellet-pellet interfaces).

and synergistic effects of stress on the cladding (a result of mechanical interaction) and the release of fission product species, the zircaloy cladding can fracture by a stress corrosion cracking (SCC) in the pellet-to-pellet interface region. Although the chemical species responsible for SCC have not been positively identified, iodine is thought to be one of the most likely candidates. The exposure of fission products to the stressed region can be called pellet-clad chemical interaction (PCCI). The SCC failure is thus due to pellet-clad interaction (PCI), which can be defined as the combined effect of pellet-clad mechanical interaction (PCMI) and PCCI. In Figure 19 these aspects of PCI are illustrated schematically. The principal mechanisms of PCI are shown in Figure 20. The SCC/PCI must be emphasized as serious. It causes the cladding to lose its ability to retain the radioactive fission products.

The propensity to PCI failure is a generic feature of all zircaloy-clad fuel. It may, however, be mentioned that this failure is attributed to severe local power ramps, defined as a first and fast power increase after extended exposure accummulated at low power (previous steady state). The power cycling — defined as repeated changes on a daily or weekly basis without exceeding the previously reached limits of the power level — to which the rod has been conditioned does not lead to PCI failure. Some facts about iodine, the species that has been identified as responsible for bringing PCCI, can be given. About 15% of the total amount of isotopes formed by fission of ^{235}U are iodine isotopes. Only one iodine isotope, ^{129}I, is stable. Its half-life is 1.6×10^7 years, while others have relatively short half-lives of 8 d or less. About 0.8% of the total amount of fission products consists of ^{129}I. The potential threat of iodine to zircaloy cladding arises because it forms volatile compounds with zirconium at the LWR working temperature, about 300°C. Not all the iodine seems to be available for reaction with the zircaloy cladding. In the first place, it should be mentioned that most of the iodine is trapped in the UO_2 pellets, because the UO_2 temperature is too low for iodine to diffuse away to the surface in any large quantity. Second, iodine reacts with other fission products, for example, with cesium, forming very stable compounds. During the power ramp, fuel fragments pressing into the cladding cause a state of stress which, combined with the iodine released when the fuel temperature rises, leads to SCC, as has been stated.

For avoiding interaction failures, two methods can be considered: (1) developing clad materials that have sufficient ductility even at high burnups and (2) designing and operating the fuel so that excessive interaction at high burnup is avoided. The first method would definitely be the most convenient, but a large number of reactors are committed to zircaloy-

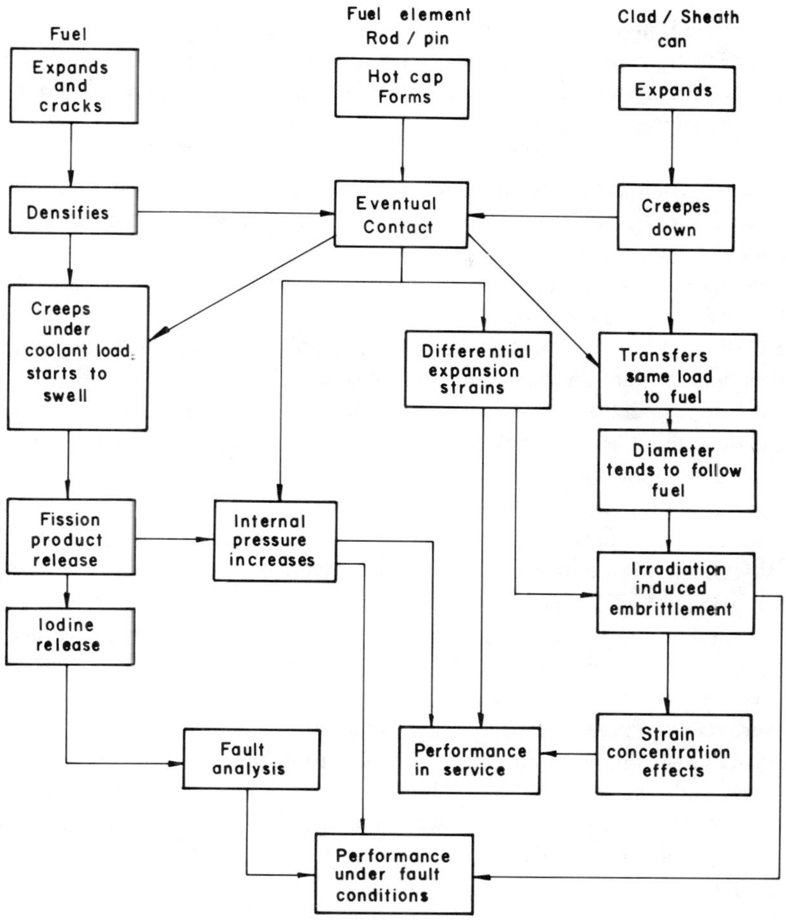

FIGURE 19. Representation of the major fuel element interactions.

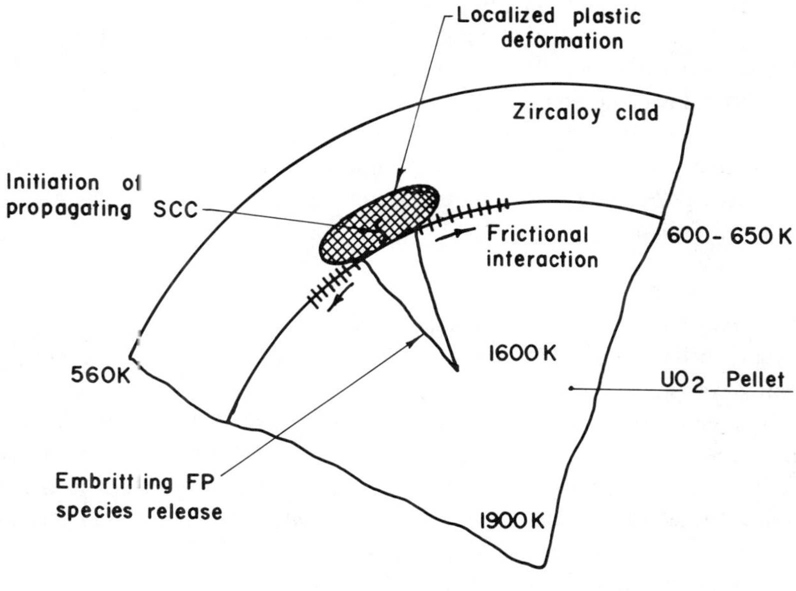

FIGURE 20. PCI failure mechanism.

2 or zircaloy-4 clad, and no large improvement in these alloys is expected. Moreover, there does not seem to be any serious competitor with the zirconium alloys as clad material. It therefore turns out that the approach to avoiding interaction failures is confined to the second method. The following are the measures taken or proposed:

1. Fuel design change has been introduced. It includes shorter chamfered pellets to reduce local clad stress. Fuel assemblies have been rearranged to contain more fuel rods of smaller diameter, which lower the power levels and peak temperatures per rod.
2. Special start-up procedures have been introduced to minimize mechanical interaction of the pellet and its cladding.
3. Use of the so-called barrier fuel has been employed. The inside cladding is coated with copper or pure zirconium to prevent attack of the zircaloy itself.
4. Control of temperature transients in the fuel during operation has been used to limit stress produced when the fuel contacts the cladding. This is accomplished by limiting the heat generation rate during operation and the rate of power increase during reactor start-up in PWRs.
5. Pressurizing the fuel rods during fabrication (helium gas at 20 to 25 atm is often used in PWRs) extends the time before the cladding creeps down into contact with the fuel, thus minimizing PCI. By reducing the amount of possible fuel-clad interaction, the stresses and strains to which the cladding will be subjected are reduced and reliability is improved. The improvement stems from two sources: delayed fuel-clad contact, and reduced cyclic stresses and strains after contact. Even a moderate level of internal pressure results in a manifold increase in the time to establish contact. Prepressurizing of the fuel also improves the thermal conductivity, resulting in lower operating temperatures and less fuel swelling. The fuels for BWRs, on the other hand, are not as subject to creep because of the lower system pressure. Only about 3 atm of helium are required inside the fuel in order to counteract the system pressure and to prevent creep-down of the cladding onto the pellet.
6. Coating the interior surface of zircaloy cladding tube with graphite is done in CANDU fuel rods (CANLUB fuel). The graphite serves as a protective layer presenting a physical barrier to the direct contact of iodine with the zircaloy and lubrication to bring a salutory effect to the frictional interaction.
7. Special pellet shapes substantially different from conventional designs have been used. One such pellet design is the annular pellet (the central portion of the fuel pellet is hollow). Compared with the standard pellet, the annular fuel pellet provides lower temperature, lower stored heat, lower fission gas release, improved resistance to pellet-clad interaction, and improved water-to-fuel ratio. Use of annular pellets does not require design alterations to the reactor and fuel assembly. The benefits cited for the annular fuel pellets have provided incentives for the development and exploitation of such a fuel design.

3. *Irradiation Growth*[41]

Irradiation-induced, stress-free deformation, called the irradiation growth of zircaloy, is of considerable technological importance. It can result in distortion of fuel assemblies and core structural components. Pressure tube elongation in CANDU reactor occurs as a result of irradiation creep and growth, with growth as the contributor, and may limit the useful lifetime of the pressure tubes. In general, growth continuously increases with fluence, growth magnitude is a function of alloy content, and growth magnitude increases as the irradiation temperature increases. The metallurgical parameters which influence elongation rate have been identified as crystallographic texture, grain shape, and dislocation density. The crystallographic texture is primarily controlled by the extrusion process, whereas the grain shape and dislocation density depend on the postextrusion thermomechanical processes.

F. Summary

Metals intended for use inside the reactor core — for fuel cladding, for example, or for pressure tubes in CANDUs — must possess low thermal neutron-absorption cross sections. The four potentially useful low-absorption metals are aluminum, beryllium, magnesium, and zirconium. Magnesium and aluminum have low melting points which restrict their utility to modest temperatures, as in Magnox. Beryllium is brittle, difficult to fabricate, and not at all cheap. Zirconium, therefore, is the most attractive: it is reasonably abundant and has good corrosion resistance and a high melting point.

The principal commercial source for industrial production of zirconium is zircon, from which, conventionally, the metal is extracted by caustic fusion-solvent extraction (using organic extractant, either TBP or MIBK) — the Kroll reduction route. The pyrochemical process involving separation of zirconium from hafnium by extractive distillation of chloride intermediate and the use of TNOA as the organic extractant in sulfuric-acid-bearing aqueous feed of zirconium and hafnium for achieving separation between the two are the two newly introduced processes in the extractive metallurgy of zirconium.

Zirconium is used in reactors only in alloyed form. There are corrosion problems associated with use of the pure metal. The present applications of zirconium alloy demand (1) low capture cross section for thermal neutrons, (2) adequate corrosion resistance, (3) good strength and ductility at service conditions, and (4) good formability to enable production of high-integrity tubes with close tolerances. The most widely used material meeting these requirements is zircaloy, which is a Zr-1.5% Sn alloy, the composition of the two important variants, zircaloy-2 and zircaloy-4, differing slightly in the amounts of minor elements. The absence of nickel from zircaloy-4 decreases the uptake of the alloy of hydrogen, and this is the only difference in their properties. The zirconium alloys, the zircaloys, and the other alloys contained with niobium (described later) have emerged as principal materials for nuclear fuel cladding, fuel channels, and pressure tubes for aqueous power reactors. The characteristics mentioned for zircaloy-4 have qualified it to become the standard alloy for PWRs. The BWRs make use of zircaloy-2. Zircaloy-2 is the most widely used pressure tube material in CANDU-PHW and CANDU-BLW reactor systems.

The following generalizations summarize the main factors of unirradiated corrosion of zirconium and its alloys. The corrosion film consists of a single oxide composition: ZrO_2. The protective oxide is hypostoichiometric in oxygen. The stoichiometric oxide tends to lose protective character. The regimes for zirconium alloy corrosion have been regarded as pre- and posttransition, where transition denotes a point on the kinetic curve where oxide loses its protective character. The process is gradual. Formation of protective, pretransition oxides is described by the kinetic equation $W = K_1t^{1/n}$. Posttransition is often described as linear: $W = K_2t$.

A special feature of zirconium-based alloys, due solely to the hcp structure of zirconium, is the formation of preferred orientation textures depending on the fabrication process and forming conditions, for instance, by hot forming in the alpha range or by cold working. It is well known that these textures have an influence on many properties, for example, hydride orientation, mechanical properties, and SCC.

In contact with the coolants, such as in a pressurized or boiling water reactor, zirconium alloy surfaces are subjected to oxidation and hydriding phenomena. The oxidation process can become very rapid when the temperature increases. For instance, about 17% of a typical fuel cladding will be oxidized within 1 h at a temperature of 1100°C, and film boiling on the surface of a fuel rod can cause oxygen embrittlement of zircaloy-4 in the temperature range of 800 to 1100°C. Therefore, the usual design goal is to keep the cladding temperature below 550°C. Hydrogen has very limited solubility in zirconium, and consequently hydrogen picked up at higher temperatures will precipitate as hydride at lower temperatures. The metal-water reaction and the radiolytic decomposition of coolant are the sources of hydrogen.

There exists a pronounced texture effect on hydride orientation. Under normal circumstances most of the hydride platelets would be aligned parallel to the circumferential direction. The least desirable alignment would be in the radial direction.

The mechanical properties of zircaloy tubing for fuel cladding play an important role in nuclear fuel designs. The mechanical properties of zircaloy tubing can be influenced by such parameters as texture, oxygen content, and Q value (defined as the ratio of the wall thickness reduction to the diameter reduction during fabrication). The Q value has a significant effect on texture, hydride orientation, and deformation behavior.

The PCI mechanism typically occurs midway to late in fuel when mechanical pellet-clad contact is established and sufficient fission products are available to initiate SCC. It has been clearly established that PCI failures are not caused by power cycles within the previous steady-state power level. They are always correlated to power ramps, i.e., the first severe (fast and large) power increase beyond the steady state. A number of remedial measures as given in the text have been taken to counteract the SCC. All measures have in principal been based on delaying the onset of and reducing or minimizing the severity of mechanical interaction.

Irradiation of zirconium alloys increases strength and reduces ductility in a fashion similar to many metals and alloys. The most significant difference is the nearly complete loss of uniform elongation in cold-worked and irradiated zirconium alloys. Irradiation may increase or decrease creep rates, depending on neutron spectrum, flux, temperature, and prior fabrication history. However, the irradiation-induced effect on creep rate is sufficiently small.

V. NIOBIUM

A. Extractive Metallurgy[42-50]

The principal commercial sources of niobium are (1) niobite-tantalite, $(Fe,Mn)(Ta,Nb)_2O_6$; (2) pyrochlore, $NaCaNb_2O_6F$; and (3) niobium-tantalum-bearing cassiterite deposits. Just as in the case of zirconium and hafnium, niobium and tantalum are rarely found apart in nature and never in the free state. The element niobium amounts to 20 ppm in crustal abundance. In comparison, tantalum is a scarcer metal with abundance of only one tenth of that of niobium.

The extraction starts with chemical processing of the resource separating niobium from tantalum and recovering suitable intermediates of both; finally a chemical reduction step liberates the metal. Figure 21 shows the various known processes for reduction to niobium. A schematic flow diagram for extractive metallurgy of niobium is shown in Figure 22. The schemes shown with continuous lines have been commercially followed, whereas those shown with broken lines have been investigated either on the laboratory or pilot plant scale or to various stages of development. Niobium resources have been treated by a variety of processes to yield finally either niobium metal or ferroniobium. All the niobium metal produced originates from niobite or niobite-tantalite concentrate, whereas pyrochlore is mainly used for the production of ferroniobium. Present resource processing technology makes use mainly of digestion with hydrofluoric acid to open the ore, followed by solvent extraction with MIBK or TBP extractant for the separation of niobium and tantalum. The solvent extraction process efficiently recovers both niobium and tantalum as separate streams which are further chemically processed individually into oxides or double fluorides with potassium. The fractional crystallization route, the process originally developed for niobium-tantalum separation, has been displaced by the solvent extraction process. Niobium metal is prepared from the oxide by carbothermic or aluminothermic reduction. Other processes, such as chlorination for opening the ores, particularly of lower-grade types; separation of chlorides of niobium and tantalum; and reduction of the separated chloride of niobium with hydrogen, have been applied on an expanding scale, though not so far on a commercial

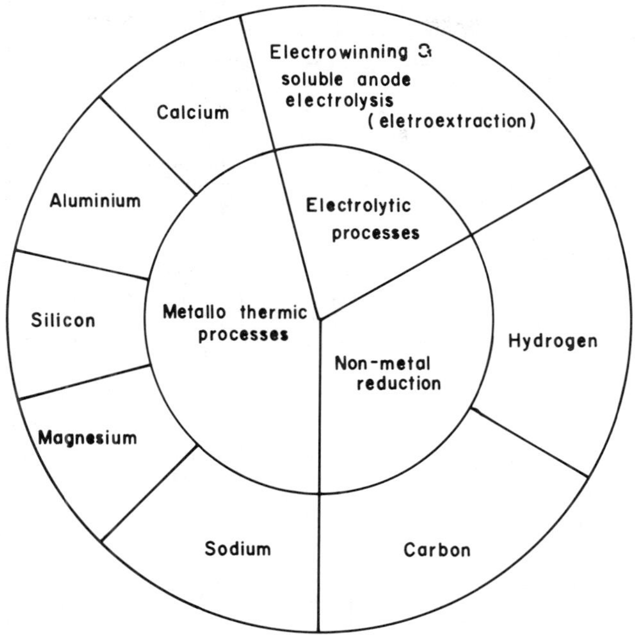

FIGURE 21. Reduction processes for preparation of niobium metal.

scale. Figure 23 presents a flowchart of niobium processing based on aluminothermic reduction starting from columbite-tantalite resource concentrate.

B. Properties[51-56]

The properties of niobium are given in Table 8. The metal is endowed with a unique combination of properties. It has a high melting point and a density lower than three of the other most important refractory metals: tantalum, molybdenum, and tungsten. Perhaps the most important property of niobium is its low thermal neutron capture cross section, which is lower than that of all other structural metals barring aluminum, magnesium, zirconium, and beryllium, which, on the other hand, have lower melting points than niobium. Niobium is considered quite corrosion resistant to most of the mineral or organic acids at ordinary temperatures. More significant is its corrosion resistance to liquid alkali metals. Though the refractory metals mentioned above also exhibit satisfactory corrosion resistance to liquid metals, niobium is preferred for nuclear applications because of its low cross section and good radiation stability. Among the refractory metals, except hafnium, niobium has the lowest work function. From the point of view of high-temperature structural applications, in spite of its high melting point and low density, the modulus of elasticity of niobium is low and decreases rapidly with temperature. The metal is too soft for structural use at about 600°C. Besides, the oxidation resistance of pure niobium is also not satisfactory. The low modulus of elasticity of niobium can be an advantage in certain applications. The combination of a low coefficient of thermal expansion, a high thermal conductivity, and a low modulus of elasticity gives niobium excellent resistance to thermal stress. Niobium is a good conductor of heat — about three times better than zirconium — and electricity — the resistivity is only eight times that of copper.

Niobium is so ductile that it is always fabricated at room temperature. It work-hardens more slowly than tungsten or molybdenum, and resists tearing and splitting more than other metals when subjected to deep-drawing.

The superconducting transition temperature is high, and its alloys with titanium or tin

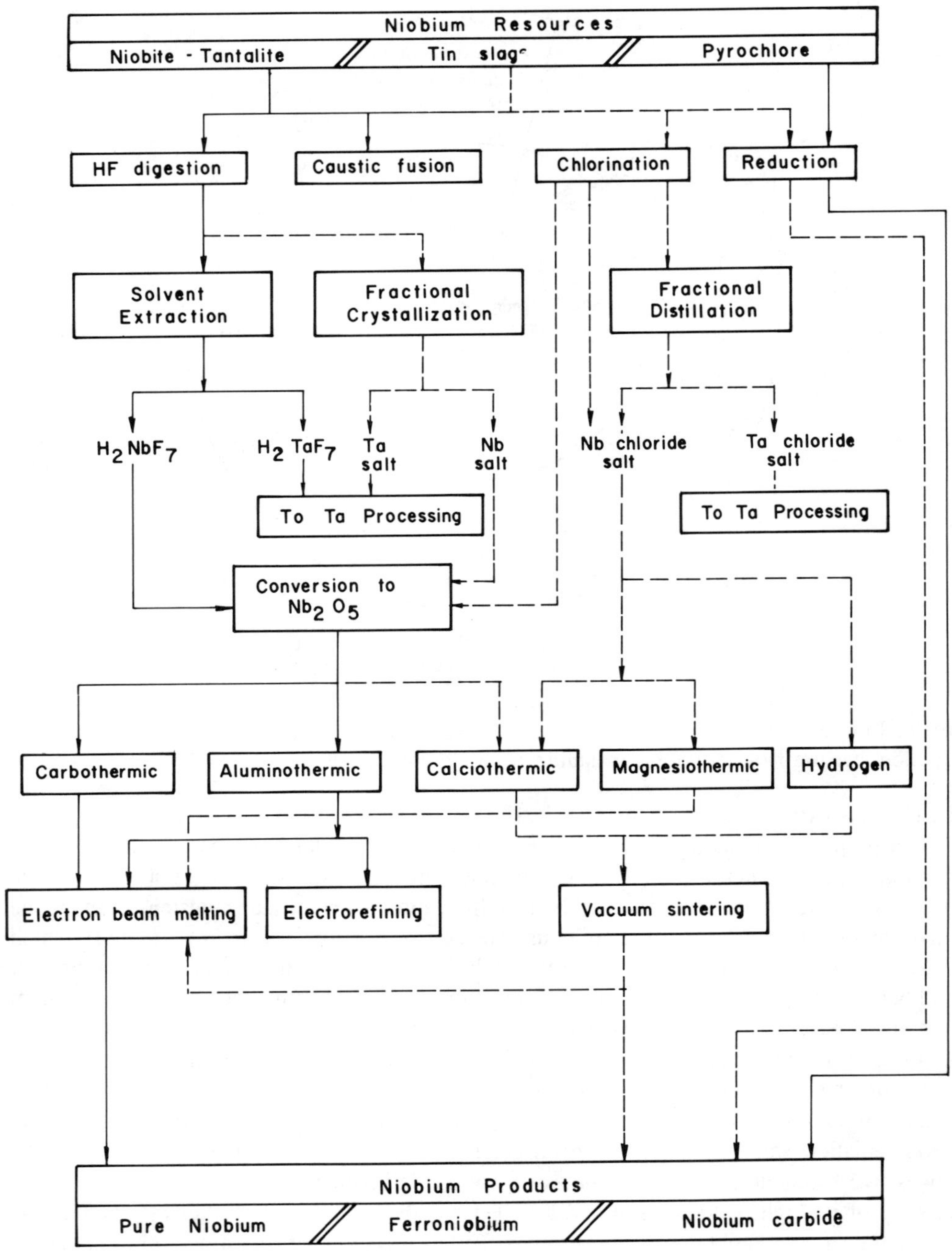

FIGURE 22. Schematic flow diagram for extractive metallurgy of niobium.

exhibit higher critical constants. Alloy development programs on niobium have been aiming at offsetting its limitations of strength and oxidation resistance.

C. Applications

The historical record of the development of zirconium for cladding and structural material in nuclear reactors shows that the development of zirconium-based alloys proceeded along two major paths. One development involved a zirconium-tin system and led to the devel-

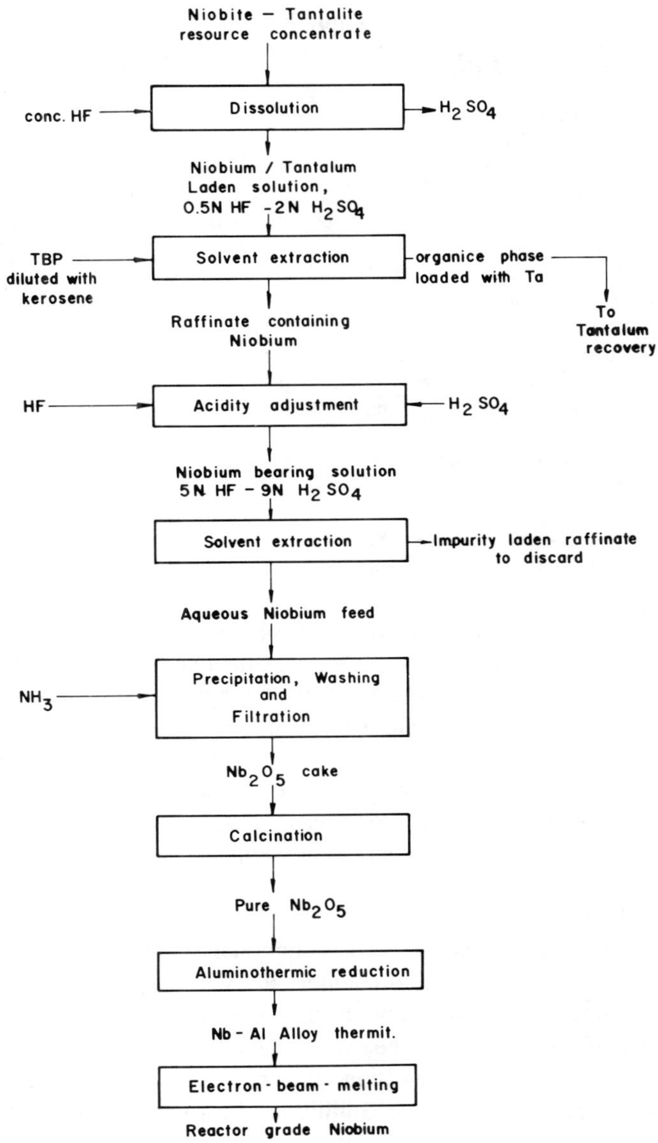

FIGURE 23. Processing route of niobite-tantalite resource concentrate to reactor-grade niobium.

opment of zircaloys, whereas the other concentrated upon zirconium-niobium materials and produced two major alloys of this system in use today: Zr-1Nb and Zr-2.5Nb. The Zr-Nb alloys, in comparison to zircaloys, are stronger, display better creep resistance properties, and show a lower tendency to pick up hydrogen. Because of their relatively high strength, the thinner-walled pressure tubes possible with niobium containing zirconium-based alloys result in desirable neutron economy.

Pressure tubes in the following alloys and metallurgical conditions are now in service in CANDU power reactors: Zr-2.5Nb, quenched and aged, and Zr-2.5Nb, cold-worked, are the current reference tube materials (Table 9). Tubes from a new alloy, Excel,[57] are now being developed. The addition of copper to the Zr-2.5Nb alloy improves the corrosion resistance in air and carbon dioxide.[58] Because of this property, the Zr-2.5Nb-0.5Cu alloy

Table 8
PROPERTIES OF NIOBIUM

Atomic number	41
Atomic weight	92.91
Crystal structure	bcc
Work function (eV)	4.0
Thermal neutron capture cross section (b/atom)	1.15
Transition temp (K)	9.25
Density (g/cm^3)	8.57
Melting point (°C)	2468 ± 10
Boiling point (°C)	4827
Coeff of linear thermal expansion (cm/cm/°C)	7.31×10^{-6}
Thermal conductivity (cal/cm^2/cm/s/°C) at 0°C	0.125
Electrical resistivity ($\mu\Omega$ cm) at 0°C	12.5
Tensile strength (MPa)	170—205
Yield strength (MPa)	75—105
Elongation (%)	30—35
Hardness (HV)	40—80
Elastic modulus tension (GPa)	105
Fatigue strength (MPa, 1 million cycles)	140

is used as a spacer in the annulus between pressure and calandria tubes in all CANDU reactors. The copper addition is known to have important effects on the physical metallurgy of Zr-2.5Nb alloy.[59] The fuel channel tubes of the Leningrad, Kursk, and Chernobyl atomic power solutions are made of Zr-2.5Nb alloys. Their fuel element claddings employ Zr-1Nb. The use of Zr-1.0Nb seems to be confined exclusively to fuel cladding in Russian reactors, of which the icebreaker 'Lenin' is perhaps the best publicized. Zirconium-based niobium alloy systems along with their application areas are summarized in Table 10. Figure 24 shows the use of zirconium-based alloys in core components of the CANDU reactor.

D. Physical Metallurgy

Next to the zircaloys, Zr-2.5Nb is the most important commercial alloy. Niobium is a strong stabilizer of the beta phase in zirconium alloys. During quenching from the cubic beta phase (above about 927°C) the alloy undergoes a martensitic transformation to a hexagonal alpha phase which, when aged below the monotectoid temperature (the addition of Nb to Zr depresses the beta-alpha transition temperature to a monotectoid at about 600°C and 17% Nb), forms precipitates of the equilibrium beta niobium phase containing about 85% niobium. If the alloy is slowly cooled through the alpha- plus beta-phase field (927 to 602°C), an equilibrium alpha phase and a metastable beta-zirconium phase are formed. The beta phase, which contains about 20% Nb, is stable at room temperature, but when heated in the temperature range of 302 to 500°C transforms initially to metastable omega and niobium-enriched beta phase. Continued heating transforms the structure to the equilibrium phases alpha zirconium and beta niobium. Thus, in terms of metallurgical structure, the Zr-2.5Nb, is a two-phase material: hexagonal alpha grains and niobium-rich beta phase at the grain boundaries. The structure of zircaloy-2 consists of alpha grains and uniformly distributed intermetallic particles of 1 μm diameter.

E. Fabrication[60]

Zirconium-based niobium, the Zr-2.5Nb, for application as pressure tube material (for CANDUs) is used either in heat-treated (quenched and aged) or in cold-worked condition. Two fabrication routes that have been used commercially to produce reactor pressure tubes are illustrated in Figure 25.

As shown in the figure, in practice the heat-treated pressure tube is extruded, water-

Table 9
CANDU POWER REACTORS USING PRESSURE TUBES OF Zr-2.5Nb

Reactor	Power output (MWe)	Metallurgical condition
Gentilly 1	250	Quenched and aged
Pickering 3 and 4	500 × 2	Cold-worked
Kanupp	125	Quenched and aged
Bruce 1—4	750 × 4	Cold-worked
Gentilly 2	600	Cold-worked
Rio Tercero	600	Cold-worked

quenched from solution treatment in the alpha plus beta phase, and then aged at 500°C, generally receiving some cold-working between the quench and age. As also shown in the figure, for the cold-worked condition, the tube is extruded and cooled fairly rapidly, then cold-drawn (typically 25%).

Fuel sheathing may be either cold-worked or annealed. One common annealing treatment is to cool at a rate of about 1 K/s from a solution temperature of about 800°C. Finally (in fuel sheathing), an additional complication of the metallurgical condition of Zr-2.5Nb arises from welding. The weld structure and some of the heat-affected zone (HAZ) has of necessity been cooled from above the beta transus; thus, the phase distribution is dependent both on the metallurgical condition prior to welding and the cooling rate below the beta transus.

Zirconium-based niobium spacer alloy in use in CANDUs is produced by quenching the alloy from the alpha plus beta phase and then overaged at about 545°C. This metallurgical condition gives the same combination of strength and ductility as the standard quenched and aged condition of the Zr-2.5Nb alloy. In this condition the spacer alloy retains excellent fracture ductility after irradiation at temperatures between 20 and 325°C, while undergoing moderate increments in strength.

F. Structure-Property Correlation

The process of making Zr-2.5Nb pressure tubes by extruding preheated hollow billets into tubes followed by cold-working to size, as described, develops a two-phase microstructure of strongly textured alpha grains and a grain boundary network of beta phase. The extrusion process determines the texture and microstructure of the finished tubes, provided that they are not treated above 600°C at any stage during fabrication. Tubes extruded by different manufacturers can have different textures and microstructures, and therefore different mechanical properties. The texture usually present in extruded tubes is that with the majority of alpha grains oriented with their basal plane normals close to the circumferential direction. However, in some tubes the majority of grains are oriented with their basal plane normals in the extrusion direction. Factors which can affect the texture and microstructure of extruded tubes are composition of the alloy, billet microstructure, temperature during extrusion, deformation mechanisms operative in the alpha grains during extrusion, die design, lubrication, and cooling rate after extrusion.

Zirconium alloys commonly pick up D_2 from corrosion in D_2O, and H_2 from corrosion in H_2O. The presence of H_2 (D_2) in excess of the solubility limit results in the formation of hydride platelets, as has been extensively described in the case of zircaloy. After reactor service for 10 years, heat-treated Zr-2.5Nb is expected to pick up H_2 (D_2) to a total concentration of 110 ppm in a 0.095-in. walled tube. Thus, some platelet precipitation should occur in heat-treated Zr-2.5Nb at operating temperatures. The presence of hydride (deuteride) platelets can lower the low-temperature ductility and raise the ductile-brittle transition temperature. Hydrides (deuterides) have some effect on the mechanical properties of the zir-

Table 10
NOMINAL COMPOSITION OF ZIRCONIUM-BASED NIOBIUM ALLOYS

Alloy designation	Chief alloying elements (wt%)								Applications
	Sn	Fe	Cr	Ni	Cu	Nb	Mo	O	
Zr-1Nb	—	—	—	—	—	1.0	—	—	Fuel cladding (Russian reactors)
Zr-2.5Nb	0.001 max	0.08	0.01	0.003	—	2.5	—	900—1300 ppm	Pressure tubes (CANDUs), fuel channels (Russian reactors)
Zr-2.5Nb-0.5Cu	—	—	—	—	0.5	2.5	—	900—1300 ppm	Garter springs or spacers (CANDUs)
Ozhennite 0.5	0.2	0.1	0.02 max	0.01	—	0.1	—	—	Russian experimental alloy
Excel	3.3	—	—	—	—	0.8	0.8	—	Pressure tube material under development

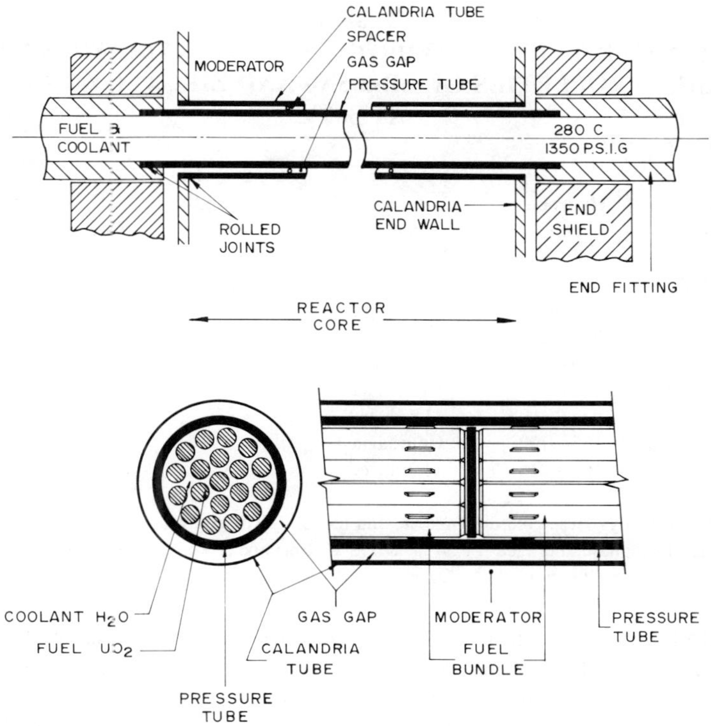

FIGURE 24. Core components of a CANDU reactor.

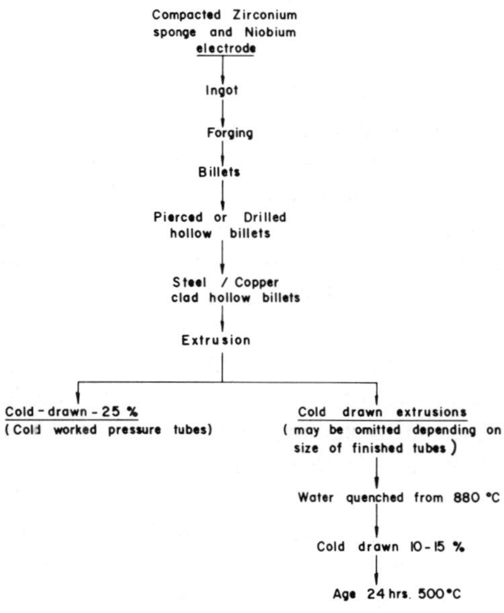

FIGURE 25. Fabrication routes for Zr-2.5Nb pressure tubes.

conium alloy pressure tubes regardless of orientation, but, from a ductility point of view, hydrides (deuterides) located perpendicularly to the major stress direction are the most serious. It thus becomes desirable to establish the precipitation characteristics of hydrides (deuterides) likely to occur in Zr-2.5Nb during reactor service and to ascertain whether or not hydrogen (deuterium) seriously affects the mechanical properties of the material. Apart

Table 11
COMBINED EFFECT OF HYDROGEN AND IRRADIATION ON THE TENSILE PROPERTIES OF ZR-2.5Nb AND ZIRCALOY-2 IN TESTS AT 300°C

Material	Hydrogen conc	Irradiation	0.2% yield strength (MPa)	UTS (MPa)	Total elongation (%)
Zr-2.5Nb	0	Unirradiated	547	620	14.8
	0	Irradiated	789	812	9.5
	220—290	Unirradiated	570	645	13.2
	250—320	Irradiated	890	820	5.5
Zircaloy-2	0	Unirradiated	291	329	12.8
	0	Irradiated	439	439	5.8
	136—190	Unirradiated	316	356	18.0
	162—216	Irradiated	453	453	8.8

Note: The zircaloy-2 was cold-drawn 20%, and the Zr-2.5Nb was water-quenched from the alpha plus beta phase and aged for 54 hr at 773 K. Specimens were irradiated at about 553 K to about 2.7 × 10^{24} n/m² (E > 1.0 MeV).

from the hydriding (deuteriding), it is also necessary to take into account the irradiation influence on Zr-2.5Nb for the purpose of its reactor service performance evaluation as a pressure tube material. Table 11 indicatively shows, for the purpose of illustration, the combined effect of the hydrogen and irradiation on the tensile properties of Zr-2.5Nb. For the purpose of comparison the data[61] for zircaloy-2 have been included in Table 11.

VI. STAINLESS STEEL[62-65]

A. General

Stainless steel represents a group of steels that contain at least 10.5% chromium, the minimum level required to form the passive film that characterizes their corrosion resistance. Other alloy elements contained in stainless steels improve their corrosion resistance and impart specific mechanical and physical properties, weldability, and such other properties required by a wide variety of applications. The passive behavior of stainless steels is due to the presence of an oxide film which protects the metal surface from attack. The film is self-healing and reforms instantly in oxidizing conditions, including even such mildly oxidizing environments as aerated water. It is not stable in reducing acids such as hydrochloric acid or in the presence of significant amounts of chloride ion, although the presence of an oxidizing agent such as nitric acid restores passivity at low chloride ion concentrations.

Of the alloy constituents present in stainless steel, it is the chromium which dominates in terms of effect on passive behavior. Steels with less than 12 to 14% chromium in practical application show borderline passive behavior. The chromium is present in the passive oxide film as a mixed oxide of uncertain composition but enriched somewhat in chromium compared to bulk alloy composition. It may be in the form of spinel, such as $Fe(Cr,Fe)_2O_4$, but all that is certain is that chromium is in the 3+ valency state. Under very oxidizing conditions (e.g., boiling HNO_3 + Cr^{6+}) the chromium can be oxidized to a soluble 6+ valency state, and the protective oxide film is disrupted.

Chromium has a strong affinity for carbon (always present as an impurity in stainless steel) and can form a chromium-rich $Cr_{23}C_6$ carbide phase. If stainless steel containing more than about 0.03% carbon is heated (e.g., during welding) and cooled slowly through the range 440 to 650°C, it is possible for a continuous matrix of $Cr_{23}C_6$ to precipitate at boundaries between the grains in the alloy matrix. This results in depletion of the chromium on the surface of the grains, with a consequent decrease in their corrosion resistance, leading to enhanced corrosion between the grains and eventual grain dropping. This phenomenon is called sensitization, and stainless steel so affected is sensitized to intergranular corrosion. Sensitization can be overcome in three ways: (1) by reheating the steel to about 1100°C to redissolve any $Cr_{23}C_6$ and allow chromium to diffuse to depleted grain surfaces, followed by rapid cooling; (2) by adding a carbon scavenger like niobium; and (3) by manufacturing very pure stainless steel with a low carbon level. The latter two methods prevent the formation of $Cr_{23}C_6$ precipitates by reducing the availability of free carbon.

B. Types and Characteristics

Depending on their microstructure, stainless steels may be grouped into five families: austenitic, ferritic, duplex (austenitic/ferritic), martensitic, and precipitation hardened.

The austenitic stainless steels are the most commonly used (basic type 18% chromium and 8% nickel). They are resistant to intergranular corrosion in the quenched condition, in which these steels are supplied. Apart from nickel, other alloying elements such as manganese and nitrogen are used in these steels to stabilize the austenitic crystal structure. Austenitic steels are distinctive in having very great toughness, excellent ductility, and a high rate of hardening by cold work. These stainless steels are generally nonmagnetic, although some of the grades become weakly magnetic after cold work. The austenitic structure provides for a somewhat lower thermal conductivity and higher thermal expansion than others in the stainless family. AISI type 304 is by far the most commonly produced and used stainless steel in all kinds of applications. It has good corrosion resistance in a wide range of moderately aggressive chemical environments, and resists pitting and crevice attack in most fresh waters at moderate temperature. It is readily weldable and provides excellent mechanical properties. When higher corrosion resistance is required, the next alloy typically considered is AISI type 316, in which 2 to 3% molybdenum has been added, along with additional nickel to assure the stability of austenite. The molybdenum strongly increases resistance to chloride pitting and crevice corrosion. Type 316 is generally similar to type 304 in its mechanical properties and fabricability. When greater corrosion resistance is needed, use is made of AISI type 317, which features 3 to 4% molybdenum, a slight increase in chromium, and again more nickel to provide for the stability of the austenitic structure. Nitrogen is an inexpensive alloying element and saves on the cost of nickel addition. Nitrogen provides improved pitting resistance for molybdenum, containing grades and increased strength for all austentic grades.

A stainless steel may be produced simply by alloying iron with chromium. However, the ferritic structure of such an alloy poses problems with toughness and weldability unless carbon and nitrogen contents are minimized. Ferritic stainless steels have generally taken a back seat to the more versatile austenitic stainless steels. With increasing costs of nickel and alloying elements, the ferritic stainless steels are receiving increased interest. There are a number of service conditions of such severe corrosivity that austenitic steels must be used in preference to the ferritics. However, there are many more applications of stainless steels which have less severe service conditions and in which the unique properties of the austenitics are not required. Additional applications exist where ferritics have property advantages over the austenitics, including improved machinability, higher thermal conductivity, lower thermal expansion, and immunity to chloride SCC. There are an increasing number of applications where the higher cost of austenitic stainless steels, compared with that of ferritics, makes the austenitics economically unsuitable. Many present applications for austenitic steels used ferritics during early development. Metallurgical objections to the use of ferritic stainless steels as engineering materials include 885°F (475°C) embrittlement, sigma phase, and notch sensitivity. Each of these lowers the ductility and toughness of the ferritic stainless steels at room temperature. These conditions are manifest to the fabricator and user in various restrictions on operations such as hot-forming, brazing, welding, and heat treating in order to minimize high-temperature exposure. The basic grade is type 430, which has a completely ferritic structure when in the annealed condition. Since the principal alloying element is chromium, the iron-chromium phase diagram describes the metallurgical structures present. The phase diagram indicates that at 16% chromium the structure would be completely ferritic at all temperatures. However, the addition of carbon and residual elements to the levels found in commercial steels increases the austenitic loop and significantly broadens the ferritic plus austenite region.

The rapid development of several new methods of lowering the carbon and nitrogen in ferritic stainless steels, notably argon/oxygen decarburization and vacuum melting, have opened up a wide range of alloy levels not previously available. The ferritic stainless steels are all very resistant (possibly immune) to chloride SCC and now range in pitting resistance from being comparable to type 304 to exceeding the resistance of the most highly alloyed austenitic stainless steels. As is the case with the austenitics, molybdenum is added for improved corrosion resistance, and there seems to be increased flexibility in the use of higher chromium content in the ferritics. At the top of the new ferritic-stainless steel family are two steels unmatched in corrosion resistance by other stainlesses: 29Cr-4Mo and 29Cr-4Mo-2Ni. They are produced by vacuum induction melting and rely on their low carbon and nitrogen contents to prevent sensitization by welding. The corrosion resistance of these grades is outstanding, permitting use in seawater and very severe chemical environments. These grades are readily weldable, but contamination by carbon and nitrogen must be avoided. These new grades, like other ferritic stainlesses, are also susceptible to the phenomenon of 885°F embrittlement pointed out earlier.

By addition of elements at a level insufficient for a fully austenitic structure, it is possible to achieve a mixture of ferrite and austenite. With limited amounts of nickel, and sometimes manganese and nitrogen, added to ferritics, the duplex structure can be obtained. Ideally, such a structure would combine the toughness and weldability of austenite with the strength, corrosion resistance, and stress corrosion resistance of ferrite. Best known of these grades is AISI type 329, sometimes known as Carpenter 7-Mo, which provides good general corrosion resistance and SCC resistance somewhat better than that of AISI 304. Sandvik 3RE60 is a similar alloy with a lower chromium content. In the use of the duplex steels, attention must be given to the final high-temperature exposure because of the tendency for chromium and molybdenum to segregate between the phases.

With lower chromium and higher carbon, it is possible to produce stainless steels that are hardenable by heat treatment. These steels are usually selected when their excellent strength is required. AISI type 410 is the most widely used of the martensitic grades. It is occasionally used for heat exchange service, but because of its lower chromium and higher carbon, its corrosion resistance is substantially less than that of AISI type 304. The straight-chromium martensitic stainless steels are resistant to chloride SCC but are susceptible to cracking when fully hardened.

The precipitation-hardening steels use an aging reaction to produce very high-strength stainlesses. The precipitation-hardening stainless steels are subdivided into martensitic, semi-austenitic, and austenitic types. The martensitic types generally come in the martensitic condition, and precipitation hardening is achieved by a simple aging treatment. The semiaustenitic types are supplied in the austenitic condition, and this austenite must be transformed to martensite by special heat treatments before precipitation-hardening. In the austenitic types, it is the austenitic that is precipitation-hardened directly.

The precipitation-hardening process is thought to involve the formation of very fine intermetallics. For the precipitation-hardening steels, the nickel content is sufficiently high to ensure fully stable austenite at room temperature. For these, precipitation-hardening is achieved by reheating the austenite to an elevated temperature at which the fine intermetallic compounds are precipitated. The compositions of the martensitic precipitation-hardening steels are so adjusted as to have the martensite start temperature well above room temperature, and a structure fully transformed to martensite is obtained at room temperature. The class of precipitation-hardening stainless steels, referred to as semiaustenitic, requires an intermediate step to transform the metastable austenite to martensite before precipitation-hardening. This can be achieved in two ways, either by subzero cooling, which is generally practiced, or by tempering at about 750°C to reduce carbon and chromium content of the austenite by the precipiataion of carbide. The reduction of dissolved carbon and chromium in austenite significantly raises the martensite start temperature. Apart from this, cold-working also raises the martensite start temperature, and this can also be used to achieve transformation of austenite to martensite.

Compositions of the various types of steel briefly described above are shown in Table 12. The AISI 300 series consists of austenitic steels. Type 304 is the general-purpose grade, widely used in application requiring a good combination of corrosion resistance and formability. Type 304 L is a lower-carbon modification of type 304 used in applications requiring welding. Types 316 and 317 contain molybdenum and have a greater resistance to pitting in marine and chemical industry environments. Types 321, 347, and 348 are grades stabilized with titanium, niobium and tantalum, and niobium, respectively. Type 348 is a nuclear-grade stainless steel having a restricted tantalum and cobalt content. Nickel-manganese-nitrogen stainless steels are the austenitic steels designated as the 200 series. The 400 series of AISI standard grades are ferritic. The 400 series of AISI with requisite compositions are martensitic types. They have an austenitic structure at elevated temperatures that can be transformed into martensite by suitable cooling to room temperature. Type 410 is the classic 12% Cr martensitic stainless steel and is the most popular martensitic grade produced today. It is used in many wrought forms, and in sand and investment castings. Type 420, with medium carbon, is the original cutlery grade. Types 414 and 431 contain nickel, which gives rise to somewhat higher toughness and corrosion resistance, and are used in aircraft fittings, pumps, and valves.

C. Applications[66,67]

The atomic energy industry is one of the main users of stainless steel. Stainless steels are attractive as structural material in fast reactors, thermal reactors (vessels and piping systems), radioactive wastes, radioisotopes (container), and other nuclear applications. The paragraphs below present a brief coverage of some of the principal areas where stainless steels are being used in the nuclear industry.

Zirconium does not offer as much oxidation protection as stainless steel, a reason why stainless steel, particularly type 304, was chosen in some LWR cores in the past. However, since the early 1960s, the designers quickly moved to the use of zirconium, primarily on account of its low neuton absorption cross section (an improvement of a factor of 15). Another reason was a rash of fuel failures in BWRs using stainless steel cladding (e.g., Dresden 1, Humboldt Bay, and Big Rock Point). Today only one small BWR (LaCrosse) and several PWRs (Connecticut Yankee and San Onofore 1) still use stainless steel cladding. By and large, the susceptibility of stainless steel to SCC has largely led to its abandonment in favor of zirconium alloys for cladding in water reactors. Stainless steels perform exceedingly well as a cladding material for advanced gas-cooled reactors (AGRs). The main interest in the use of stainless steel cladding is now for the sodium-cooled fast reactor, where the operating temperatures are high enough to result in seriously reduced strength and corrosion resistance for most of the alternative cladding materials. Type 316, 20% cold-worked stainless steel is the reference material used for most liquid metal fast breeder reactor (LMFBR) in-core cladding and structural applications. The choice has been made on the basis of classical properties such as creep properties, corrosion behavior in liquid sodium, and nuclear data which influence the neuton absorption and the breeding ratio.

Table 12
COMPOSITIONS OF SOME STAINLESS STEELS

Designation (AISI grades)[a]	Composition (%)[b]								
	Cr	Ni	C	Mn	Si	P	S	N	Other
300 Series (Austenitic)									
304	18—20	8—10.5	0.08	2.0	1.0	0.045	0.030	—	—
304L	18—20	8—12	0.03	2.0	1.0	0.045	0.030	—	—
316	16—18	10—14	0.08	2.0	1.0	0.045	0.030	—	Mo 2.0—3.0
317	18—20	11—15	0.08	2.0	1.0	0.045	0.030	—	Mo 3.0—4.0
321	17—19	9—12	0.08	2.0	1.0	0.045	0.030	—	Ti 5 × C[c]
347	17—19	9—13	0.08	2.0	1.0	0.045	0.030	—	Nb + Ta 10 × C[c]
348	17—19	9—13	0.08	2.0	1.0	0.045	0.030	—	Nb + Ta 10 × C[c]; Ta 0.1; Co 0.20[d]
200 Series (Austenitic)									
201	16—18	3.5—5.5	0.15	5.5—7.5	1.0	0.060	0.030	0.25	—
202	17—19	4—6	0.15	7.5—10	1.0	0.060	0.030	0.25	—
205	16.5—18	1—1.75	0.25	14—15.5	1.0	0.060	0.030	0.40	Mo 1—1.75
400 Series (Ferritic)									
405	11.5—14.5	—	0.08	1.0	1.0	0.040	0.030	—	Al 0.10—0.30
430	16—18	—	0.12	1.0	1.0	0.040	0.030	—	—
446	23-27	—	0.20	1.5	1.0	0.040	0.030	0.25	—
400 Series (Martensitic)									
410	11.5—13.5	—	0.15	1.0	1.0	0.040	0.030	—	—
420	12—14	—	0.15 (min)	1.0	1.0	0.040	0.030	—	—
414	11.5—13.5	1.25—2.50	0.15	1.0	1.0	0.040	0.030	—	—
431	15—17	1.25—2.50	0.20	1.0	1.0	0.040	0.030	—	—

				Duplex Stainless Steels					
329	28.0	6.0	0.10	2.0	1.0	0.040	0.030	—	Mo 1.5
3 RE 60	18.5	4.5	0.02	1.5	1.6	0.01	0.01	—	Mo 2.5
				Precipitation-Hardening Stainless Steels					
17-4 PH (martensitic)	16.50	4.25	0.04	0.4	0.50	—	—	—	Cu 3.6 Nb + Ta 0.25
17-7 PH (semi-austenitic)	17.00	7.00	0.07	0.70	0.40	—	—	—	Al 1.15
17-10 P (austenitic)	17.0	10.50	0.12	0.75	0.50	—	—	—	P 0.28
				Highly Alloyed Austenitic Materials[e]					
Alloy 800	21	33	0.03 (max)	—	—	—	—	—	Co 0.10 (max), Ti 0.60 (max), Fe 38
Inconel alloy 600	15	76	0.04	—	—	—	—	—	Co 0.03, Ti 0.25, Fe 8
Inconel alloy 625	22.0	62.5	0.05	—	—	—	—	—	Nb 3.5, Mo 9.0, Fe 3.0
Hastelloy alloy C-276	16.0	59.0	0.02	—	—	—	—	—	W 4.0, Mo 16.0 Fe 5.0
Hasterlloy alloy G	22.0	45.5	0.03	—	—	—	—	—	W 0.5, Cu 2.0, Nb 2.0, Mo 6.5, Fe 19.5
Incoloy alloy 825	20.0	42.0	0.03	—	—	—	—	—	Cu 2.0, Mo 3.0, Fe 30

[a] *Source Book on Stainless Steels,* American Society of Metals, 1976; various suppliers' literature.
[b] Balance Fe; single values are maximum values unless otherwise indicated.
[c] Minimum.
[d] Maximum.
[e] These generally contain less than 50% Fe. It is convenient to divide these highly alloyed austenitic materials into two groups: (1) Ni-Cr-Fe alloys (alloy 800 used in West Germany for steam generator tubing, Inconel alloy 600 used in the U.S. for steam generator tubing); (2) Ni-Cr-Fe (Mo, Cu, Nb) alloys used in application requiring corrosion resistance to reducing acids, pitting, and crevice corrosion; Inconel alloy 625 and Hastelloy alloy C-276 are among the most resistant austenitic alloys currently available for severe aqueous environments.

Table 13
CORE REFERENCE FUEL ELEMENT STRUCTURAL MATERIALS FOR PROTOTYPIC AND COMMERCIAL FAST BREEDERS

Country	Reactor	Cladding material	Subassembly wrapper
DEBENE[a]	SNR 300	1.4970	1.4981
France	Phenix	AISI 316	316 L
	Superphenix	316 Ti	316 Ti
Japan	Monjou	316	316
U.K.	PFR	M 316 (FV 548)	Nimonic PE 16 (AISI 321)
	CDFR	316	PE 16
U.S.	FFTF	AISI 316	AISI 316
U.S.S.R.	BN 350	OCr16Ni15Mo3B	1Cr-18Ni-10Ti
	BN 600		

[a] Deutschland, Belgium, and the Netherlands.

In Table 13 a list of reference cladding and subassembly wrapper materials is given for the different fast breeder projects. Typical analyses of these materials (in weight percent) are collected in Table 14. With the exception of Nimonic PE 16, which is already a Ni-based alloy, all the other materials are austenitic stainless steel with good sodium compatibility and high creep strength at the operational temperature of 400 to 650°C. As can be seen, some of these steels are unstabilized (AISI 316 and M 316); types 1.4981, 0Cr16Ni15Mo3B, and FV548 are niobium-stabilized, and types 1.4970 (12R72HV), AISI 321, and 1Cr18Ni10Ti are titanium-stabilized. In addition there exist titanium-modified versions of AISI 316. The question of whether or not commercial fast breeders with much higher neutron damage can be used by using reference austenitic stainless steels as core structrual materials is still open. Research programs are on all over the world for new cladding and wrapper materials. Potential candidates as replacements are high-nickel alloys (such as 10Cr25Ni, Inconel 706, and Nimonic PE 16) for both cladding and wrapper, dispension-strengthened ferritic steels for the cladding only, and martensitic materials for the wrappers only.

An effect of considerable concern found in fast reactor cladding materials is fast-neutron-induced swelling. Since an appreciable number of neutrons in the fast reactor have energies in excess of the (n,p) or (n,α) threshold for the steel constituents, hydrogen and helium are produced in the cladding. At the typical operating temperature, the hydrogen diffuses fairly readily out of the material, but the helium, on account of its low solubility, precipitates in the form of small bubbles, principally at dislocation sites and at grain boundaries. It is found, however, that the total volume of bubbles produced in the cladding is much greater than can be explained by the helium produced in this way and that bubbles are largely empty voids. The explanation appears to be in the accummulation of lattice vacancies at the nucleation sites of the helium bubbles. The average energy of a neutron in the fast reactor is above 100 keV, greatly in excess of the 25 keV or so which is required to displace an atom from a lattice site. Neutron collisions, followed by multiple cascade processes, therefore lead to a high density of vacancies and interstitials. The interstitials have a higher mobility than the vacancies and tend to be more rapidly absorbed at grain boundaries and dislocation where they lose their identity. The surplus vacancies are then available for the formation of voids at nucleation centers of helium produced by the (n,α) reactions. The target burnup for a fast breeder reactor is around 10,000 MWd/t (10% heavy atom burnup), which corresponds to a fast neutron fluence of the order of 5×10^{23} n/cm^2. The fast-neutron-induced voidage for this fluence can result in substantial dimensional changes in the cladding size.

Table 14
ANALYSES (WT%) OF MORE COMMON REFERENCE CLAD AND WRAPPER STRUCTURAL MATERIALS FOR PROTOTYPIC AND COMMERCIAL FAST BREEDERS

Denomination	C	Si	Mn	Cr	Ni	Mo	Ti	Nb	N	B
AISI 316 (1.4919)	0.04—0.08	≤0.75	1.5—2.0	16—18	12—14	2—2.5	—	—	≤0.01	≤0.001
M 316 (FV 548)	0.03—0.06	≤0.6	1.5—2.0	15.4—17.5	13—14	2—2.75	0.05 max	0.05 max	≤0.02	0.001
1.4981	0.04—0.1	0.3—0.6	≤1.5	15.5—17.5	15.5—17.5	1.6—2.0	—	0.4—1.2	0.01	≤0.001
0Cr16Ni15Mo3B	0.05—0.09	0.6	≤0.8	15—17	14—16	2.5—3	—	0.6—0.9	—	0.005
1.4970	0.08—0.12	0.3—0.75	≤2.0	14.5—15.5	14.5—15.5	1—1.4	0.33—0.55	—	≤0.015	0.003—0.008
Nimonic PE 16	0.05—0.1	≤0.3	≤0.29	15—18	42—45	2.5—4	1.1—1.5	Al 1.1—1.5	0.011	0.001

This has led to changes in fast reactor design, including initial clearances between fuel pins to allow for swelling, fuel assemblies with provision for expansion, and the adoption of periodic rotation of fuel elements in regions of high flux gradient to reduce differential swelling. Since the requirement for high start-of-life clearance between the fuel pins results in an increase in the coolant-to-fuel volume ratio, and thus a neutron degradation of neutron spectrum with a consequent reduction in breeding ratio, there is an obvious incentive to develop cladding materials with minium voidage production. Also, in addition to the voidage effect described, the helium production leads to embrittlement and loss of ductility.

It is found that cold-working of stainless steels reduces the fast-neutron swelling by producing greater dislocation density and thus more sinks for their mutual annihilation of vacancies and interstitials. It may be noted that irradiation in steels in thermal reactors also reduces embrittlement, although in this case the main contributor of helium is the (n,α) reaction of slow neutrons with boron impurity in the steel. Because of the high thermal absorption cross section of boron, however, this burns out at fairly moderate exposures, leading to the saturation effect. This is in contrast to the fast-neutron case, where the reaction cross sections are much lower and helium content increases steadily with increasing fluence.

Carbon steels lose strength rapidly with increasing temperature and are not particularly resistant to corrosion. For this reason, the inside surface of all primary system components must be covered with a stainless steel cladding. A few exceptions exist. The head of a BWR vessel which is in contact with steam rather than water does not have to be clad with stainless steel. The stainless steels most common in reactor system applications are the 300 series of the AISI. The main types are 304, 316, 317, and 304 L. The L following the number indicates a steel with low carbon content, typically 0.03% compared to 0.08% for normal stainless steel.

The dominant material in primary coolant piping of operating BWR plant is 304 stainless steel; types 304 L, 316, and 347 stainless steel are also used in smaller quantities, as is Inconel 600. The low-carbon variety of stainless steels have an enhanced capacity to resist corrosion, and for this reason types 304 L, 316 L, and 347 are introduced increasingly in applications, particularly where 304 stainless steel has shown susceptibility to SCC. These steels have shown superior resistance — nearly over a factor of 20 compared to 304 stainless steel — to intergranular SCC. Ferritic steels are commonly used in steam lines, feedwater lines, and so on, in forms including seamless, welded, and small amounts of cast pipes. PWR piping materials are mostly 304 and 316 stainless steels. In many cases ferritic steels clad with 308 L austenitic stainless weld overlay are used.

An example of the LMFBR can be used to provide an idea of application ranges of stainless steels in the reactor system. The LMFBR vessel of the Clinch River Breeder Reactor (CRBR) has a diameter of about 6 m and a relatively thin wall (about 75 mm), since the internal pressure is quite low (about 1 MPa or 145 psi). It is constructed entirely of stainless types 304 and 316. The 304 type is selected for piping in LMFBRs, although in some designs type 316 is selected for the hot leg regions, where the higher temperature exists. The 316 stainless steel is more expensive than 304 but has superior strength properties. Welds or fills are made with type 308, 316, and 16/8/2 (Cr/Ni/Mo) materials.

The high-temperature gas-cooled reactors (HTGRs) operate at even higher temperatures. The helium coming out from the core has a temperature in the range of 680°C (1260°F) and a pressure of about 7.25 MPa (1050 psi). Stainless steel used in cladding materials and other core internals must have good strength characteristics at these elevated temperatures. High-alloy austenitic steels of the type 18Cr-8Ni are required for satisfactory strength up to about 705°C (1300°F); a number of superalloys have been developed for high-temperature service (N-15, S-90, Refractory 26, and Refractory 80). In the context of materials for high temperature, reference may be drawn to a few others. Nickel-based and cobalt-based alloys have excellent high-temperature strength characteristics and corrosion resistance. They include the nickel-chromium series known as Inconel alloys; the nickel-chromium-molybdenum alloys like the Hastelloys, Cosmoloy F, GMR-235, Inconel 713 C, and so on; the nickel-chromium-cobalt alloys (Nimonic 90 K-42-B); and the nickel-chromium-cobalt-molybdenum series (InCO 700, M-252, Rene 41, Udimet 500, Waspaloy, etc.). The cobalt-based alloys include the original alloy of this group known as stellite or HS-21, a cobalt -chromium-nickel-tungsten type (HS-31), and alloy HS-25 suitable for application up to 1095°C (2000°F). Stellite is used in nuclear reactors as a hard facing material on walls and on control rod drive mechanisms. It has excellent wear resistance and corrosion properties. Its principal components are 27Cr-3Ni-64Co. The cobalt, however, can cause activation problems in the primary coolant systems.

Nuclear fuel reprocessing activity sizably consumes stainless steels. The conditions in the dissolver are hot, corrosive, and highly radioactive both in the liquid phase and in the vapor space above it. Dissolver vessels are constructed using very high standards from a suitably resistant material, usually a stainless steel. The stainless steel used to date for such service has been stabilized 18Cr-13Ni-Nb steel developed specifically for plant service in the nuclear reprocessing industry. The dissolver in a reprocessing plant is a primary vessel where failure due to corrosion could seriously disrupt plant operations. The material used thus far has now been superseded for dissolver service by 25Cr-20Ni nitric acid grade (NAG, containing low levels of C, P, S, and Si impurities) stainless steel, which corrodes at a significantly lower rate than 18Cr-13Ni-Nb steel and will permit vessels of lighter construction to be used with an attendant improvement in the ease of fabrication. The dissolver with 18-Cr-13Ni-Nb material is about 35-mm thick, and the niobium addition has an effect of making steel difficult to weld. The solvent extraction plant operation is the next operation following fuel dissolution. The material of choice for the plant construction, because of its satisfactory resistance to a variety of process reagents, is stainless steel, usually an 18Cr-8 to 10Ni grade. In the waste management there are requirements for tanks for storage of highly active liquor containing a majority of fission products and waste actinides. The heat arising from radioactive decay of the fission product requires continuous cooling by a system of water-cooled coils. The conditions to which the tanks are exposed, therefore, are potentially similar to those encountered in a fuel dissolver, and a similar approach is adapted to prevent any corrosion which would compromise the integrity of the tanks. Tanks in current use are fabricated from the same grade of stainless steel (18Cr-13 Ni-Nb), as has been used for dissolver service.

VII. REACTOR PRESSURE VESSELS[68-71]

The pressure vessel contains most materials of the primary components of a nuclear fission reactor. They are nuclear fissile and fertile fuels; structural materials; reflector, moderator, or blanket materials; control element; coolant; shielding; and safety system materials. For instance, in an LWR the vessel itself is made of structural materials. Figure 26 shows the plan cross section of a typical PWR pressure vessel containing the primary reactor components.

A. Pressure Vessel Types and Materials

Types of pressure vessels depend on different types of nuclear reactors. Table 15 shows the major reactor concepts. Since fuel elements perform the most important function, these reactor systems must be completely addressed with some details of fuel element design features. Table 16 presents some typical data of the fuel elements of some of the reactor systems among those listed in Table 15. A consideration of pressure vessels for these major reactor systems leads to the following types of pressure vessels: (1) heavy-walled pressure vessels for LWRs (PWRs and BWRs), (2) the pressure tube reactor (CANDUs), (3) pressure

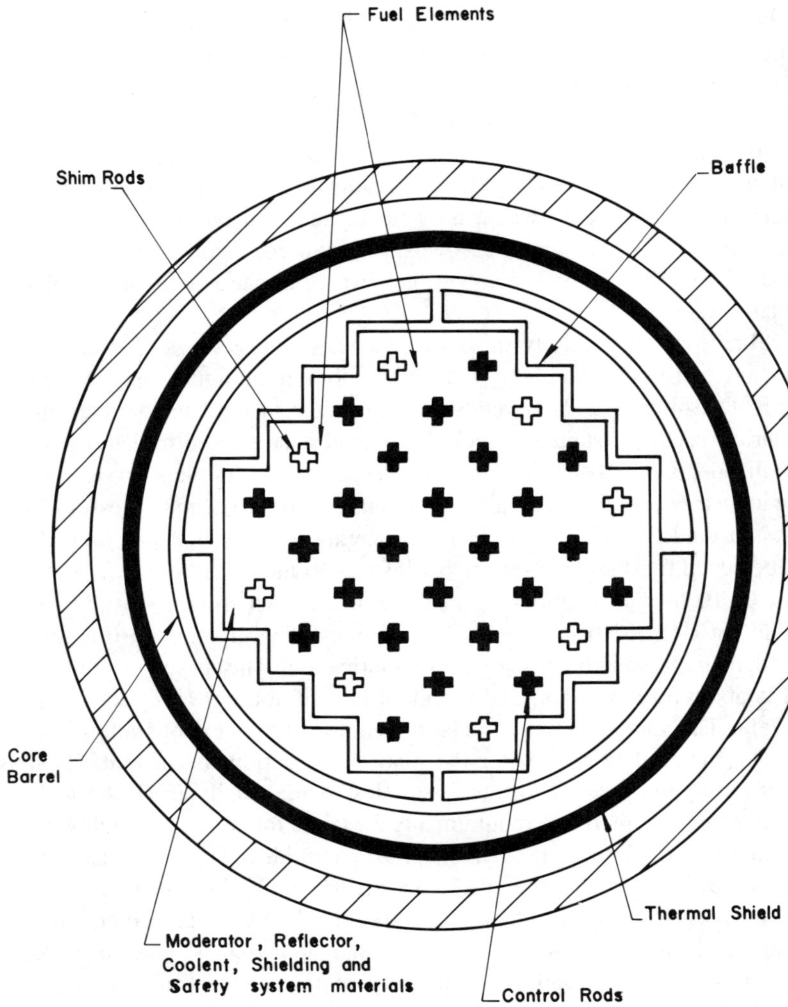

FIGURE 26. Cross section of typical PWR pressure vessel and primary components.

vessels (LMFBRs), and (4) prestressed concrete reactor pressure vessels (PCRVs) for gas-cooled reactor systems. These are the four general types of pressure vessel which have found engineering acceptance. According to the construction materials, the vessels fall into two general categories. The first is metallic reactor vessel (MRV), in which the first three are grouped. The other category is PCRV. According to the relative power density of a given power level, the relative reactor vessel size and selected material for various types of nuclear power reactors are given in Table 17.

For a PWR, operating at high power density and high pressure in the core, a thick-walled medium-size reactor vessel is desirable. The BWR has relatively low power density, medium operational pressure (about half of the operational pressure of PWR), and larger size. The reactor vessel is concurrently a steam generator of a BWR power plant.

Steels containing a carbon of 0.10 to 0.30% and other alloying elements are widely used as a base metal for pressure vessels for both PWRs and BWRs as well as for other large pressure vessels such as pressurizers. Typical materials are A 508/class 2 (these designations are according to the American Society for Testing and Materials, or ASTM) forgings and A 533/grade B class 1 plate. The first is a manganese-molybdenum steel modified with

Table 15
SOME OF THE MAJOR POWER REACTOR CONCEPTS

Reactor abbreviation	Characteristics
LWR	Light water reactor; light water moderates neutrons and removes heat from slightly enriched UO_2 fuel rods.
PWR	Pressurized water reactor; circulating water at about 140 atm and 300°C transfers heat from the fuel elements to steam generators.
BWR	Boiling water reactor; circulating water transfers heat from fuel rods, boils at top reactor, and furnishes steam at about 300°C and 70 atm directly to turbines.
LWBR	Light water breeder reactor; pressurized water provides cooling and moderation; core consists of ThO_2 rods containing up to 6% ^{233}U; blanket consists of ThO_2 rods containing up to 3% ^{233}U.
HWR	Heavy water reactor.
PHWR-CANDU	Pressurized heavy water reactor Canada deuterium uranium; heavy water in a calandria moderates neutrons; heavy water at 100—150 atm, circulating in pressure tubes, transfers heat from natural UO_2 fuel rods to steam generators.
BLWR-CANDU	Boiling light water reactor Canada deuterium uranium; pressure tubes in a tank of heavy water; the tubes are vertical, unlike PHWR-CANDU, where tubes are horizontally placed; light water is the coolant; obviously, this involves some neutron penalty, but natural UO_2 can still be used to fuel the reactor; Canadian Gentilly reactor is a typical example; like the BWR, these reactors dispense with boilers by allowing in-core boiling of the water and separating steam for the turbines by means of steam drums
SGHWR	Steam-generating heavy water reactor; similar to PHWR and BWR; it is actually the U.K. version of the BLWR-CANDU; light water at 100 atm boils in the upper part of reactor, supplying steam directly to turbines.
GCR	Gas-cooled reactor.
Magnox	Graphite-moderated neutrons provide substantial part of core structure; circulating CO_2 transfers heat from natural uranium metallic fuel rods to steam generators at about 400°C; name is derived from Mg-Al alloy cladding developed for fuel rods.
AGR	Advanced gas-cooled reactor; graphite moderator and CO_2 heat transfer similar to Magnox; stainless steel cladding for slightly enriched UO_2 fuel rods permits steam generation at over 500°C.
HTGR	High-temperature gas-cooled reactor (HTR in U.K.; AVR, THTR in Germany); graphite moderator provides core structure, and moderation circulating helium at 700—1000°C and 50—80 atm transfers heat to gas turbines or heat to chemical processes.
FBR	Fast (neutron) breeder reactor.
LMFBR	Liquid metal fast breeder reactor; liquid sodium at 400—600°C and 2—3 atm transfers heat from stainless-steel-encased $(U,Pu)O_2$ or (U,Pu)C fuel rods to intermediate heat exchanger from which sodium transfers heat to steam generator via another heat exchanger; the system is the most developed among the FBR systems.
GCFR	Gas-cooled fast reactor; helium at 300—600°C transfers heat from vented and pressure-equalized U-Pu oxide (or carbide) fuel rods to steam generators; the system is in the development phase.
FFTF[70]	Fast flux test facility; it is a 400-MW sodium-cooled fast reactor operated by Westinghouse Hanford Company for the U.S. Department of Energy to conduct fuels and materials testing in support of the U.S. LMFBR program; heat removal is by sodium-to-air heat exchangers rather than steam generators; nickel reflectors surround the core in lieu of a breeding blanket to intensify the core neutron flux; the net result is a plant that is relatively quick and easy to start up and operate, and that provides the flexibility needed to fulfill its test objectives; the reference driver fuel loading comprises about 15,000 pins of co-precipitated U-Pu mixed oxide fuel of 22.5 and 26% Pu; each fuel subassembly contains 217 pins spaced by wire wrap and identified by Xe-Kr gas mixture to detect possible cladding failure; fuel cladding is type 316 stainless

Table 15 (continued)
SOME OF THE MAJOR POWER REACTOR CONCEPTS

Reactor abbreviation	Characteristics
	steel (20% cold work), 5.842 mm O.D. by 0.381 mm wall thickness; technical parameters are thermal power 400 MW, reactor vessel inlet operating temp 360°C, reactor vessel outlet operating temp 503°C, nominal core temp 167°C, primary sodium flow capability 43,500 gpm; primary sodium dynamic head 152.4 m, core diam 121.92 cm, core height 91.44 cm, peak fast flux 7 × 10^{15} n/cm^{2}s, decay heat removal by natural circulation.
MSBR	Molten salt breeder reactor; molten Li, Zr, Be, U fluoride salt circulates through channels in structural graphite moderator and a Hastelloy N heat exchanger; Li, Zr, Be fluoride salt transfers heat at about 600°C to steam generators; U-bearing fuel and thorium-bearing blanket salt processed continuously or in batches.
TRIGA	General atomic training, research, and isotope production reactor; cooled and shielded by light water circulating at ambient pressure; core consists of rod containing Zr-U hydride clad with Al or SS.
Swimming pool reactor	A pool of water provides cooling, moderation, and shielding; core consists of plates of aluminum-enriched uranium alloy clad with Al.

small amounts of nickel and chromium (gases, H_2 and N_2, sulfur, and phosphorus are maintained as low as possible in A 508 steels). The second has a similar composition but much reduced chromium. Both materials are first heated to a temperature of 870 to 900°C where austenite is formed, quenched, and finally tempered to about 650 to 675°C for over an hour. These materials are optimized for high strength and toughness and to ensure weldability.

The LMFBR has a very high power density, very low vapor pressure of sodium or sodium-potassium, and very small vessel size because of no moderator in the reactor core. In the primary system design of LMFBR, there are two types of reactor pressure vessles: the pipe loop (piped-loop) type and the tank (pot or pool) type. In the pipe type the pressure vessel contains only the reactor core, control elements, and liquid-metal coolant. The heat exchangers, pumps, and other components of the primary systems are placed outside the pressure vessels, in a design similar to a PWR power plant. In the tank type, the pressure vessel contains the entire primary system (such as the core, control elements, heat exchangers, pumps, etc.), which is submerged in a large sodium tank by taking advantage of the basic sodium properties of low vapor pressure and high thermal conductivity. Stainless steels 304, 304 L, and 308 are the basic materials of the vessels.

As for the CANDU-type heavy water reactor, pressure tubes, instead of a pressure vessel, are employed. The tubing material is zircaloy-2, or Zr-2.5Nb, similar to the cladding material of BWR fuel elements.

An alternative to the steel reactor pressure vessels is the PCRV. Several gas-cooled reactors use PCRVs. Basically, the PCRVs consist of a thick-walled concrete shell fabricated at a side and containing steel reinforcing bars and ducts. The ducts serve as guides for the tendons, either stranded or single-wire, located axially and circumferentially in the vessel walls. These tendons are tensilely loaded by jacketing so that the entire vessel is placed in compression. In essence, it is necessary to exceed the yield strength of the tendons by internal overpressurization before cracking of the concrete wall occurs. There exist now a good many gas-cooled reactors that use PCRVs. Most are thick-walled right cylinders. A few, however, are roughly spherical in shape. The Wylfa PCRV is a 30-m sphere, while Bugey is an 18-m-diameter by 40-m-height right cylinder. Wall thickness may be as much as 628 m. Operating gas pressure reaches 600 psi in current designs, and gas temperature may exceed 540°C. The thick walls permit incorporation of steam generators and gas circulators so that

Table 16
TYPICAL FUEL ELEMENT DESIGN DATA FOR SOME MAJOR REACTOR SYSTEMS

Characteristic	PWR	BWR	CANDU/PHWR	HTGR	LMFBR	GCFR
Fuel material	UO_2 pellets	UO_2 pellets	UO_2 pellets	UC_2ThC_2 pellets	UO_2-PuO_2 pellets	UO_2-PuO_2 pellets
Fuel density (g/cm^3)	10.3 (94%)	10.3 (94%)	10.6 (96.7%)	0.58	9.90 (91%)	10.15 (93.3%)
Pellet diam (mm)	9.29	12.4	14.25	15.70	4.94	6.21
Fuel portion length (mm)	3650	495	500	793	900	1360
Fuel initial enrichment (%)	3.0	2.5	0.7 (natural)	93	15 (Pu)	15 (Pu)
Fuel peak temp (K)	2530	2690	2270	1520	2470	2470
Fuel av burnup (MWd/kg)	24.4	27.5	9.0	100	100	100
Fuel max burnup(MWd/kg)	35	40	—	700	150	150
Fission gas release (%)	1—4	1—4	1—4	10^{-3}	80—100	100
Fuel-clad gap, diameteral (mm)	0.18	0.25	0.08	0.076	0.18	0.15
Cladding material	Zr-4	Zr-2	Zr-2	Graphite	20% c-w 316 SS	20% c-w 316SS
Cladding wall thickness (mm)	0.62	0.81	0.38	0.17/0.16	0.375	0.370
Cladding peak temp (K)	620	569	583	1520	920	998
Cladding fast-neutron fluence (0.1 MeV) n/cm^2	10^{22}	10^{22}	10^{22}	8×10^{21}	3×10^{23}	2.3×10^{23}
Fast-neutron flux (0.1 MeV av) n/cm^2s	2×10^{14}	2×10^{14}	4.5×10^{13}	2.4×10^{14}	5×10^{15}	3.8×10^{15}
Number of pins per element	204	49	28	132	271	331
Number of elements per channel	1	1	12	8	1	1
Av core rating (W/g)	34.8	22.0	22.1	76	170	132
Peak heat rating (W/cm^2)	171.2	134	126.1	42	230	105
Linear heat rating, peak (W/cm)	577	600	603	210	475	500
Av power density (kW[t]/ℓ)	90	50	10	8.4	300	250

Note: From among the reactor systems listed in Table 15.

Table 17
REACTOR PRESSURE VESSELS

Reactor type	Relative power density	Relative vessel size	Vessel type
PWR	High	Thick, medium	MRV
BWR	Medium	Large	MRV
LMFBR	Very high	Small	MRV
CANDU	Medium	Pressure tube	MRV
GCR, HTGR	Very low, low	Very large	PCRV

the entire primary system is inside the vessel. The steel tendons are particularly critical in a PCRV, since they are expected to operate for nearly 40 years without significant loss in properties. All told, PCRV represents a well-tested concept.

B. Irradiation Effects[72-76]

The effects of neutrons are an increase in yield stress and the hardening effect of radiation, accompanied by loss of ductility. An important effect of neutron irradiation on the low-carbon structural steels in reactor pressure vessels is a shift in the transition temperature at which the material goes from ductile to brittle and a reduction in upper-shelf energy (Charpy impact energy-vs.-temperature curve). The shift of both these quantities means an embrittlement effect of neutron irradiation on reactor pressure vessels.

The neutron exposures on the LWR vessels range from about 5×10^{18} to 5×10^{19} n/cm^2. The important factors governing radiation embrittlement of pressure vessel steels are the sensitivity of the steel to embrittlement, the neutron fluence and energy spectrum, and the irradiation temperature. The main criterion used in specifying the operational limitation of the pressure vessel steel is ductile-brittle transition temperature (DBTT), which in the irradiated steel must not exceed 33°C below the lowest operating temperature. It has been observed that minor amounts of impurities, particularly Cu and P, have profound influence on the radiation embrittlement at elevated temperature. Fine-grained vacuum-deoxidized steels with low impurity content (Cu 0.01%, P and S 0.012%) provide a material that shows remarkably good resistance to radiation embrittlement.

Radiation effects in LMFBRs are far more severe than in LWRs. A much larger amount of power is generated per unit core volume in a LMFBR. The neutron flux is higher by about a factor of 100 than in an LWR. In such severe environments additional damaging processes take place. In LMFBRs austenitic stainless steels are the preferred choice for the present generation of reactors. Apart from the general effects of radiation on stainless steels, there is an additional problem of significant changes in volume at high neutron fluences. These volumetric changes are attributed to agglomeration of vacancies generated in the high neutron flux. Furthermore, the helium and hydrogen generated by (n,α) and (n,p) reactions of neutrons with metals such as nickel are assumed to stabilize these agglomerated vacancies. At a fluence approaching 10^{23} n/cm^2 there is a 10% increase in volume. Several means of modifying the alloys to minimize swelling have been developed. These include (1) controlling (n,α) helium generators such as boron, nitrogen, titanium, vanadium, sulfur, calcium, oxygen, and magnesium; (2) developing a high dislocation density by cold-working to counteract void formation and swelling, balancing cold work against loss of ductility in stress rupture; (3) introducing a finely dispersed precipitate as a stable phase to serve as nucleation sites for bubbles and to prevent grain boundary bubble segregation. Other factors requiring control include (1) minimizing tantalum, niobium, and molybdenum because of their capture of fast neutrons; (2) minimizing the tendency for intergranular failure in stress rupture by decreasing grain size and nickel content, and stabilizing with titanium or niobium or doping with boron (stabilization also minimizes carbon migration from alloy to sodium); and (3) minimizing nickel content to prevent excessive corrosion in sodium.

REFERENCES

1. **Wilkinson, W. D. and Murphy, W. H.,** *Nuclear Reactor Metallurgy,* D Van Nostrand, New York, 1958.
2. **McIntosh, A. P. and Heal, T. J., Eds.,** *Materials for Nuclear Engineers,* Interscience, New York, 1960.
3. **Hasner, H. H. and Schumer, J. F., Eds.,** *Nuclear Fuel Elements,* Reinhold, New York, 1959.
4. Materials in Nuclear Applications, ASTM STP-276, American Society for Testing and Materials, Philadelphia, 1960.
5. Irradiation Effects on Structural Alloys for Nuclear Reactor Applications, ASTM STP-484, American Society for Testing and Materials, Philadelphia, 1970.
6. **Heal, T. J.,** Magnesium, in *Materials for Nuclear Engineers,* McIntosh, A. G. and Heal, T. J., Eds., Interscience, New York, 1960.
7. Fuel Manufacturing Technology at BNFL Springfields, BNFL, Risley, Warrington, 1981.
8. **Gupta, C. K.,** Zirconium: atomic apparel, *Sci. Today,* p. 70, September 1984.
9. **Jamrack, W. D.,** *Rare Metal Extraction by Chemical Engineering Techniques,* Pergamon Press, Elmsford, NY, 1963.
10. **Lustman, B. and Kerze, F., Jr.,** *The Metallurgy of Zirconium* (National Nuclear Energy Series), Division 7, Vol. 4, McGraw-Hill, New York, 1955.
11. **Miller, G. L.,** *Zirconium,* Academic Press, New York, 1954.
12. *Zirconium and Zirconium Alloys,* American Society for Metals, Cleveland, 1953.
13. **Hampel, C. A., Ed.,** *Rare Metals Handbook,* Reinhold, New York, 1954, chap. 28.
14. **Murthy, T. K. S., Koppiker, K. S., and Gupta, C. K.,** Solvent extraction in extractive metallurgy, in *Recent Developments in Separation Science,* Vol. 3, Li, N. N. and Navratil, J. D., Eds., CRC Press, Boca Raton, FL, 1986, 1.
15. **Takahashi, M., Miyazaki, H., and Katoh, Y.,** New solvent extraction process for zirconium and hafnium, paper presented at the 6th Int. Conf. Zirconium in the Nuclear Industry, Vancouver, Canada, June 28 to July 1, 1982.
16. **Skagga, R. L., Rogers, D. T., and Hunter, D. B.,** Review of Anhydrous Zirconium-Hafnium Separation Techniques, Inf. Cir. No. 8963, U. S. Bureau of Mines, 1983.
17. **Spink, D. R.,** Extractive metallurgy of zirconium and hafnium, *Can. Min. Metall. Bull.* November 1, 1977.
18. **Robertson, J. A. L.,** Zirconium - an international nuclear material, *J. Nucl. Mater.,* 100, 108, 1981.
19. **Thomas, D. E.,** Aqueous Corrosion of Zirconium and Its Alloys at Elevated Temperatures, Proc. Int. Conf. Peaceful Uses of Atomic Energy, Vol. 9, United Nations, New York, 1956, 407.
20. **Videm, K.,** Properties of zirconium base cladding materials — corrosion and hydrogen pickup, *Nucl. Eng. Des.,* 21, 200, 1972.
21. **Pickman, D. O.,** Properties of zircaloy cladding, *Nucl. Eng. Des.,* 21, 212, 1972.
22. **Stehle, H., Kaden, W., and Manzel, R.,** External corrosion of cladding in PWRs, *Nucl. Eng. Des.,* 33, 155, 1975.
23. Zirconium in Nuclear Applications, ASTM STP-551, American Society for Testing and Materials, Philadelphia, 1974.
24. Zirconium, *Can. Metall. Q.,* 11(1), 1972.
25. **Pickman, D. O.,** Design of fuel elements, *Nucl. Eng. Des.,* 21, 303, 1972.
26. **Kallstrom, K.,** Zircaloy canning tubes production and mechanical properties, *Scand. J. Metall.,* 4, 65, 1975.
27. **Northwood, D. O. and Kosasih, U.,** Hydrides and delayed hydrogen cracking in zirconium and its alloys, *Int. Met. Rev.,* 28(2), 92, 1983.
28. **Tyson, W.,** Hydrogen in metals, *Can. Metall. Q.,* 18, 1, 1979.
29. **Steiner, A. A. S.,** Mechanical interaction between fuel and cladding, *Nucl. Eng. Des.,* 21, 237, 1972.
30. **Garzalli, F., Jan Von, R., and Stehle, H.,** The main cause of fuel element failure in water-cooled power reactors, *At. Energy Rev.,* 17(1), 31, 1979.
31. **Syrett, B. C., Cubicciotti, D., and Jones, R. L.,** The origin of variations in the iodine stress corrosion cracking susceptibility of commercial zircaloy-2 tubings, *J. Nucl. Mater.,* 92, 89, 1980.
32. **Hofmann, P. and Spino, J.,** Chemical interactions between simulated fission products and zircaloy-4 at elevated temperatures, including influence on ductility and time-to-failure, *J. Nucl. Mater.,* 102, 117, 1981.

33. **Gngloff, R. P. and Coffin, L. F.,** Simulated Fission Product Embrittlement of Zircaloy Fuel Cladding, Report No. 79CRDO16, General Electric, Schenectady, NY, 1979, 18.
34. **Batey, W. and Bagley, K. Q.,** Fuel/clad reactions in irradiated oxide fuel pins, *J. Br. Nucl. Energy Soc.,* 13(1—4), 49, 1974.
35. **Holzer, R., Knodler, D., and Stehle, H.,** Fuel Clad Interactions: Experience, Testing and Evaluation, A KWU Review, ANS Topical Meeting on Water Reactor Fuel Performance, St. Charles, IL, May 9 to 11, 1977.
36. **Weinberg, J. G.,** Iodine SCC of Zircaloy-4 Tubing, Rep. WAPD-TM 1048, 1974.
37. **Rosenbaum, H. S.,** The interaction of iodine with zircaloy-2, *Electrochem. Technol.,* 4, 153, 1966.
38. **Cox, B. and Wood, J. C.,** Iodine induced cracking of zircaloy fuel cladding — a review, in *Corrosion Problems in Energy Conversion and Generation,* Electrochemical Society, 1974, 275.
39. **Wood, J. C.,** Interaction between stressed zirconium alloys and iodine at 300°C, *Nucl. Technol.,* 23, 63, 1974.
40. **Rolstad, E. and Knudsen, K. D.,** Studies of Fuel/Clad Mechanical Interaction and the Resulting Interaction Failure Mechanism, Proc. 4th Int. Conf. Peaceful Uses of Atomic Energy, International Atomic Energy Agency, Geneva, 1972, 257.
41. **Leger, M. and Fleck, R. G.,** Irradiation growth in Zr-2.5 wt% Nb; **Willard, H. J., Jr.,** Irradiation growth of zircaloy; **Tucker, R. P., Fidleris, V., and Adamson, R. B.,** High fluence irradiation growth of zirconium alloys at 645—725°K; **Williams, J., Darby, E. C., and Minty, D. C. C.,** Irradiation growth of annealed zircaloy-2; **Fleck, R. G., Price, E. G., and Cheadle, B. A.,** Pressure tube development for Candu reactors, papers presented at the 6th Int. Conf. Zirconium in the Nuclear Industry, Vancouver, Canada, June 28 to July 1, 1982.
42. **Gupta, C. K.,** Extractive metallurgy of niobium, tantalum and vanadium, *Int. Met. Rev.,* 29(6), 405, 444, 1984.
43. **Krishnan, T. S. and Gupta, C. K.,** Application and process: metallurgical aspects of niobium, tantalum and hafnium, in *Industrial Applications of Titanium and Zirconium,* Vol. 4, Young, C. S. and Dusham, J. C., Eds., ASTM-SP 917, American Society for Testing and Materials, Philadelphia, 1986, 99.
44. **Zelikman, A. N., Krein, O. E., and Samsonov, G. V.,** *Metallurgy of Rare Metals,* Israel Programme of Scientific Publications, Jerusalem, 1966.
45. **Songina, O. A.,** *Rare Metals,* 3rd ed., Israel Programme of Scientific Publications, Jerusalem, 1970.
46. **Miller, G. L.,** *Tantalum and Niobium,* Butterworths, London, 1959.
47. **Sisco, F. T. and Epremium, E., Eds.,** *Columbium and Tantalum,* John Wiley & Sons, New York, 1963.
48. **Gonser, B. W. and Sherwood, E. M., Eds.,** *Technology of Columbium (Niobium),* John Wiley & Sons, New York, 1958.
49. **Dennis, W. E. and Adamson, A. F.,** Reduction of Niobium Compounds by Metal Reductants, Tech. Note No. 2, UKAEA, 1954.
50. **Kamat, G. R. and Gupta, C. K.,** Open-aluminothermic reduction of niobium pentoxide and purification of the reduced metal, *Met. Trans.,* 2, 2817, 1971.
51. **Tietz, T. E. and Wilson, J. W.,** *Behavior and Properties of Refractory Metals,* Stanford University Press, Stanford, CA, 1965.
52. **Hampel, C. A., Ed.,** *The Encyclopedia of the Chemical Elements,* 2nd ed., Reinhold, New York, 1968.
53. **Douglas, D. L. and Kunz, F. W., Eds.,** *Columbium Metallurgy,* Interscience, New York, 1961.
54. **Douglas, D. L.,** The metallurgy of zirconium, *At. Energy Rev.,* 1971.
55. *Metals Handbook,* desk ed., ASM, Metals Park, OH, 1985.
56. *Metals Handbook,* 9th ed., ASM, Metals Park, OH, 1980.
57. **Ellis, C. E. and Evans, W.,** The pressure tubes in the CANDU power reactors, *CIM Bull.,* 74, 105, 1981.
58. **Dalgaard, S. B.,** Corrosion and Hydriding Behaviour of Some Zr-2.5 Wt% Nb Alloys in Water, Steam and Various Gases at High Temperature, Report AECL-1513, Atomic Energy of Canada Ltd., Chalk River, Canada, 1962.
59. **Jones, A. W.,** The Effect of Heat Treatment on the Properties of Some Zirconium-Niobium-Copper Alloys, Report UPC/R 374, United Power Co., 1965.
60. **Cheadle, B. A. and Evans, W.,** The Fabrication and Metallurgical Evaluation of $3^1/_4$ In. Diameter Cold-Worked Zirconium-$2^1/_2$ % Niobium Pressure Tubes, Report AECL-2652, Atomic Energy of Canada Ltd., Chalk River, Canada, 1962.
61. **Ells, C. E.,** Deformation of irradiated zirconium-niobium alloys, in *Zirconium in Nuclear Applications,* ASTM STP-551, American Society for Testing and Materials, Philadelphia, 1974, 311.
62. **Davison, R. M. et al.,** A review of worldwide developments in stainless steels, *Mater. Des.,* 7(3), 111, 1986.
63. **Sedriks, A. J.,** *Corrosion of Stainless Steels,* John Wiley & Sons, New York, 1979.
64. **Kaltenhauser, R. H.,** Improving the engineering properties of ferritic stainless steels, *Met. Eng. Q.,* p. 41, May 1971.

65. **Bond, A. P.,** Mechanisms of Intergranular Corrosion in Ferritic Stainless Steels, *Trans. AIME,* 245, 2127, 1969.
66. **Steele, D. F.,** Corrosion control in nuclear fuel reprocessing plants, *Atom,* 353, 5, 1986.
67. **Shaw, R. D. and Elliot, D.,** Stainless Steels in UK Reprocessing Plants, Proc. Conf. Stainless Steels, September 1984, Cotherburg, Sweden.
68. **Bush, S. H.,** Pressure vessels for nuclear reactors, in *The Science of Materials Used in Advanced Technology,* Parker, E. R. and Columbo, U., Eds., John Wiley & Sons, New York, 1973, 399.
69. **Roberts, J. T. A.,** *Structural Materials in Nuclear Power Systems,* Plenum Press, New York, 1981.
70. **Waldo, J. B., Franz, G. R., and Loika, E. F.,** FFTF operating experience 1982—1984, *Atom,* 335, 3, 1984.
71. Proc. ICE/BNES Conf. Prestressed Concrete Pressure Vessels, London, March 1967, CONF-670, BNES, Institute of Civil Engineers, London, 1967, 301.
72. **Steele, L. E.,** Neutron Irradiation Embrittlement of Reactor Pressure Vessel Steels, STI/DOC/10/163, International Atomic Energy Agency, Vienna, 1975.
73. **Clark, D. S. and Varney, W. R.,** *Physical Metallurgy for Engineers,* D Van Nostrand, New York, 1962.
74. **Glasstone, S. and Sesonske, A.,** *Nuclear Reactor* Engineering, D Van Nostrand Reinhold, New York, 1967.
75. **Ma, B. M.,** Heat generation and temperature distribution in cylindrical reactor pressure vessels, *Nucl. Eng. Des.,* 11, 1, 1970.
76. **Ma, B. M.,** Irradiation swelling, creep and thermal fatigue behaviours of LMFBR pressure vessels, *J. Franklin Inst.,* 299, 127, 1975.

Chapter 2

MODERATOR AND REFLECTOR MATERIALS

I. INTRODUCTION[1-2]

Neutrons released during the fission process possess a relatively large amount of energy in the form of velocity. The average energy ranges between 1 and 2 MeV, and in this energy range the neutrons are not very effective in sustaining a chain fission reaction in a natural-uranium-fueled reactor. Because of their speed, neutrons in this energy range — fast neutrons, as they are called — miss most nuclei and then disappear harmlessly away from the surface of the fissile material. In order to establish a chain reaction in natural uranium, it is necessary to slow down fission neutrons to thermal energy (in the energy of 0.01 to 0.1 eV, generally taken typically to be about 0.025 eV). Because of their reduced energy, the neutrons in this energy range — thermal neutrons, as they are called — cause the fission cross section of ^{238}U to be much larger than the total capture cross section in ^{235}U and ^{238}U. The possibility is thus established for a chain reaction to occur in natural uranium with thermal neutrons.

The process of scaling down the energy of the fission-produced neutrons from an average of 2 MeV to values of 0.025 to 1 eV, which correspond to the thermodynamic equilibrium with the environment in the reactor and favor ^{235}U fission in thermal reactors, is called neutron moderation, and the materials in the reactor core that bring about this effect are, accordingly, known as moderator materials. The process of how a moderator is capable of slowing neutrons can be easily understood by comparing a high-speed neutron to a fast-moving billiard ball. Now, if this billiard ball were to stike head on another billiard ball of the same mass which is at rest, the striking ball would leave the scene of collision at considerably less speed than it had prior to collision. Of course, the second billiard ball is also now moving, having acquired some of the kinetic energy of the first ball. On the other hand, if the fast-moving billiard ball collides with a much heavier ball, say, a bowling ball, a very small amount of energy will be transferred to the latter, and the billiard ball will simply bounce in a different direction at almost the same speed it was traveling when it hit, while the bowling ball itself would scarcely move. This shows that moving matter can be slowed fastest by colliding with objects of equal mass. The more nearly equal the mass of the struck object is to the mass of the colliding object, the more kinetic energy or speed the colliding object will lose. Thus, for a neutron moving at high speed, the best way to slow it down is to have it collide with another particle whose mass is as close to that of a neutron as possible.

In estimating a figure of merit for moderators, certain specific quantities that are introduced below are of importance:

1. From a practical point of view, it is advantageous to characterize moderators by means of their macroscopical scattering (Σ_s) and absorption cross sections (Σ_a). These cross sections are related by $\Sigma = \sigma N$, where σ is the microscopic cross section and N is the number of atoms in the unit volume given by $N = (N_A\rho)/A$, where N_A is Avogadro's number, and ρ is the density of the material.
2. The typical elastic scattering process by which moderation results can basically be taken as the neutron colliding with the nucleus of one of the atoms of the medium (assumed to be at rest before the collision) and rebounding at an angle, say, θ to its original direction of travel with reduced speed, while the struck nucleus is set in motion with the transferred energy or, in other words, with a kinetic energy equal to that lost by the neutron. It can be proved that kinetic energy lost by the neutron is independent

of the initial energy of the neutron, but dependent upon the atomic mass, A, of the moderator and the angle, θ, through which the neutron scatters at collision. The relationship is given by

$$E_1/E_2 = 1/2\ [(1 + R) + (1 - R)\cos\theta]$$

where E_1 and E_2 are neutron energies before and after collision, respectively, and R is equal to $[(A - 1)/(A + 1)]^2$. We can define a quantity known as the average or mean logarithmic energy decrement per collision, ξ, which is given by $\ln(E_1/E_2)$. By averaging $\ln(E_1/E_2)$ — where E_1/E_2 is given as related with θ, the angle by which the neutron is scattered — over all possible values of the angle θ, one obtains the magnitude of the mean logarithmic decrement as a function of only the atomic mass of the moderator $\xi = 1 + R\ln R/(1 - R)$. For materials with $A > 10$, this may be approximated by $\xi = 2/(A + 2/3)$. For a substance consisting of several varieties of nuclei (1, 2, . . . , n), the mean value of ξ is obtained as

$$\bar{\xi} = (\xi_1\sigma_{s1} + \xi_2\sigma_{s2} + \ldots.. + \xi_n\sigma_{sn})/(\sigma_{s1} + \sigma_{s2} + \ldots.. + \sigma_{sn})$$

The mean logarithmic energy decrement may be used to calculate the mean collision number (MCN) that is required for a neutron resulting from fission (E_1 = 2 MeV) to scale down to the energy level of the thermal neutrons (E_2 = 0.025 eV).

$$\text{MCN} = \ln(E_1/E_2)/\xi = 18.2/\xi$$

3. The effectiveness of a moderator in slowing down neutrons depends not only on the value of ξ but also on the probability per unit value that scattering does take place. Quantitatively, then, the effectiveness may be described by the product $\xi\Sigma_s$, where Σ_s is the macroscopic scattering cross section of the moderator. This product is known as slowing-down power or moderating power (MP).
4. The slowing down is not in itself a sufficient indication of the suitability of the material as a moderator. It is necessary to take into account the absorption cross section. Boron, for example, which has good slowing-down power, is impractical as a moderator because of its very high absorption parameter for thermal neutrons. One can obtain an overall index of suitability by what is called the moderating ratio (MR), obtained by dividing the slowing-down power by the absorption cross section ($MR = \xi\Sigma_s/\Sigma_a$). The moderating ratio in essence enhances the information provided by the moderating power by taking into account the absorption cross section of the moderator. It is a measure of the relative slowing-down power and absorbant capability of a substance.
5. A moderator parameter of importance is mean free path λ, which is equal to $1/\Sigma$. It is defined as the mean distance traveled between events, e.g., scattering mean free path $\lambda_s = 1/\Sigma_s$.
6. The moderating density (MD) is yet another parameter taken to describe the suitability of the moderator. In a special situation where there is no neutron absorption and consequently no losses, then MD equals the number of neutrons that result from fission and consequently remains independent in their energy. In effect, there are two distinct stages of the moderation process. First is the neutron slowdown in which neutron distribution is inversely proportional to neutron energy. The second stage is the thermalization in which the energy distribution of the neutrons is a Maxwell-Botzmann one. In the first stage the moderation process can be described by applying the classical neutron-nucleus interaction, while in the second stage, use is made of interaction model based on quantum mechanics.

7. The time that is required for slowing down of a neutron from the fast to the thermal neutron stage is crucial because of the need to have the neutron in a form in which it can cause fission before it is either captured or disappears altogether from the reactor core. One can express this by $t = L[(2/(M + 0.677)](2m)^{1/2}(\frac{1}{E_s^{1/2}} - \frac{1}{E_f^{1/2}})s$, where L is the mean free path of the collisions between the neutrons and the nuclei of the moderator, M is the number of atomic mass units of the moderator used, m is the mass of the neutron (1.67×10^{-27} kg), E_f is the kinetic energy of the fast neutron, and E_s is the kintetic energy of the neutron after it has been slowed down to the desired energy content. The time which a neutron can spend inside the moderator between thermalization and capture is given by the diffusion time. This is given by dividing the mean free absorption path length by the average velocity of a slow neutron with kinetic energy of 4×10^{-21} J, which corresponds to a speed of 2200 m/s. Apart from diffusion time, another parameter in dealing with moderating properties of materials is the slowing-down time, which is defined as the time required for thermalization of neutron.
8. The Fermi age of neutrons is related to the mean square distance traveled while slowing down. The square root of the age is called the slowing down length, which, for thermal neutrons, is essentially a measure of net reactor distance traveled from their formation as fission neutrons to their attainment of thermal energy. The sum of the square of the diffusion length of thermal neutrons L^2 and of the age τ is called the migration area $M^2(M^2 = L^2 + \tau)$, and the square root is referred to as the migration length $M(M = \sqrt{L^2 + \tau})$. Migration length for thermal neutrons is related to the total distance covered between generation as a fission neutron and capture as a thermal neutron. The migration length has a direct bearing on the size of a critical reactor.

The foregoing list has provided the definition and physical significance of moderator parameters. A good moderating material has the following qualities which make it suitable for this role:

1. The material should be an element of low mass number so that large energy loss per collision results.
2. The nuclei of the material should have a fairly large cross section for scattering neutrons, which implies frequent collisions per unit time, thereby providing the neutron with a better chance of being moderated before capture or loss through leakage.
3. The material should have a small cross section for capturing neutrons so that the probability of neutrons being lost in that way is small.
4. The material should present itself providing a relatively short mean free path, or distance for a fast neutron to travel during slowing down, thereby reducing the leakage and escape.
5. In order that the scattering should be the dominant process at all energies, including the ^{238}U resonance region, the quantity of moderator should be much greater than the quantity of fuel.
6. Moderators must have attributes of the structural materials. They must have adequate mechanical strength, thermal and irradiation stability, fabricability, and corrosion resistance. Reasonably good heat transfer characteristics are also required for moderators, to deposit and dissipate the energies lost by neutrons while slowing down in thermal reactors.

In a self-sustaining nuclear chain reaction that proceeds in a nuclear reactor, most of the neutrons which are produced during the fission process are used in the reactor (1) to sustain the chain reaction, (2) for control purposes, or (3) for breeding purposes. A certain number of neutrons found near the surface of a nuclear core, however, tend to escape from the region of the chain reaction and hence can be put to no useful purpose. This leakage of neutrons outside the assembly impairs the neutron economy, and consequently an additional amount of fissionable material must be added as fuel. This is costly and often requires that the core of the reactor be somewhat larger. To cut down on leakage losses, radial and axial reflector materials are used surrounding the reactive region. The purpose is to scatter back escaping neutrons into the core in both thermal and fast reactors without absorbing them. The ratio of neutrons scattered back from the surface of the reflector to the amount reaching the surface is called the albedo. A reflector basically is nothing more than a low-cross-sectional material. which permits neutrons to collide with its own nuclei without neutron absorption. When a neutron collides with any low-cross-section nucleus, its direction of motion is changed. This, in a sense, is a neutron getting reflected. A material having a low cross section surrounding the core of a reactor acts as a neutron reflector. The choice of material, of course, depends upon several factors. In a thermal reactor the reflector can be made of a moderating material. In this case, when a neutron is reflected back into the reactor core, it has given up much of its kinetic energy to the reflector-moderator and hence is more useful to the chain reaction than if it had been reflected with all its original kinetic energy. In a fast reactor, slowing down of neutrons is not desirable, and many materials that scatter neutrons strongly can be used as reflectors. The reflector should be an element of high mass (a nonmoderator) so that the neutrons will be reflected back into the core with most of their original kinetic energy. They can thus be of maximum benefit in helping to maintain the fast chain reaction. In fast reactors, therefore, the role of reflector may only be played by heavy materials such as iron, lead, bismuth, thorium, uranium, etc., that possess a high-scattering cross section for fast neutrons.

The reference to the uses particularly of uranium and thorium in the fast reactors as given above suggests the need to identify another class of materials. The fast reactors are designed to operate without the need for high reflectors. Commonly, neutrons which are escaping from the chain reaction in the reactor are intercepted and used for secondary purposes. This is precisely the case with the fertile materials, such as natural thorium (^{232}Th) or depleted natural uranium (^{238}U) provided in fast breeder reactors surrounding the cores both radially and axially. New fuels are bred by the absorption of leading or escaping neutrons in these fertile materials. When a material is thus placed around the core of a reactor for the purpose of converting some of this material into another element, the material is known as a blanket. It may be pointed out that every collision between the neutron and the core of a blanket atom may not result in capture of the neutron. The neutron may just enter into an elastic collision with the blanket nucleus and may bounce back into the reactor core. Blanket materials, then, almost always act as reflectors. The extent of reflection depends on the material used in the blanket and the energy range of neutrons.

In a thermal reactor the nuclear requirements are the same for moderators and reflectors. Applying the basic criteria for moderating materials, *prima facie,* the fact that the nucleus of a hydrogen atom contains but a single proton and that the mass of a proton is virtually identical to that of a neutron would make the hydrogen nucleus an ideal collider for neutrons. Hence, hydrogen would be an ideal moderator if it were not for the fact that hydrogen nuclei have a certain affinity for neutrons and often tend to capture a colliding neutron. Although the probability of neutron capture by a hydrogen nucleus is relatively small, this probability is still large enough to be undesirable from the point view of neutron economy. The capture by hydrogen nuclei can result in too many neutrons being lost from the chain reaction, and on this basis one rules out the use of hydrogen as a moderator, except in those few cases where there are enough surplus neutrons so that one can afford neutron capture by hydrogen nuclei. In that situation, the high moderating power of hydrogen can be realized.

Aside from hydrogen, the present discussion can be extended to the next lightest nucleus as a potential moderator. Such a nucleus is that of deuterium or heavy hydrogen, which is composed of one neutron and one proton, making it only twice the mass of the neutron and thus still an effective slowing-down material for the neutron. Unlike the hydrogen nucleus, the deuterium nucleus possesses little affinity for a neutron; and, on account of this, it is extremely unlikely that a neutron will be captured by a deuterium nucleus and thus lost to the chain reaction. In fact, deuterium is an outstanding moderator because it has, as stated, such a low capture cross section. Deuterium and its compounds have the added nuclear advantage that deuterium undergoes (γ,n) reaction in the pile that contributes some neutrons to the chain reaction. Its widespread use has been limited in the past only because of its scarcity and high cost. Although hydrogen and deuterium excel in service as moderators, they cannot be used as gaseous moderators because the atomic density is too low. Consequently, moderators will be liquid or solid. Hydrogen (and deuterium) are used more frequently in the forms of their oxides (ordinary and heavy water). However, there are many compounds such as hydrides, hydrocarbons, and deuterocarbons, which have high concentractions of hydrogen (deuterium) in their molecules, and hence they can be used as moderators. Other light materials such as beryllium and carbon or their compounds are the best solid moderators. The atoms of lithium and boron have the necessary small mass requirements for good moderators, but they are excluded from consideration because the nucleus of each of these elements has an exceedingly high affinity for neutrons, so that the probability of elastic scattering, in preference to capture, is relatively small. The use of elements of a mass number greater than carbon has not drawn attention simply because of the fact that they have considerably lower moderating ability. When carbon is used, it is usually in the form of graphite. The use of beryllium is either in the metallic form or as beryllium oxide.

A summary is now in order. The purpose or chief function of a moderator is to moderate or slow down fast fission neutrons from high energy to thermal energy in a thermal reactor. On the other hand, the purpose of the reflector is to reflect leaking neutrons back into the core in either thermal or fast reactors. From a nuclear consideration, the primary requirements of moderator and reflector materials are the same: (1) low neutron absorption, (2) high neutron scattering, and (3) low mass number (large energy loss by neutron per collision). For solid moderators and reflectors, desirable characteristics include (1) fabricability, (2) structural strength, (3) corrosion resistance, (4) high density of moderators or reflectors, (5) high thermal conductivity, and (6) good thermal and radiation stability. For fluid moderators or reflectors, desirable characteristics include (1) melting point below room temperature, (2) low vapor pressure at high temperature, (3) being noncorrosive to structural materials, (4) high density of moderating or reflector atoms, (5) good heat transfer properties, and (6) good thermal and radiation stability. Table 1 shows the major moderator-reflector materials along with their requirements. Their leading characteristics are presented in Table 2. Among the major moderator and reflector materials, in a thermal reactor, ordinary water (H_2O) and heavy water (D_2O) not only perform the functions of moderator and reflector simultaneously, but also function as coolant material. Because of its abundance and economics, ordinary water has been used extensively as the coolant and working substance in LWRs. Of all the elements, beryllium and carbon are the only attractive solid moderators. They can simultaneously perform the function of moderator, reflector, and structural materials. Graphite has long been used as moderator, reflector, and structural material in the gas-cooled, graphite-moderated reactor or high-temperature gas-cooled reactor. The first nuclear fission reactor, CP-1, in 1942 was a natural uranium-graphite reactor. Table 3 presents moderators employed in commercial thermal reactor types along with their required fuels.

II. ORDINARY WATER

Fairly small neutron absorption cross section, excellent slowing-down power, and small

Table 1
MODERATOR-REFLECTOR MATERIALS IN THERMAL REACTORS

Major nuclear requirements	Large energy loss by neutron per collision Low absorption of neutron High neutron scattering
Candidate materials	Light water Heavy water Beryllium, beryllium oxide Graphite

Table 2
CHARACTERISTICS OF MODERATING MATERIALS[1,2]

	Light water	Heavy water	Beryllium	Graphite
At. (or molecular) wt.	18.0	20	9.01	12.0
Density(g/cm^3)	1.0	1.10	1.84	1.57
Atoms (or molecules) per cm^3	3.3×10^{22}	3.3×10^{22}	1.2×10^{23}	8.1×10^{22}
Microscopic scattering cross section epithermal (σ_s b)	49	10.5	7.0	4.8
Microscopic absorption cross section thermal (σ_a b)	0.66	0.0026	0.0095	0.0045
Macroscopic scattering cross section epithermal (Σ_s cm^{-1})	1.64	0.35	0.74	0.39
Macroscopic absorption cross section thermal (Σ_a cm^{-1})	22×10^{-3}	85×10^{-6}	1.1×10^{-3}	0.03×10^{-3}
Average logarithmic energy loss per collision (ξ)	0.95	0.51	0.2078	0.1589
Mean collision number to thermalize (MCN, 2 MeV—0.025 eV)	19.6	35.7	86.4	115
Slowing-down or moderating power (MP, s cm^{-1})	1.28	0.18	0.11	0.063
Moderating ratio (MR, $\xi\Sigma_s/\Sigma_a$)	70	2100[a]	150	170
Slowing-down time (s)	1.02×10^{-5}	4.50×10^{-5}	6.7×10^{-5}	1.5×10^{-5}
Diffusion time (s)	2.1×10^{-4}	6.9×10^{-2}	3.7×10^{-3}	1.3×10^{-2}
Slowing-down length (cm)	5.6	11.0	9.2	18.7
Diffusion length (cm)	2.76	100	21	64.2
Migration length (cm)	6.2	101	23	57

[a] Pure heavy water has a moderating ratio of 12,000. The figure given corresponds to 99.8 at. % heavy water (0.2% is light water).

migration length for thermal neutrons are among the important neutronic characteristics of ordinary water. These, along with excellent specific heat properties and, above all, cheapness and abundant availability of water, have conferred it a specific place in nuclear power generation. In fact, ordinary or light water has been used as the moderator, reflector, coolant, solvent, shielding (for neutrons) material, and working substance for the steam turbine generator system in the LWRs. It is estimated that light water is presently in use in these roles in some two thirds of all nuclear power plants in operation throughout the world. In LWRs, the moderator is ordinary (though highly purified) water, which is contained under pressure and serves as the coolant. Hydrogen has a relatively high absorption cross section for neutron capture.

A critical system in this case can be attained with light water as moderator only if uranium enriched in the fissile isotope is used as fuel. With enriched material and with the small migration length of thermal neutrons in water, it is possible to design a compact reactor

Table 3
MODERATORS EMPLOYED IN COMMERICAL THERMAL REACTOR TYPES

Reactor type	Moderator	Fuel required
Thermal Gas-Cooled Reactors		
Magnox	Graphite	Natural uranium metal
AGRs	Graphite	UO_2: 2.5% ^{235}U in ^{238}U
HTGRs or HTRs	Graphite	UO_2: 10% ^{235}U in ^{238}U
Thorium high-temperature reactors (THTRs)	Graphite	U-Th oxide: 10% ^{235}U in ^{232}Th
Water-Cooled Reactors		
PWRs	Light water	UO_2: 3% ^{235}U in ^{238}U
BWRs	Light water	UO_2: 2.2% ^{235}U in ^{238}U
CANDUs	Heavy water	Natural UO_2
SGHWRs	Heavy water	UO_2: 2.3% ^{235}U in ^{238}U
Fast Reactors		
LMFBRs	None	UO_2: 20% ^{235}U in ^{238}U

core. This has been realized in the most widely used commercial nuclear reactor, the light water reactor (LWR), in which light water has established itself as the most well-proven moderator. The subvarieties of LWRs include the pressurized water reactor (PWR), in which the heat is transferred from the coolant in a steam generator to a separate steam-water circuit driving the turbine, and the boiling water reactor (BWR), in which the coolant is allowed to boil as it passes through the core and the steam formed therein is led directly to the turbine. The primary coolant water circuit, which also acts as moderator in the LWR, is different from the heavy water. In it the functions of moderation and cooling are separated. The smaller value of the slowing-down power of heavy water, about 0.18 cm^{-1} as opposed to 1.28 cm^{-1} for light water, necessitates a small fuel-to-moderator ratio in the reactor lattice. This in turn leads to low power densities and large-volume cores. Since heavy water cores need not be as closely packed as those using ordinary water as moderator, it is possible to utilize the pressure tube approach. This avoids the problem of large pressure vessels inherent in systems in which ordinary water is the moderator-coolant. Unfortunately, the use of water immediately raises some problems too. It is corrosive, a superb solvent, and undesirably affected by irradiation. One of the main drawbacks to the use of water as a moderator is its comparatively low critical temperature (647 K). This sets a limit to the highest steam temperatures that can be achieved, and this in turn limits the plant thermal efficiency. It also requires a high pressure in the primary system, which poses a challenge to structural materials. Pressures in the vininity of 2000 psi are used in PWR, although the values are somewhat lower in boiling-water systems.

The following sections outline the chief physical and technological aspects pertaining to the use of water in nuclear reactors.

A. Activation

Nuclear reactions leading to radioisotope formation take place under the influence of high energy neutrons, photons, and γ radiation. The oxygen isotopes of water give rise to the main activation products. Among the nuclear reactions that take place, the most important ones are $^{16}O(n,p)^{16}N$ and $^{18}O(n,\gamma)^{19}O$. In the former case, ^{16}O is found in significant amounts, and the beta and gamma radiation emissions from ^{16}N are quite strong. In the latter case, the neutron capture cross section is important, and the ^{18}O proportion is by no means negligible. Both cases are characterized by short half-life, and consequently the saturation activity is quickly attained. The resulting activity is relatively high, and therefore it is necessary for water to follow long enough routes prior to coming into contact with the operating personnel.

Apart from water activation, as mentioned above, impurities that are either ordinarily in water or occur as a result of corrosion process (such as iron, aluminum, sodium, etc.) are strongly activated by neutrons. In the case of iron, irradiation induces the occurrence of radioactive isotopes ^{59}Fe and ^{60}Co with their respective half-lives of 4×10^6 s and 1.67×10^8 s, which allows the water activity to decrease to an acceptable limit only over some weeks after irradiation has ceased. The situation is not difficult to this extent for the case of aluminum. In water, even in small amounts, the aluminum activation product — an isotope with a short half-life of 120 s — contributes significantly to the activation of the moderator. However, a shutdown taken for a brief period of a few hours is enough for the water activity to disappear completely. In PWRs, where special steels are used, the impurities that are of concern as far as activation is concerned are tantalum, cobalt, sodium, and manganese. Apart from radioactivity, the impurities contribute to the consumption of thermal neutrons in the reactor core, and to corrosion acceleration.

B. Radiolysis[3]

Upon exposure of water to nuclear radiations, it gives rise to hydrogen, hydrogen peroxide, plus the two free radicals. The situation can be represented according to the reaction $3H_2O \rightarrow H_2 + H_2O_2 + H^* + OH^*$. If the water is perfectly pure, the reactions $H_2 + OH^* \rightarrow H_2O + H^*$ and $H_2O_2 + 2H^* \rightarrow 2H_2O$ take place. This means that one achieves status quo, provided the water is thoroughly pure. If, however, there are any impurities in the water that are capable of reacting with H_2O_2, H^*, or OH^* produced during the radiation treatment of the water, such reactions then occur, and the residual substances are left in the water. For example, traces of bromide ions are able to eliminate both the H^* and the OH^* by the chain reactions $Br^- + OH^* \rightarrow Br^* + OH^-$, $Br^* + H^* \rightarrow Br^- + H^+$, and so forth. This means that hydrogen peroxide builds up in the irradiated water and hydrogen is evolved. As a general rule, it can be stated that the presence of acidic ions in water is accompanied by the formation of hydrogen gas under conditions of radiation, leaving hydrogen peroxide behind. The quantities of hydrogen which are usually formed are of the order of 25 to 30 $\times 10^{-6}$ mol/l reactor volume, assuming normal water-cooled reactor conditions. The formation of peroxide is inhibited at temperatures in excess of 250°C, as H_20_2 tends to split into water and free oxygen. Formation of hydrogen in irradiated water can be reduced by the addition of nitrates. In general it can be said that in any reactor using either ordinary or heavy water as moderator, reflector, or coolant there would be some decomposition into hydrogen (or deuterium) and oxygen due to the action of various radiations. Provision must therefore be made for the removal or recombination of these gases.

In PWRs, a high temperature (and hence a considerable rate of recombination) and also accumulation of hydrogen prevent water hydrolysis. In BWRs, the radiolysis proceeds more vigorously than in PWRs. Recombination is more sluggish in the steam phase than in the condensed phase. Moreover, in compliance with the water-steam distribution ratio, the radiolytic hydrogen goes mainly to steam and is discharged from the system through the condenser ejector. Hydrogen does not accummulate in the system. Intensive water radiolysis takes place in the BWR core zone. BWR primary coolant contains 0.2 to 0.4 ppm oxygen during normal operation. These levels are inherent in the operating characteristics of a BWR; the oxygen content is the direct result of radiolysis. Radiolysis cannot be controlled in the core, but is minimized to prevent accelerated corrosion of the system. Since the BWR is an open system, hydrogen cannot be easily introduced into the system to scavenge oxygen. The pH of the coolant water in a BWR is close to neutral between 6 and 8 at 25°C. Unlike

Table 4
PHYSICAL PROPERTIES OF D_2O AND H_2O

Property	D_2O	H_2O
Molecular mass (amu)	20.03	18.016
Density (kg/m^3)	1107	1000
Boiling point (K)	377.5	373.1
Freezing point (K)	276.96	273.15
Critical temp (K)	644.6	647.3
Temp corresponding to max density (K)	284.35	277.13
Vaporization heat (kJ/kg)	2073.4	2232.9
Thermal conductivity (W/mk)	1.96	2.12
Viscosity (kg/ms)	1.2514×10^{-3}	1.005×10^{-3}
Refractive index	1.3283	1.3326

a PWR, boron and lithium are not added to the coolant. More details regarding water chemistry are given in Chapter 3 of this volume.

III. HEAVY WATER

Heavy water (deuterium oxide) differs from the vastly more abundant light water (H_2O) in having the higher isotope of hydrogen, ^2H, in place of ^1H. The physical properties of heavy water and light water are given in Table 4. Their specific weight differs by about 11%. Other properties are close to those of ordinary water. It freezes at 3.96°C and boils at 104.5°C, the values being a few degrees above the corresponding temperatures for ordinary water. High-temperature reactor systems involving heavy water, like those using ordinary water, would therefore require application of high pressures. The density of heavy water is about 1107 kg/m^3 at room temperature. Heavy water is the most valuable material in the nuclear power industry for a number of reasons. From data provided in Table 2, some important characteristics may be mentioned. It has a remarkably low absorption cross section for neutrons, which is about 0.15% of that of water, which in itself is not very large. Its slowing-down power is satisfactory (not as high as that of water). Its moderating ratio is the highest among all moderators. All of these characteristics combined make the heavy water an excellent moderator and reflector.

It is entirely feasible that a heterogeneous reactor with natural uranium as fuel and with heavy water as moderator can be made critical. This route has been the principal line of Canadian reactor system development. This has been extensively referred to in an earlier chapter. The economic incentive for the use of heavy water as moderator has not yet been established. However, use of natural uranium fuel in a heavy-water-moderated system is the noteworthy advantage. The economics of heavy water reactors are determined by the costs of isolating deuterium relative to those of enriching uranium, the inventory charges for the large amount of expensive heavy water required, and a number of other factors that could be affected by emerging technological developments. For example, if a plutonium-fueled fast breeder reactor system becomes important in nuclear technology, the role of heavy water would be adversely influenced. Each MWe installed in the most widespread heavy water reactors — the CANDU PHWRs — requires 850 kg of moderator. Each unit in Bruce A station in Canada with 745-MWe capacity has a 568-t heavy water inventory. Of the approximately 3400 t/year worldwide capacity for producting heavy water, about 75% is located in Canada, often sharing the same site with a CANDU reactor plant.

In most ways, the chemistry of heavy water is similar to the chemistry of light water, except that velocities of reaction of compounds in which deuterium is present are usually lower, and in some cases a good deal lower, than those in which protium (^1H) is present.

This means that under similar conditions corrosion reactions are somewhat slower with heavy water than they would be with light water. However, care is still to be taken that there are no corrosive conditions within reactors.

A. Irradiation

The heavy water radiolysis phenomenon follows the same pattern as that of light water. For heavy water, the deuterium in the gaseous state has to be recovered and D_20 regenerated by a reforming process. As for light water, the radiolysis rate may be lowered by an initial purification of water and subsequent removal of corrosion products. Formation of radioactive products is an important effect of irradiation in heavy water. Unlike light water, heavy water gives a nuclear reaction from which tritium emerges. Tritium is a beta radioactive emitting a radiation of 0.018 MeV and having a half-life of 12.4 years. This leads to what is called heavy water tritiation, causing a radioactivity, and so there arises a need to submit to a tritium scrubbing process. Besides this, another characteristic reaction is $^2H(\gamma,n)^1H$ that leads to permanent depletion of deuterium in heavy water, by hydrogen formation. Elimination of hydrogen that has entered the moderator in various ways is carried out in heavy water reconcentration facilities, based on a vacuum distillation process.

B. Production of Heavy Water[4-8]

Deuterium occurs in nature in all hydrogen-containing compounds. The deuterium found in the surface water is in the form of HDO. The deuterium content is about 0.050 atom%, which varies slightly because of a slight isotopic fractionation caused by evaporation and condensation of water vapor in the atmosphere (1 out of 6666 molecules of water is heavy water). The credit for first isolation of deuterium in relatively pure form goes to Urey and co-workers at Columbia University in 1931, and nearly pure D_2O was prepared by G. N. Lewis shortly thereafter by electrolysis. As far as heavy water production processes are concerned, a large number of them have been proposed, but only a few of them proved technically reliable and economically sound. General process characteristics for heavy water production bear specialities. The very low value of deuterium-to-hydrogen ratio in nature of about 130 ppm is the main factor responsible for the high cost of heavy water. It is necessary to process at least 8000 mol of feed per mole of product for all processes. Reactor-grade heavy water is 99.7 mol% D_2O. Thus, the overall concentration ratio from feed to product is about 3×10^6. This means that hundreds of separative elements in series must go from natural water to reactor-grade heavy water. The combination of very large feed flow and a very large number of separative elements means that heavy water plants are very large in relation to other chemical plants. As a consequence, heavy water is an extremely capital-intensive product. The combination of a dilute feed and a large overall concentration ratio means that cascading confers very important advantages.

We shall briefly digress here to provide an elementary understanding of some of the terminologies used in isotope separation plant practice. A unit in plant practice is recognized as the smallest element capable of performing some separation of the process material (a single diffusion barrier, one stage of a mixer-settler, one plate of a distillation column, a gas centrifuge, and one electrolytic cell are examples of a single separating unit). An ensemble of separating units in parallel constitutes one stage. The parallel-connected separating units are fed with material of the same composition and yield partially separated product streams of the same composition. Often a single unit serves as a stage, like a plate of a bubble-plate column. However, separation methods involving low-capacity units (e.g., an electrolytic cell) require use of many units in parallel. In case the degree of separation achieved by a single stage is less than the intended or desired degree of separation between product and waste, it is essential to connect stages in series.

A cascade is such a series-connected group of stages (e.g., a complete distilling column,

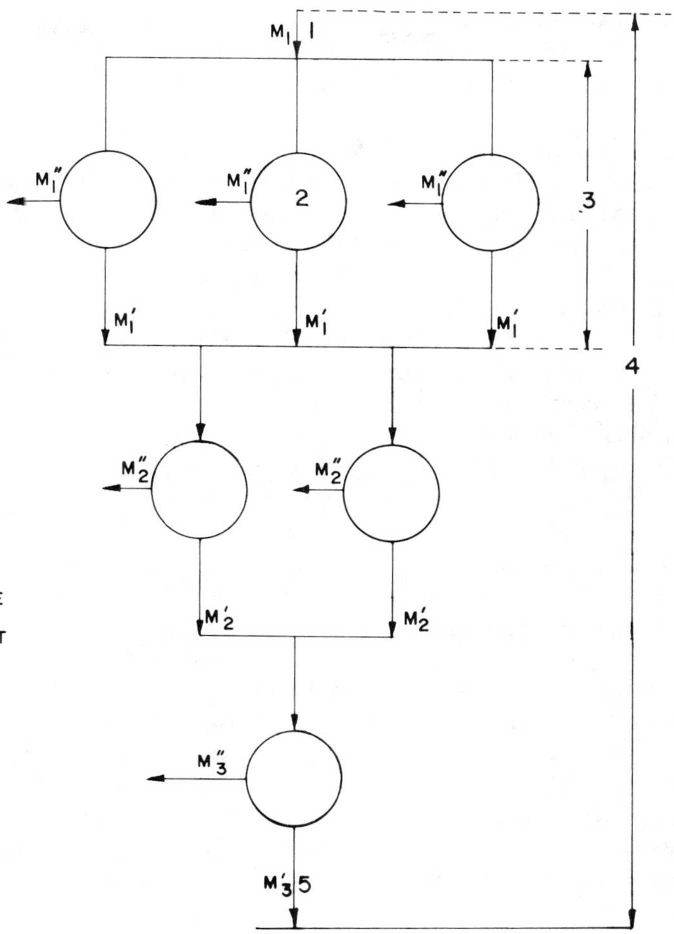

FIGURE 1. Illustration of the relation among unit, stage, and cascade.

a battery of solvent-extraction mixer-settlers). The connection that exists between unit, stage, and cascade is shown in Figure 1. Each unit of this cascade might represent, for example, one gas centrifuge. The group of parallel-connected units, each of which separates feed of composition M_1 into a partially enriched stream of composition M_1^1 and a partially depleted stream of composition M_1^{11} constitutes the first stage of this cascade. The cascade is the entire group consisting of series- and parallel-connected centrifuges. A simple cascade is one in which no attempt is made to reprocess the partially depleted tail streams leaving each stage. The feed for one stage is the head fraction from the next lower stage of the cascade. In instances where partially depleted stage tails have sufficient value, a countercurrent recycle cascade may be applied. A gaseous diffusion cascade, a bubble-plate distillation column, and a battery of series-connected solvent-extraction mixer-settlers are illustrative of such an approach. The arrangement essentially comprises the two portions of the cascade, one between the feed point and product end, called the enriching section, and the second between the feed point and waste end, called the stripping section. The aim of the enriching section is to make material of product composition. The stripping section objective is to enhance the recovery of the desired isotope from feed. The essentiality of the enriching section lies in making product of the desired grade. The stripping section, on the other hand, is employed only to obtain a reduction of the amount of feed required to make a given amount of product. When the feed has no value, as with water feed for a deuterium plant, there can be a total elimination of the stripping section.

Major methods of heavy water production fall into three categories: distillation (hydrogen distillation and water distillation); electrolysis (water electrolysis); and chemical exchange (exchange between hydrogen sulfide and water, hydrogen and water, and hydrogen and ammonia).

To inject a historical note, it may be mentioned that the real stimulus for the development of industrial processes for the production of heavy water came about only after the recognition in 1940 of the fact that deuterium as heavy water has nuclear properties that make it a highly desirable moderator and coolant for nuclear reactors fueled by uranium of natural isotopic composition. It was between 1940 and 1945 that three heavy water plants based on water distillation were designed, built, and operated by the DuPont company in the U.S. These plants were located at Morgantown, WV; Childersberg, AL; and Dana, IN. Among these, the Morgantown plant was the smallest and most efficient, and produced 500 lb of D_2O per month, with a deuterium recovery of 2.8%. All the plants used electrolysis for final concentration after accomplishing the initial separation. During this period another plant located in Canada at Trail, British Columbia, was operated. The Trail plant used chemical exchange between hydrogen gas and steam for the initial isotope separation followed by electrolysis for final concentration. The four plants produced a total of about 32 metric tons of 99.8% pure D_2O before they were shut down in 1945. Construction of a truly large-scale facility was initiated at the Savannah River site near Aiken, SC in 1950 to produce heavy water for the nuclear reactors operated there. The plant used the dual-temperature exchange of deuterium between hydrogen sulfide and water for isotopic enrichment of deuterium. The plant design capacity was to produce 450 metric tons of heavy water per year, and a plant of similar capacity was constructed at Dana, IN. Water electrolysis is the earliest technique of deuterium separation, and, in fact, until 1943 all commerical production of heavy water was by electrolysis. The largest single producer of heavy water was the Norsk Hydro Co., which operated the world's largest electrolytic hydrogen plant at Rajukan, Norway. In 1942, this plant was producing about 1.7 short tons of heavy water per year as a by-product of the production of 6.4 lakhs std ft^3 of electrolytic hydrogen per hour, used for synthesis of ammonia.

1. Distillation

a. Hydrogen Distillation

Distillation of liquid hydrogen as a method for separating deuterium received early consideration. The cryogenic temperatures and the requirement that the large hydrogen feed required be extremely pure have been deterrents in the past to the use of this process. It may also be mentioned that, since hydrogen is costly, it is not economical to make hydrogen solely to serve as feed for a distillation plant. A hydrogen distillation plant, therefore, can work economically by taking a stream of hydrogen-rich gas produced for other purposes, extracting most of the deuterium, and returning the impoverished hydrogen to the host plant.

A typical hydrogen distillation flowchart is shown in Figure 2. Natural hydrogen containing 0.0258% HD at slightly above atmospheric pressure is cooled and admitted into the primary distillation tower which fractionates it into an overhead product substantially free of HD and a bottoms product with 5 to 10% HD. The bottoms from this tower are fed into the upper half of a secondary tower. In this tower, fractionation into a bottoms product of nearly pure HD is completed. This HD is heated to room temperature in a heat exchanger and taken through a catalytic exchange reactor where its disproportionation into an equilibrium mixture of H_2, HD, and D_2 is catalyzed. The exchange reaction product is cooled to liquid hydrogen temperatures in the heat exchanger and admitted to the bottom half of the secondary tower in which it is fractionated into an overhead product of HD plus H_2 and a bottoms product of pure deuterium. This is warmed to room temperature in the heat exchanger and constitutes the plant product. The HD and H_2 overhead from the bottom of the secondary tower is fed to the top of the secondary tower for recovery of HD.

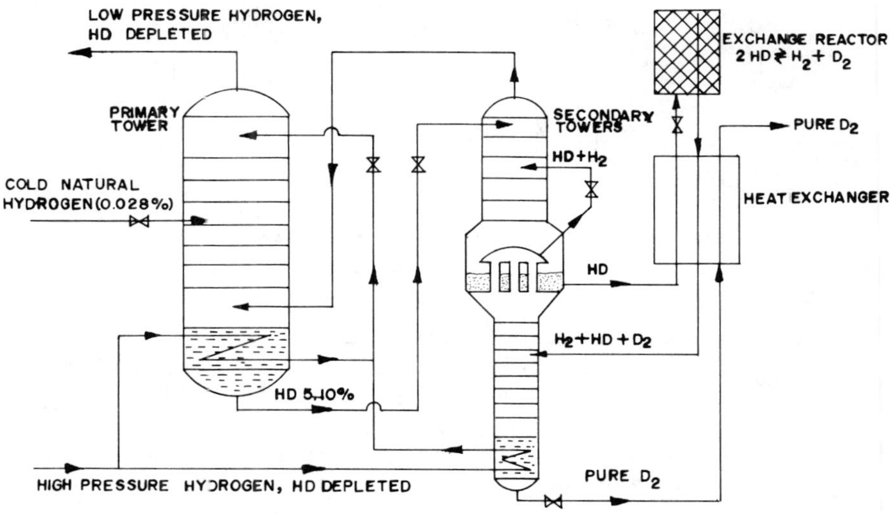

FIGURE 2. Production of deuterium by distillation of liquid hydrogen.

Hydrogen distillation plant production capacity is limited by that of the hydrogen source. The hydrogen needs careful purification to eliminate paramagnetic impurities such as oxygen. It should be brought below about $10^{-4}\%$ in order to avoid the ortho- to parahydrogen conversion. Extraction of heat released during this transformation consumes a high amount of work. The heat release is, however, strictly necessary in order to avoid a quick release of liquid from the storage vessels. One feature of deuterium production through hydrogen distillation is the exclusive enrichment in the HD species, and this is so because equilibration among the H_2, HD, and D_2 molecules takes place only in the presence of a specific catalyst. As a result, the maximum molar deuterium concentration that may be achieved is 50%. Further deuterium concentration is, preferably, done by vacuum distillation or by the electrolysis of water obtained by burning the hydrogen product from the distillation installation. The overall specific energy consumption in this process is high, reaching up to 5500 kWh/kg D_2O.[9]

b. Water Distillation

It is apparent from the historical note provided earlier that vacuum distillation of water was the first method used for the extraction of deuterium on a large scale. Water includes three molecular species — H_2O, HDO, and D_2O — that are in equilibrium according to the reaction $H_2O(l) + D_2O(l) \rightleftharpoons 2HDO(l)$. In concentrating heavy water by distillation, the deuterium separation factor is given by the ratio of the atomic ratio of deuterium to hydrogen in the liquid to the corresponding ratio in the vapor. In terms of the mole fractions of individual compounds in the liquid a and vapor b, the separation factor α is

$$\alpha = \left(\frac{a_{HDO} + 2a_{D_2O}}{2a_{H_2O} + a_{HDO}}\right)\left(\frac{2b_{H_2O} + b_{HDO}}{b_{HDO} + 2b_{D_2O}}\right)$$

It has been shown that in all D:H ratios, and on the conjecture that (1) the equilibrium constant of the reaction ($H_2O + D_2O \rightleftharpoons 2HDO$) has the value 4.0, (2) the equilibrium is maintained in the liquid, (3) liquid and vapor phases form ideal solutions, and (4) HDO vapor pressure is the geometric mean value of the vapor pressures of H_2O and D_2O, α may be expressed in terms of the vapor pressures P of H_2O and D_2O by

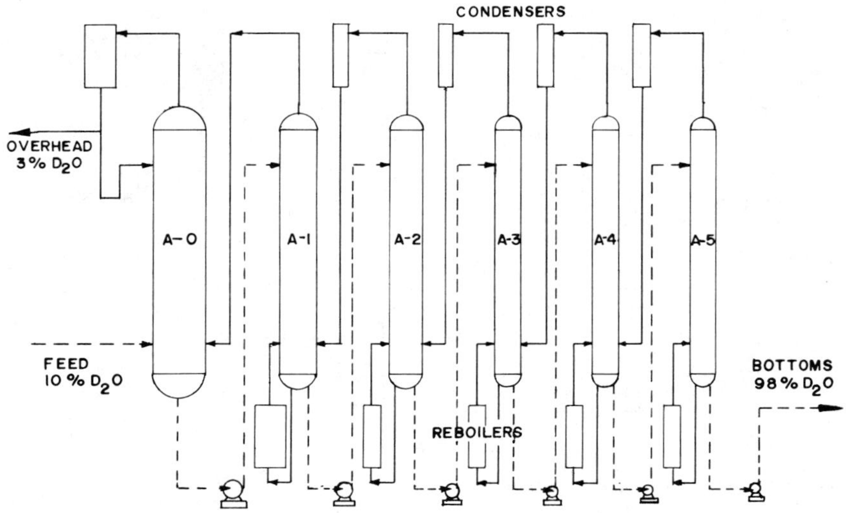

FIGURE 3. A water distillation cascade.

$$\alpha = \sqrt{\frac{P_{H_2O}}{P_{D_2O}}}$$

The lower the pressure, the higher the separation factor, and this provides a strong incentive for the operations of a water distillation facility to be conducted at subatmospheric pressures. For all that, the magnitude α remains rather small, approximately 1.05. Using this value, a rough idea of the requirements or the water distillation process may be derived which projects that the plant will need a large number of towers and will consume a large amount of steam in comparison with its output. Despite these severe requirements, the water distillation process has drawn industry interest because it is simple and straightforward, and above all has established itself as a proven method technologically.

Typical water distillation plant features are shown in Figure 3. In this type of plant, feed enters A-O tower and progressively traverses through a sequence of towers to yield an enriched water containing 98% D_2O. In each tower the enriched water comes out from the bottom and is pumped to the next tower. The vapor coming out from the top is fed to the preceding tower through a condenser. A part of the vapor is fed back to the same tower through a reboiler.

The water distillation process for the primary concentration of deuterium is no longer considered to be economically sound. This is because of the huge amounts of water to be vehiculated and because of exceedingly high heat demands. However, due to the rather straightforward and simple nature of the process and the good safety record, vacuum distillation is successfully applied for upgrading heavy water to 99.8% D/(D + H), using as feed the primary concentrated product at (5 − 30)% D/(D + H). These facilities are operated by using columns filled inside with corrugated phosphorus bronze wire nets folded in packs.

2. *Water Electrolysis*

In electrolytic decomposition of water into hydrogen and oxygen, the deuterium content of the hydrogen produced at the cathode is substantially lower than that of the water remaining in the cell. With continuation of electrolysis, the water gets enriched progressively in deuterium. The gas produced in the latter stages of electrolysis is richer in deuterium than the initial water. This gas, if desired, may be burned, the water condensed and returned to the cell.

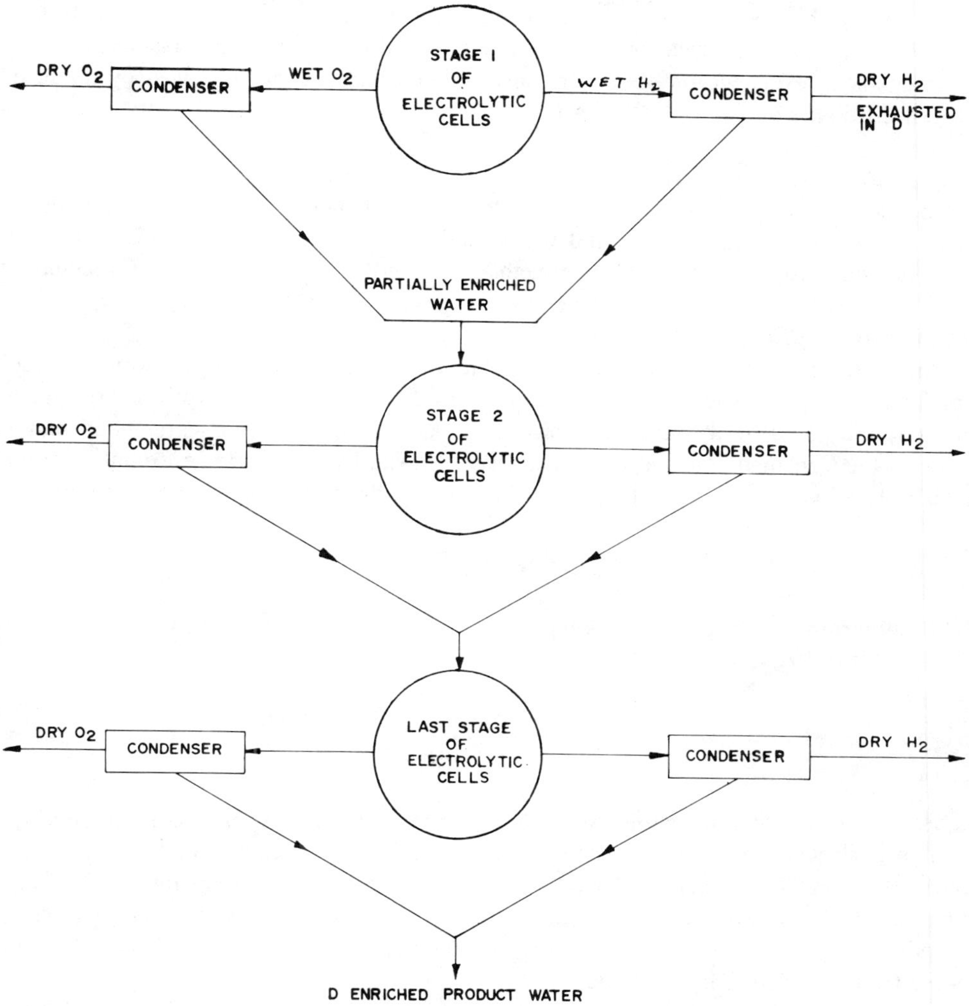

FIGURE 4. Electrolytic production of heavy water.

An electrolytic plant for enrichment of heavy water consists of electrolytic cells arranged in a cascade. The hydrogen depleted in deuterium leaves the cell and is directed to the preceding one. The remaining water that is richer in deuterium goes to the next cell. The cells use diaphragms to prevent mixing of hydrogen and oxygen. A schematic flowchart of an electrolytic plant featuring the stages being connected in a series cascade, without recycle of partially enriched hydrogen and the cascade operated in steady flow, is shown in Figure 4. The isotope separation factor in the electrolysis of water is between 7 and 8. The magnitude of the separation factor depends on the cathode material, quality of electrode surfaces, temperature, electrolyte purity, etc. The separation factor decreases significantly with increasing temperature of the electrolytic bath.

Because of the high separation factor compared to other processes and the ease with which it can be operated on a small scale, electrolysis has been the method of choice for further enrichment of partially concentrated heavy water. The usefulness of the process for production of heavy water from natural water is limited by (1) the large amounts of water that must be handled, (2) the high costs of electricity, and (3) the low recovery. The power consumption for the production of 1 kg of heavy water is very high, about 125 MWh, and consequently the costs are high as compared to other processes. Water electrolysis is, therefore, mainly being used for the final upgrading of heavy water that has resulted from other chemical exchange processes, but also for preconcentration, followed by liquid hydrogen distillation.

3. Chemical Exchange Processes

Chemical exchange processes involving isotopic exchange reactions between hydrogen sulfide and water, hydrogen gas and water, and hydrogen gas and ammonia provide the basis for the most efficient large-scale methods known for the concentration of deuterium.

a. Hydrogen Sulfide and Water

Among the processes for separating deuterium by chemical exchange, the most widely employed is the exchange between hydrogen sulfide and water. The world's larger plants of today are based on this exchange processes, and as typical examples mention may be made of the Canadian plants at Glace Bay, Port Hawkesbury, and Bruce Complex, and an Indian plant located at Kota. The process is based on the reversible reaction set out as

$$H_2O(1) + HDS(g) \rightleftharpoons HDO(1) + H_2S(g)$$

The equilibrium of the reaction is temperature dependent. The equilibrium constant of the reaction given by

$$K_{gl} = \frac{(HDO)_1(H_2S)_g}{(H_2O)_1(HDS)_g}$$

varies inversely with temperature. When temperature decreases, K_{gl} increases, and the equilibrium is shifted toward the forward direction. The values of equilibrium constants at 30 and 130° C are $K_{30} = 2.18$ and $K_{130} = 1.83$, respectively. The temperature effect upon equilibrium is used to accomplish heavy water production in the water-hydrogen sulfide system in the form of a dual-temperature or bithermal exchange process. It is also known as the Gieb and Spevack (GS) process.

In the bithermal H_2O-H_2S process, exchange of deuterium between $H_2O(l)$ and $H_2S(g)$ is carried out at pressures of 20 atm. At elevated tempratures deuterium tends to displace hydrogen (^1H) in the hydrogen sulfide and thus concentrates in the gas. At lower temperatures the deuterium concentrates in H_2S in contact with water in the liquid phase. The deuterium exchange reactions take place in the liquid phase without the necessity for a catalyst. The bithermal feature of the process is illustrated in Figure 5. As shown in the figure, the basic element of the GS process is a pair of gas-liquid contacting towers, each containing a number of sieve or bubble-cap plates. The cold tower operates at a temperature of 30°C, and the hot tower at 120 to 140°C. Hydrogen sulfide circulates between the towers, carrying deuterium as D_2S. Water entering the system flows downward through the cold tower and then through the hot tower counter current to a stream of hydrogen sulfide gas at 1896 kPa (275 psig). The water is progressively enriched in deuterium as it passes through the cold tower and is progressively depleted as it passes through the hot tower, eventually leaving the hot water at a concentration below that at which it entered the system. The HDO and HDS that build up within the process are withdrawn from the base of the cold tower and the top of the hot tower, respectively, by withdrawing a fraction of the water and gas flow. These enriched fractions are fed to a succeeding stage for further concentration. The hydrogen sulfide gas, which acts as a vehicle for transporting deuterium, circulates in a closed loop within the several stages of the processes. A plant might typically produce about 20% of the deuterium in natural water economically. For securing additional concentration of the

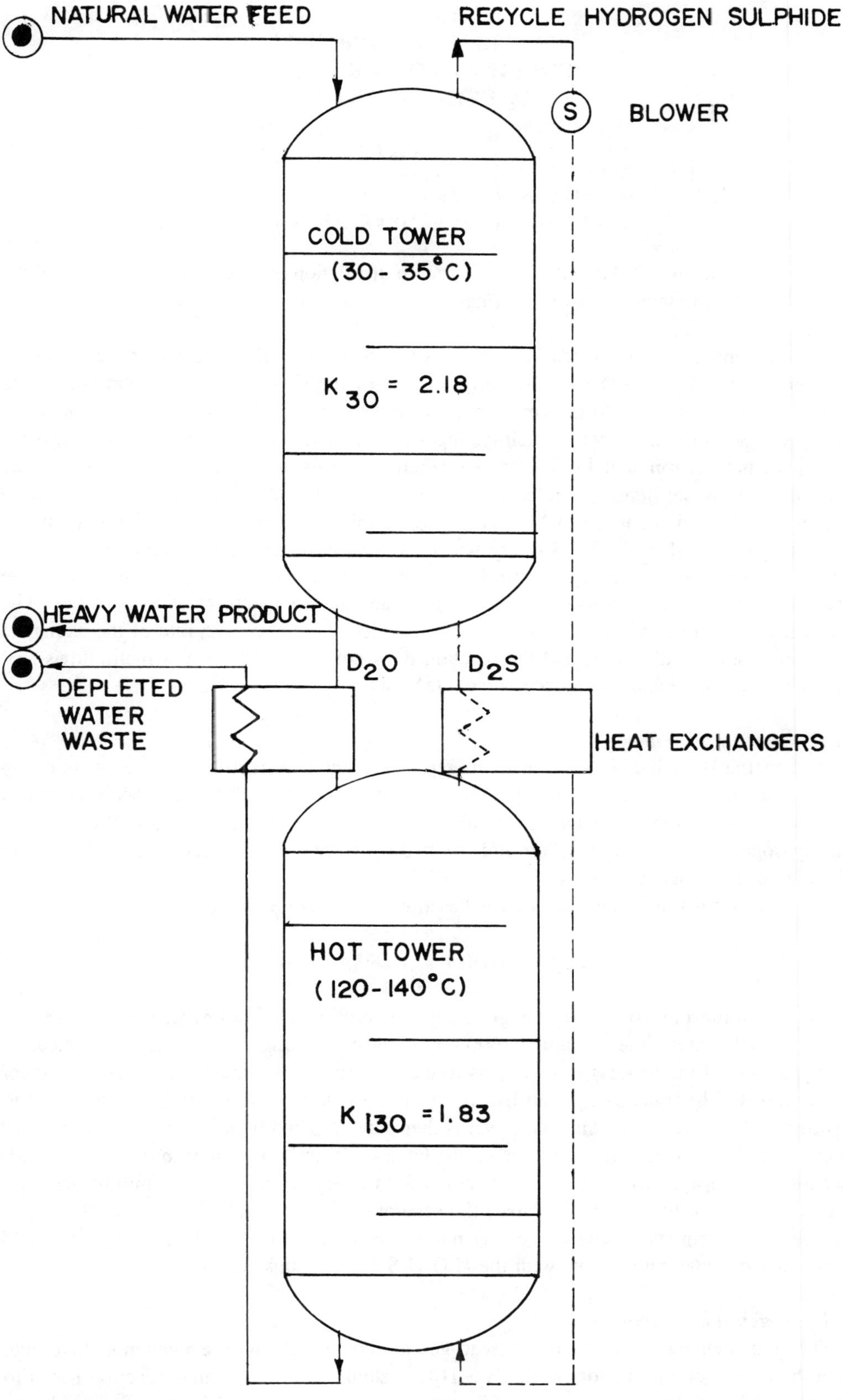

FIGURE 5. Simplified flow diagram for bithermal H_2O-H_2S chemical exchange process (GS).

Table 5
MATERIAL AND ENERGY REQUIREMENTS FOR A 600-T D_2O/YEAR PLANT[10]

Process water	35 t/kg D_2O
Cooling water	10 t/s
Thermal energy	600 MW (23 GJ/kg D_2O)
Electrical energy	67 MWe (700 kWh/kg D_2O)

GS product from about 20 to 90% D_2O, a vacuum distillation process is used. An electrolytic process then completes the work to finally produce reactor-grade concentrate of 99.75% D_2O.

Material and energy demands in an H_2O-H_2S heavy water plant are rather high, as may be seen from Table 5.[10] Due to the corrosiveness of the H_2S aqueous solution and of the humid D_2S, and also due to the very large quantities of toxic gas that are found practically in a pure state in a heavy water facility, high-quality materials and special precautions for personnel and environmental protection are required. Particular care is devoted to recovering D_2S from the waste before its release into rivers, so that residual concentration should be around 10^{-5}%. It is believed that the H_2O-H_2S chemical exchange process will continue to stand in good stead as the chief heavy water production process on a number of grounds. The deuterium source is cheap and readily available in large quantities, and therefore there are no constraints due to availability of raw materials on the production capacity. The operating pressure is not very high; it is of the order of 20 atm. The rate of the exchange reaction is sufficiently high, and the reaction does not need a catalyst, which allows employment of simple and large-size contact elements.

b. Hydrogen and Water

The chemical exchange between hydrogen and water is attractive because of its cheap deuterium source and the noncorrosive character of the process. The main drawback of this process is a low exchange rate, which calls for the use of catalysts. Two specific systems, the hydrogen-liquid water reaction and the hydrogen-water vapor reaction, are known in this chemical exchange process.

The hydrogen-water vapor reaction is based on the reaction given as

$$H_2O(g) + HD(g) \rightleftharpoons HDO(g) + D_2(g)$$

Since the equilibrium constant of the given reaction is around 2.8 at low temperatures (80°C) and around 1.3 at high temperatures (600°C), a dual-temperature scheme seems applicable. The principle of the process is analogous to the GS process. In this case, in the cold tower deuterium will be transferred from hydrogen to steam and in the hot tower from steam to hydrogen. The steam from the cold tower is therefore enriched in deuterium content to 1.19 times feed and that from the hot tower is depleted in deuterium content to 0.81 times feed. For a special case in which 1 mol of steam is fed to the hot reactor and 1 mol to the cold, for every mole of hydrogen circulated, the separation factor is 1.19/0.87 (or 1.47). A full economic assessment of this process is not yet possible, and the chances that it would eventually become competitive with the H_2O-H_2S process look remote.

c. Hydrogen and Ammonia

The fundamental parameters for the hydrogen-ammonia exchange are much more favorable than the corresponding factors in the H_2S-H_2O system, but the exchange reaction needs to be catalyzed in order to achieve a usefule rate of exchange. The finding that the amide ion

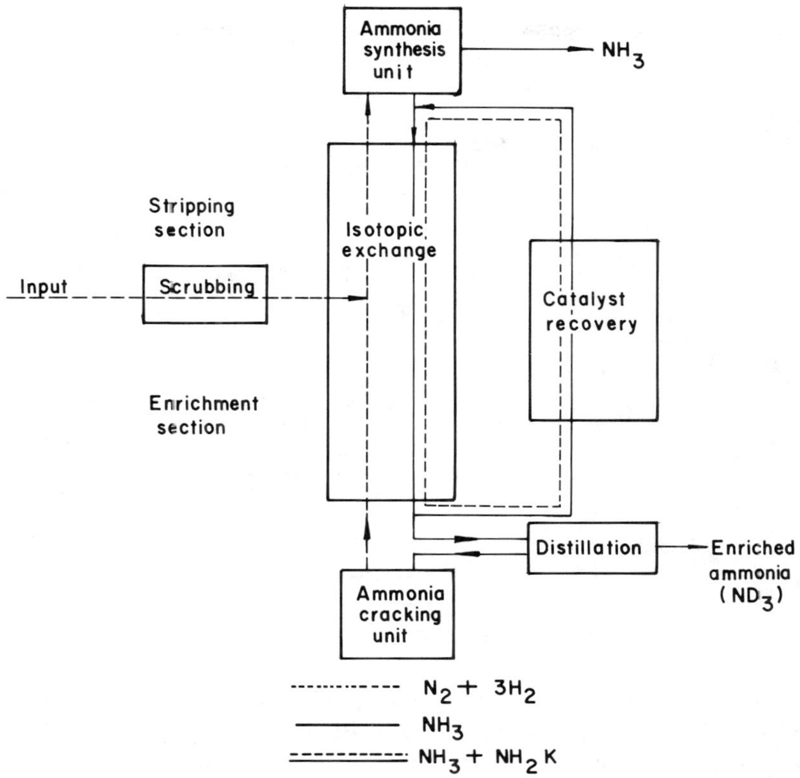

FIGURE 6. Flowchart of the process of monothermal ammonia-hydrogen isotopic exchange.

NH_2^- produced by addition of alkali metal to liquid ammonia is an efficient catalyst for the NH_3-H_2 exchange has provided intense interest in this system. In practice, the hydrogen gas and ammonia exchange uses the hydrogen employed in the ammonia synthesis as a deuterium source and potassium amide (KNH_2) dissolved in the liquid phase in a 1 to 2% ratio as a catalyst. The catalyst has a number of drawbacks in industrial use: (1) its price is relatively high, and it must therefore be recovered and recycled; and (2) it is very reactive and combines with CO_2, CO, H_2O, and O_2 to form solid impurities and with O_2 to form KN_3, which is explosive; concentrations of these impurities in the feed gas must be reduced to less than 1 ppm. Besides these drawbacks, general agreements against the H-NH exchange process rest (1) with the limitations on the availability of sufficient amounts of ammonia for plant feed (even a plant producing 1000 t ammonia per day would yield sufficient feed only to permit production of about 70 t D_2O per year); and (2) with sophistication of the exchange plate, which must provide for special ejectors for the purpose of increasing the reaction interface.

The process is based on the following reaction:

$$NH_3(1) + HD(g) \underset{}{\overset{\text{catalyst}}{\rightleftharpoons}} NH_2D(1) + H_2(g)$$

The process may be carried out either monothermally or bithermally. The process is most readily adapted to monothermal operation because the heat of formation from H_2 and N_2 is relatively low. A schematic of the monothermal process is shown in Figure 6. The process is fed by ammonia synthesis gas (N_2 + $3H_2$). The ammonia exchange liquid is not only cracked in the phase converter at the bottom of the exchange column, as shown in the figure, but is also synthesized for reflux at the top of the column; in this form the monothermal

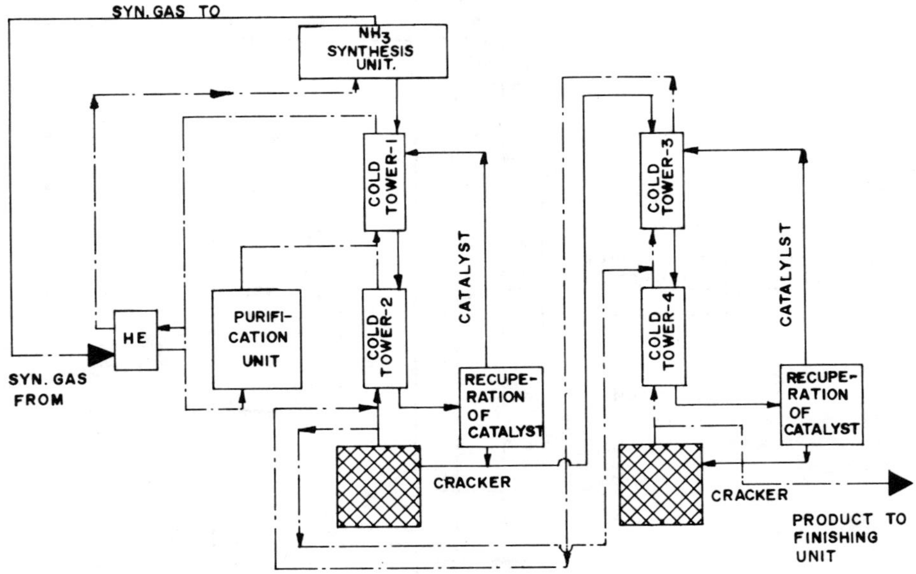

FIGURE 7. Schematic flow diagram for HWP Baroda.

flowchart is completely analogous to a distillation process. France has successfully operated a 20 t/year pilot plant using such a monothermal NH_3-H_2 exchange process at Mozingarbe. Two Indian plants, one at Baroda, and the other at Tuticorin, are also operating this monothermal exchange process.

The flowchart employing the monothermal NH_3-H_2 exchange process for heavy water production at the Baroda plant is simplified as shown in Figure 7. The plant has been designed to produce 67 t of heavy water. Synthesis gas from the fertilizer plant at a pressure of about 650 atm is first cooled in a heat exchanger by the outgoing cold gas from the top of cold tower 1 and then propelled to a purification unit. In this unit the gas is freed from impurities (H_2O, CO, CO_2, and O_2) and is saturated with ammonia. The cooled and purified synthesis gas then enters the bottom of cold tower 1, where it transfers its deuterium to a countercurrent stream of liquid ammonia fed from the top of the column. The enriched ammonia from the bottom of cold tower 1 is fed to the top of cold tower 2. The liquid gets further enriched in this column by coming in contact with the cracked gases from the cracker. The liquid from the bottom of cold tower 2 enters the catalyst recuperation unit. Here the catalyst (potassium amide) is separated from ammonia by distillation and recycled at the top of cold tower 1. Pure liquid ammonia is drawn from this unit and is divided into two streams. One is fed to the top of cold tower 5 for further enrichment, and the other is fed to the cracker. As the reaction that takes place in the cracker is endothermic, heat has to be supplied from outside, which in this case is carried out by burning of natural gas. The cracked gas from the cracker is divided into two streams. One is fed to the bottom of cold tower 2, and the other to the bottom of cold tower 3. The liquid ammonia from the bottom of cold tower 3 is fed to the top of cold tower 4, where it is further enriched to the final concentration of 99.8% ND_3 by contacting it with cracked gases from the second cracker. The deuterated ammonia from the bottom of cold tower 4 is fed to the second catalyst recovery unit, where the catalyst is separated from the deuterated ammonia and recycled to the top of cold tower 3. Pure liquid ND from this unit is fed to the second cracker. A portion of the cracked gases is taken out as produce and sent to the finishing unit. Here it is burned with air to produce heavy water. The rest of the cracked gases are sent to the bottom of cold tower 4. The synthesis gas coming out from the top of cold tower 1 is heated by the incoming feed gas,

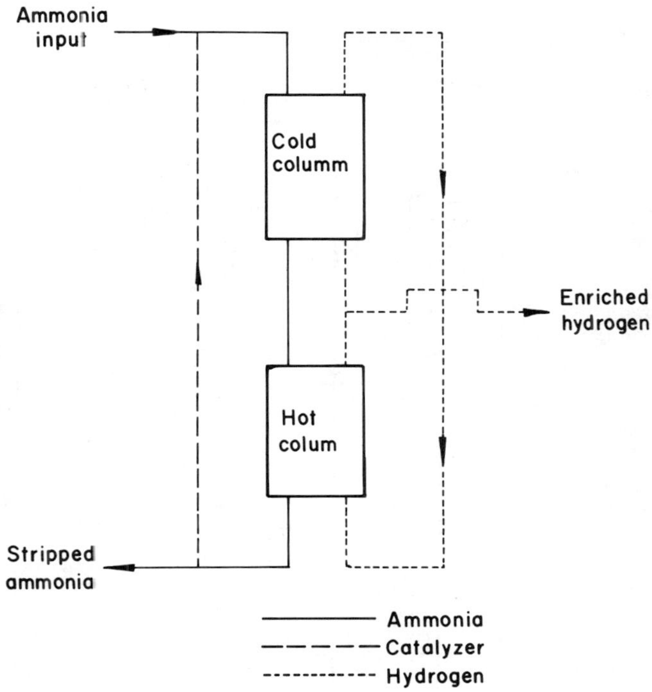

FIGURE 8. Schematic of a bithermal process for NH_3.

as shown in the flow scheme, and is sent to the ammonia synthesis unit. Here a portion of the synthesis gas equivalent to the amount of ammonia cracked in the crackers is converted to ammonia and fed back at the top of cold tower 1. The rest of the depleted synthesis gas, equivalent to the feed gas supplied by the fertilizer plant, is sent back to the fertilizer plant.

A bithermal process, as schematically shown in Figure 8, employs hot and cold columns. A hot column at 70°C (single-stage separation factor 2.9) and a cold column at 40°C (single-stage separation factor 5.9) would have an effective separation factor of 2.0, which could permit extraction of 50% of the deuterium from the ammonia feed. Catalysis of the exchange by potassium amide is sufficiently effective even at −40°C to attain equilibrium in reasonably sized exchange columns.

A plant in India located at Talcher can be taken as an example for the purpose of elaboration of plant practice of the bithermal version of NH_3-H_2 exchange process. A simplified flow diagram of the plant is shown in Figure 9. The plant is designed to produce 62.5 t of heavy water per year. In essence the plant can be divided into three sections: (1) the purification and transfer section, (2) the enrichment section, and (3) the final concentration section. In the purification and transfer sections, the synthesis gas from the fertilizer plant is first sent to a booster to enhance its pressure to take care of the pressure drop in the heavy water plant. The gas is then passed through a deoxo unit followed by a purification tower, where it is successively washed by ammonia and an ammonical solution of potassium amide. During this operation, removal of small quantities of oxygen, water vapor, and carbon monoxide associated with the gas takes place, and the gas gets saturated with ammonia. The purified gas then passes to a transfer column, where it is contacted with depleted ammonia obtained from the enrichment section. Here, transfer of deuterium from the synthesis gas to liquid ammonia takes place. The depleted synthesis gas from the transfer column goes back to the fertilizer plant, while enriched liquid ammonia is fed to the enrichment section for further enrichment. The enrichment section consists of a three-stage cascade with a

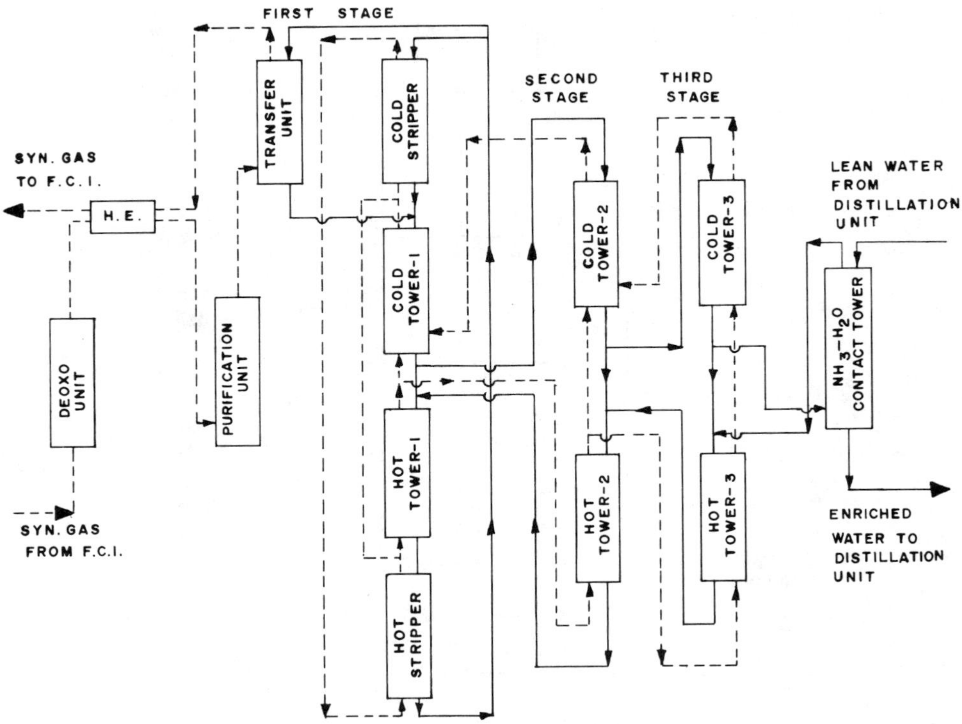

FIGURE 9. Schematic flow diagram HWP Talcher.

variation that the first stage is superimposed by two stripping columns, one cold and the other hot. Incorporation of this variation results in the increase of deuterium. Each stage consists of a pair of cold and hot columns. The deuterium concentration is maximum in liquid ammonia at the base of the cold column and in synthesis gas at the top of the hot column. Part of these streams is passed to the subsequent stage for further enrichment. The highest concentration of deuterium in this cascade is in liquid ammonia at the base of the third-stage cold column. A part of this stream is withdrawn as a product of the enrichment section and sent to the final concentration section to produce nuclear-grade heavy water. In the final concentration section, the enriched liquid ammonia from the enrichment section is first distilled to separate it from the potassium amide catalyst. The catalyst-free liquid ammonia thus obtained is then contacted with water. Part of the deuterium in ammonia gets transferred to water. The ammonia, which has now given up a part of its deuterium to water, is recovered and after mixing with the separated catalyst is recycled to the enrichment section at an appropriate point. Deuterium-enriched water after its separation from ammonia, is distilled to produce nuclear-grade heavy water. It may be mentioned here that an important feature of this plant is its adaptability to the use of amine-hydrogen exchange.

A variant of the NH_3-H_2 chemical exchange process is known which uses alkyl amines in place of ammonia. It starts from the reaction

$$CH_3NH_2(l) + HD(g) \underset{}{\overset{\text{catalyst}}{\rightleftharpoons}} CH_3NHD(l) + H_2(g)$$

which is catalyzed with potassium methylamide (CH_3NHK). Hydrogen exchange catalyzed by amide is generally faster with alkyl amines than ammonia, and a dual-temperature flowchart for an H_2-CH_3NH_2 process has been developed. However, this process does not seem to have been demonstrated on an industrial scale.

Table 6
COMPARISON OF DIFFERENT HEAVY WATER PRODUCTION PROCESSES

Process	Merits	Demerits
H_2O-H_2S exchange	Independent; noncatalyzed; conventional equipment; good separation factor	Corrosive; toxic
H_2O-H_2 exchange	Independent; noncorrosive; nontoxic; very good separation factor	Catalyzed; demanding on equipment
NH_3-H_2 exchange	Noncorrosive: very good separation factor	Dependent; catalyzed; demanding on equipment; toxic
H_2O distillation	Independent; noncatalyzed; conventional equipment; noncorrosive; nontoxic	Very low separation factor
H_2 distillation	Noncatalyzed; noncorrosive; nontoxic; good separation factor	Dependent; demanding on equipment

4. Summary

The high cost of heavy water and its importance as a reactor moderator have generated active competition for the best production methods. Processes as described include water distillation, hydrogen distillation, water electrolysis, and chemical exchange. The first chemical exchange is the GS process. The three other chemical exchange processes are water-hydrogen, ammonia-hydrogen, and amine-hydrogen. In each case the gas is hydrogen, and the other component is the liquid. A comparison of the major processes appears in Table 6.

Production of heavy water, by country, is the U.S., 59%; Canada, 35%; Norway, 4%; and India and France, 1% each. The world capacity (2500 t/year) shows an even greater dependence on the GS process (98%). Most of the heavy water (95%) is produced in Canada. For the projected demand of heavy water the world over, most of the production will be by the GS process. The major capacity will be located in Canada. India, which also employs heavy water reactors, will likely be the next largest producer. The dominance of the GS process largely results from the fact that it is the only practical process that is fed by water as a deuterium source. The only other source of sufficiently large amounts of deuterium is synthesis gas, produced from natural gas or petroleum.

IV. GRAPHITE[11-15]

Carbon in the form of graphite has the unique distinction of being the first reactor moderating material ever used This use was in the first nuclear reactor build by Enrico Fermi in 1942. From then until the present time graphite has rendered peerless service and grown into a major material on the nuclear scene. It has been used exclusively as a moderator in the Magnox-type and the AGR-type reactor systems in the U.K. Several reactors based on graphite moderation gas cooling were installed in France until that country began adopting the LWR. Graphite has played a very significant role in the development and growth of the high-temperature reactor (HTR) systems. The concept of this reactor system was born at Harwell in the mid-1950s, differing from other gas-cooled reactors by having an all-refractory core of fuel dispersed in a graphite moderator, higher working temperatures, and inert coolant. Among the important examples that can be given as far as this reactor system is concerned, special mention may be made of the Dragon reactor at Winfrith, U.K., Peach Bottom (decommissioned) and Fort St. Vrain reactors in the U.S., and AVR in the Federal Republic of Germany. Besides these illustrations of graphite moderation with gas cooling, a family of reactors moderated with graphite and cooled with water is worth a reference. These have been developed in the U.S.S.R.; the first nuclear power station in the world, Obnisk, of 5 MWe, was of this type.

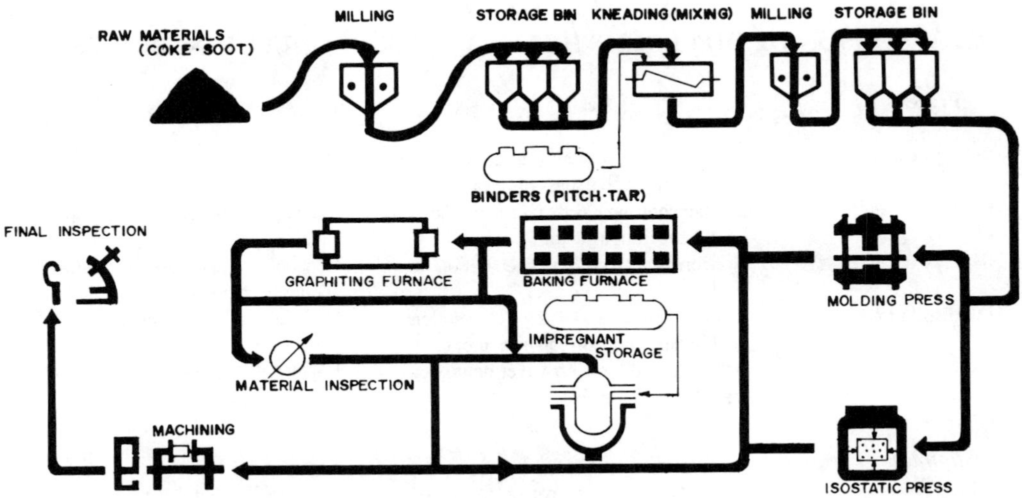

FIGURE 10. Flowchart of a carbon product manufacturing process.

The intrinsic properties of reactor-grade graphite, impermeable graphite, or pyrolytic carbon serving as moderator-reflector structure material in the thermal reactor are its low neutron absorption cross section and its good moderating properties (resulting from a low atomic weight and a high scattering cross section). To these must be added a relative ease of manufacture in a state of high purity — a prerequisite to maintaining a low absorption cross section — excellent thermal properties, acceptable strength, excellent machining properties, abundance, and economics. It is a refractory material. The high-temperature properties of graphite are outstanding and emphasize its value at high temperatures in a nonoxidizing atmosphere. The limitations in the usage of graphite arise from its low oxidation resistance and relatively low impact strength.

A. Preparation of Nuclear Graphite[16]

Graphite occurs in nature mainly in conjunction with the older crystalline rocks such as gneiss, marble, quartzite, schist, and granite. Though graphite occurs naturally, most deposits are very impure and have no reactor application without extensive purification.

The present source of reactor-grade graphite is normally a manufactured product from a high-purity petroleum coke (called the filler) and a pitch binder. A flowchart of the carbon product manufacturing process is shown in Figure 10. The procedure for making graphite as outlined in the figure may be regarded as fairly typical. Special care is taken in the selection of raw materials so that they do not contain impurities with a high neutron absorption cross section. The coke prepared by polymerization and distillation of a heavy oil is calcined to drive off volatile gases. It is then crushed, ground, and blended to give a specific particle-size distribution. The coke blend is mixed with a coal tar pitch binder, and extruded or molded to the required shape. The shape given by either the molding or the extrusion process is subjected to a baking treatment at about 1000 to 1400°C. Further pitch impregnation, followed by rebaking, may be carried out in order to achieve higher densities. The product resulting from this stage is extremely hard, is not machinable, and contains significant quantities of impurities. The baking process essentially breaks down the binder molecules and sets the shape. The baking process is followed by graphitization which is done by heating electrically the baked shapes in a packing of petroleum coke at a temperature of about 2800 to 3000°C for some hours. This period is characterized by crystal development with the formation of a typical graphite layer structure. This brings about a marked change in properties. One witnesses considerable improvement in machinability and thermal conduction. The graphitization also brings improvement in purity, and this happens because many impurities volatilize at the high temperatures used in the graphitization process.

Purification via volatilization does not, however, take place for all the impurities. In the list of such impurities, special mention may be made of boron and vanadium. They form stable carbides of high melting point and defy purification by volatilization. A chemical process is adopted to remove boron, vanadium, and similar such carbide-forming impurities from graphite. The process involves passing a mixture of chlorine and fluorine gases over the graphite at a temperature of about 2400° C. This removes boron as boron trifluoride and vanadium as vanadium pentachloride, both of which are volatile compounds. The final properties of the graphite depend upon whether molding or extrusion was adopted in shaping. Reference to this is made in Section IV. B, dealing with properties. The structure of artificial graphite as finally produced is a framework of graphitized coke particles bonded together with a graphitized pitch filler.

The so-called impermeable graphite, actually a kind of impervious artificial graphite with a low permeability to gases, is attracting interest as a special-purpose reactor material. For the preparation of impermeable graphite, pure commercial graphite is first impregnated under high pressure with a carboneous liquid, such as sugar solution or furfuryl alcohol. Following impregnation, the graphite is heated to about 1000° C to carbonize the carboneous liquid and then further heated to a higher temperature for graphitization. When the graphitization process is concluded, the resulting impermeable graphite acquires not only low permeability to gases, but also greater tensile, bending, and shearing strengths than the original graphite. Fuel element cladding in AGRs and thermal insulation between the coolant gas and ordinary graphite moderator are among the familiar applications of impermeable graphite.

Pyrolytic carbon, or pyrocarbon, is a highly oriented crystal structure of carbon (graphite). Pyrolytic carbon results as a deposit on a heated surface by thermal decomposition (temperature in the range of 1500 to 2400°C) of a hydrocarbon gas, e.g., methane (CH_4), benzene (C_6H_6), etc. The deposition temperature, the quality of the hydrocarbon gas, and other factors determine the crystal structure and actual density of the pyrocarbon. The properties have been observed to be highly anisotropic. It has been found that the properties in the plane parallel to the deposition surface are those of a metal with high thermal conductivity and tensile strength, whereas in the perpendicular direction across the deposition surface the properties are those of ceramics, with low thermal conductivity and tensile strength. For all practical purposes pyrolytic carbon is regarded as impermeable, even in thin layers, to gases, and this enables the applications of pyrolytic carbon to be essentially the same as those of impermeable graphite, as pointed out above. In addition, as has been mentioned in an earlier chapter, pyrolytic carbon is a coating material in the form of fine particles for ceramic fuel, $(U,Th)O_2$ or (U,Th)C, to retain the fission gases in the fuel. Pyrolytic carbon is used, for example, as a buffer and an isotropic coating (depicted in Figure 11).

B. Properties of Graphite[17,18]

Carbon is known as one of the nonmetallic elements. Carbon is more correctly a transition element, exhibiting properties of both metals and nonmetals. It exists as three allotropes: diamond, graphite, and amorphous carbons. Diamond has a tetrahedral crystal structure and is quite different in properties from the other two allotropes. Graphite and amorphous carbon both have a hexagonal crystal structure. Graphite is the more perfect crystalline material. The crystal structure having the hexagonal symmetry is sometimes regarded as orthorhombic and consists of flat layers of carbon atoms stacked parallel to one another. The layer of the hexagonally arranged atoms has strong binding. Because the distance between layers is relatively large, the binding between the planes is looser, and so the material presents little resistance against slippage of the parallel layers. This property makes graphite powder a good lubricant.

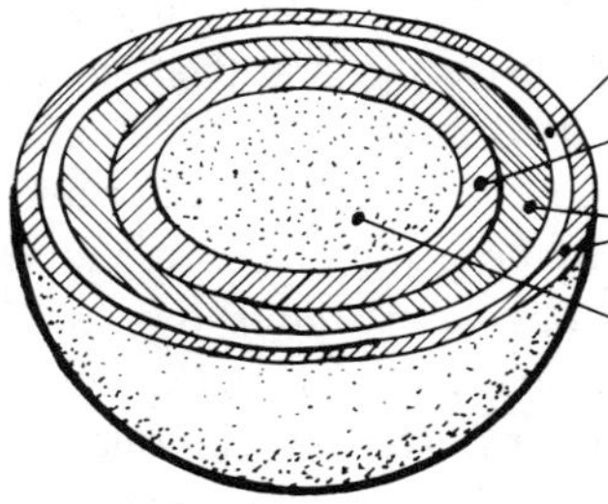

FIGURE 11. Ceramographic section of coated particles.

Carbon will not melt at temperatures below 3600K At higher temperatures vaporization at measurable rates will occur. Unlike metallic materials, melting and seizing of graphite components will not occur. Graphite is, in contrast to most nonmetallic elements or compounds, a fairly good heat conductor. This property is advantageously utilized in combination with the nonmelting and chemically inert properties not only for electrical and mechanical applications but also for metallurgical applications, particularly as molds and dies. Graphite has a low coefficient of thermal expansion. This low expansion rate contributes to dimensional stability in parts subjected to temperature changes. Graphite's excellent resistance to thermal shock is aided by its low expansion coefficient as well as its high thermal conductivity. Carbon and graphite are both well known as chemically inactive materials. It is far easier to count up chemical substances which can attack carbons than to make a list of chemicals which will not attack them. This brief account of the thermal and chemical properties undoubtedly places graphite as one of the widely used technological materials.

The theoretical density of graphite is 2.27 g/cm^3, and this value is approached by natural graphite. The density of reactor-grade graphite lies within 1.65 to 1.75 g/cm^3, the difference being accounted for by the porosity of the artificial graphite, which makes up about 20 to 30%. As a rule, pores communicate with one another, and for this reason graphite passes gases.

Reactor-grade graphite exhibits directional properties due to anisotropy resulting from the alignment of the coke particles brought about during preparation of extrusion or molding. In the extrusion process, coke particles tend to line up with the longitudinal direction along the axis of extrusion. This graphite generally possesses higher thermal and electrical conductivities, and greater mechanical strength in the direction of extrusion. In the molding process, the particles tend to line up with their largest dimension perpendicular to the molding force. This yields a lower ratio of thermal and electrical conductivities and mechanical strength in directions between parallel (longitudinal) and perpendicular (transverse) to the grain. Some properties of the fine-grained reactor-grade graphite, extruded and molded graphite, are shown in Table 7. It generally shows that the two processes employed on graphite have considerable effects on its physical, thermal, and mechanical properties. Extrusion can produce a greater amount of anisotropy and directional differences than molding.

Graphite has a high thermal conductivity which differs little from that of many metals. The flow of heat in graphite takes place along the basal planes located predominantly parallel to the axis of extrusion. The thermal conductivity increases with the density of the graphite. The variation of thermal conductivity and bulk density is shown in Figure 12. As is to be expected, crystalline graphite shows marked anisotropy in electrical resistivity. The resistance of highly oriented natural graphite perpendicular to graphite planes is 100- to 1000-fold greater than that in the parallel direction. In addition to orientation, the resistivity of polycrystalline graphite is also dependent on density and crystallite size. It has been shown that

Table 7
SOME PHYSICAL, THERMAL, AND MECHANICAL PROPERTIES OF EXTRUDED AND MOLDED GRAPHITE

Property	Extrusion		Molding	
	Longitudinal	Transverse	Longitudinal	Transverse
Actual density (g/cm^3)	1.65	1.65	1.75	1.75
Thermal conductivity (cal/cm-sec-°C)	0.41	0.31	0.5—0.3	0.3—0.2
Electric resistance (mΩ-cm)	0.86	1.62	0.96	1.32
Thermal expansion coefficient (10^{-7}/°C)	11	41	19	32
Compressive strength (kg/cm^2)	420	400	400	380
Tensile strength (kg/cm^2)	140	130	132	125
Flexural strength (kg/cm^2)	356	340	368	350
Elastic modulus (kg/cm^2 × 10^3)	128	56	98	65
Modulus of rigidity (kg/cm^2 × 10^3)	50	22	40	23

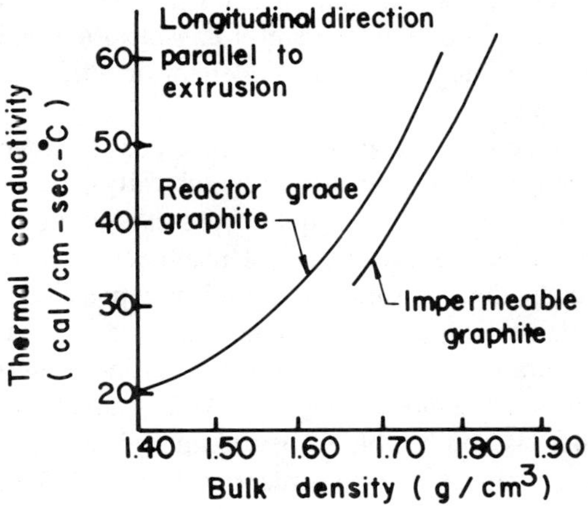

FIGURE 12. Variation of thermal conductivity of graphite with bulk density.

for nominally identical graphite blocks having different densities, the resistivity is inversely proportional to the fourth power of the density. Though electrical resistivity has no direct application for reactor use, the correlation between it and thermal conductivity, as pointed out, indicates its relative importance.

The effect of temperature on thermal conductivity of reactor-grade graphite and pyrolytic carbon in the longitudinal direction of extrusion or deposition is shown in Figure 13. The thermal conductivity of reactor-grade graphite decreases with temperature. The pyrolytic carbon with more metallic characteristics exhibits a decreasing trend in thermal conductivity appreciably different from that of the reactor-grade graphite. It thus turns out that the thermal conductivity of the pyrolytic carbon is relatively high. The thermal expansion of the reactor-grade graphite and pyrolytic carbon of crystalline graphite is particularly anisotropic. Measurements using X-ray methods indicate that the coefficient of thermal expansion along the a axis of the lattice is negative up to about 400°C. Over the range 0 to 150°C, it is -15×10^{-7}/°C and becomes more positive with increasing temperature. On the other hand, the expansion along the c axis has a large positive value of 286×10^{-7}/°C over the temperature range of −196 to 1118°C. The bulk expansion of polycrystalline graphite shows little relationship to the individual crystal behavior given above, except that anisotropy is still

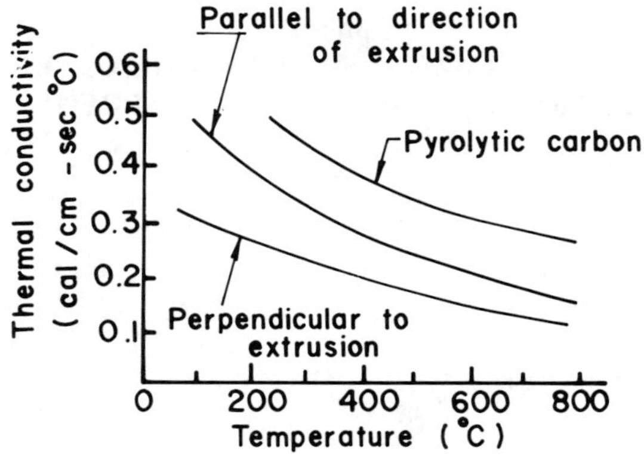

FIGURE 13. Variation of thermal conductivity of reactor-grade graphite and pyrolytic carbon with temperature.

noted in material with a preferred orientation. The temperature effect on thermal conductivity and thermal expansion of graphite requires careful consideration and evaluation related to heat conduction and structural integrity in the moderator or reflector for fission, fusion, or hybrid fission-fusion reactors.

As far as the nuclear properties of graphite are concerned, reference must be made to its thermal neutron absorption cross section. Spectrographically pure graphite has a value of 3.5 mb. Reactor-grade graphite of production quality possesses a slightly higher cross section. This, however, depends on the precise method of manufacture and the quality of the raw materials that go into production. A value of 4.0 mb is representative of high-purity production material, and a value of up to 4.5 mb represents average reactor-grade graphite. It is of interest to present a comparison of graphite with other moderators: water, heavy water, and beryllium. On the basis of the moderating ratio (MR), graphite, though not approaching the power of heavy water, is comparable to beryllium and much superior to light water. With migration length taken into account, the order is reversed. Graphite has a long migration length. It has a low moderating power (MP). Therefore, the average number of collisions that the high-energy neutrons undergo for thermalization is high. For these reasons, graphite systems are inevitably large and require large quantities. In uranium-graphite reactors, graphite is in use for moderation in the form of blocks built up to provide a core structure through which run the fuel element channels. The main requirement is, therefore, that it must have enough compressive strength to enable each block of graphite to support the load of graphite above it. Resistance to creep is high at temperatures below 1500°C. Deterioration takes place at higher temperatures. The resistance of graphite to thermal shock and the thermal stresses is good. These attributes make the materials particularly suitable for high-temperature application such as the environment of an HTGR. It may be mentioned that a critical system cannot develop with graphite moderation and natural uranium in the form of uranium dioxide as a fuel.

C. Compatibility[19,20]

The compatibility problems of a graphite moderator are almost entirely those arising from reaction with the coolant. The most popular and promising gases for cooling reactors are air, carbon dioxide, and helium. In the early days of gas-cooled reactors, air was utilized as coolant in direct contact with graphite at low operating temperatures below 200°C. Above 250°C, oxidation corrosion of graphite by air occurs, and the oxidation rate increases with operating temperature. In the temperature range 300 to 500°C, instead of air, carbon dioxide has been used as coolant in direct contact with graphite for Calder Hall-type gas-cooled reactors and AGRs. At temperatures above 500°C under neutron irradiation, oxidation and radiation corrosion of graphite by carbon dioxide become serious. Hence, helium is employed as coolant in contact with pyrolytic carbon and moderator graphite for HTGRs.

It will be of interest to detail to some extent the corrosion aspects of graphite in a CO_2 environment. Two types of graphite corrosion can occur: thermal and radiation-induced oxidation. Thermal oxidation predominates at 700°C and above temperatures of HTGRs. It is insignificant at the lower temperatures of Magnox and AGR moderators. In the CO_2 coolant, the reaction which dominates graphite moderator behavior is a radiolytically induced reaction: C(graphite) + $CO_2 \rightsquigarrow$ 2 CO(A). The overall reaction A results from the decomposition of the CO_2 coolant under irradiation to provide extremely reacting species: (O*) − $CO_2 \rightsquigarrow$ CO + O*. These species have a very short lifetime of less than 10^{-7} s before recombining with CO to reform CO_2 − CO + O* $\rightsquigarrow CO_2$. Thus, the species produced in the bulk coolant are neutralized before traveling more than a few microns, so that CO_2 appears stable under ionizing radiation in the absence of graphite. In the reactor core, therefore, graphite and CO compete for the oxidizing species. If the species reaches the graphite surface, a carbon atom may be gasified. C(graphite) + O* → CO. Since the species has high chemical reactivity, its range relative to the graphite is very short; only those species produced in gas within the graphite pores, which is about 10% of the graphite volume, cause significant corrosion. Radiolytic corrosion is thus said to take place mainly within the pores rather than the exposed surfaces of graphite. In CO_2 the corrosion rate is proportional to the open pore volume, and as this increases with corrosion, so the rate of corrosion increases. The rate of attack or rate of graphite oxidation is a function of the number of oxidizing species produced within the pores, which is proportional to the volume of the pores as stated, to CO_2 coolant gas pressure, and to intensity of radiation, but is substantially independent of temperature. It is also very dependent of the gas composition, and it is only this which can be altered without affecting reactor output.

It is necessary to inhibit the radiolytic corrosion reaction (overall reaction A). The reaction is irreversible in terms of graphite. Once a carbon atom has been removed from the lattice, it cannot be replaced. The radiolytic graphite-CO_2 reaction of inhibited by CO, which is present in the coolant as a product of the corrosion reaction, but it is not a sufficiently powerful inhibitor for AGRs. It has been found that methane at a concentration of 900 to 1000 vpm in a CO_2 + 1V/o CO coolant gives adequate inhibition. Unfortunately, the irradiation-induced decomposition of methane (methane under irradiation produces CO and H_2O) can also give rise to carbonaceous deposition on the surface of fuel pins. It is therefore necessary to strike a balance between the beneficial effect of methane on graphite and the possible detrimental consequences for the fuel.

At this point, a brief reference is drawn to the disposition of graphite in the liquid metal coolant field. In this field, by far the greatest attention rests on sodium as a coolant. The problems in the use of liquid sodium as a coolant for a graphite moderated reactor are twofold: (1) mass transfer of carbon and (2) penetration of sodium into the pores. The transfer of carbon into the sodium may arise due to reaction with the graphite or due to solution of carbon in sodium. If carbon by any mechanism can enter into sodium, mass transfer will be maintained only if a suitable sink for the carbon exists in the circuit. A material like stainless steel which is susceptible to carbonization or a nonisothermal system presenting a scope for the dissolved carbon to precipitate out in the cooler regions can provide such a sink. The conditions of mass transfer are, however, by no means well defined. The physical effects resulting from contact between sodium and graphite do exist. At temperatures in excess of 450°C, sodium is absorbed into the pores of normal reactor graphite, causing the material to swell.

D. Irradiation Effects[12,21,22]

The irradiation induces an increase in the strength, hardness, and elastic modulus of graphite, especially at normal temperatures. The important considerations pertaining to reactor moderator and reflector designs are, however, (1) dimensional changes, (2) decrease in thermal conductivity at high temperatures, and (3) stored energy.

Dimensional changes and instability, apparent as distortion and expansion, are probably related, in part at least, to the highly anisotropic nature of graphite. Neutron irradiation can considerably induce dimensional changes and instability in a highly anisotropic graphite, i.e., elongation in the longitudinal direction parallel to extrusion and contraction in the transverse direction. After prolonged exposure, it has been observed that a longitudinal contraction follows the initial elongation. This contraction continues to proceed and removes to a certain extent the initial elongation; then elongation rises up again as the irradiation time and neutron fluence increase. The dimensional changes and instability of graphite diminish with increasing temperature. Many specimens even exhibit a contraction of volume (or negative dimensional change) at higher temperatures above 350°C.

The effect of neutron irradiation on graphite is very pronounced at relatively low irradiation temperatures of up to about 200°C. At neutron fluence of the order of 10^{19} to 10^{20} n/cm^2, the thermal conductivity of the graphite samples is reduced by a factor of 40 to 50. At higher temperatures, the thermal conductivity of the irradiated graphite gradually recovers due to irradiation annealing.

An important and well-known effect of graphite irradiation is the stored energy and its consequences, also known as the Wigner effect. If the graphite temperature is sufficiently high (e.g., above about 300°C), the radiation damage is small. This is, presumably, because the mobility of the carbon atoms in the crystal lattice permits recovery to keep pace with the radiation-induced displacements. Below a temperature of about 350°C, at which annealing of defects cannot take place, radiation damage increases with time, causing the accumulation of the stored energy, sometimes called the Wigner energy, in the graphite crystal lattice. If the metastable material (loaded with stored energy) is suddenly transformed to the stable form, the excess stored energy can be released at once, accompanying a large increase in temperature. An integrated flux of 10^{19} nvt at ordinary temperature can result in a total stored energy of 400 cal/g of graphite, which, upon release, is sufficient to raise the temperature to approximately 1000°C. It is the practice in graphite-moderated reactors operating at fairly low temperatures to permit a controlled and gradual increase of temperature so that annealing of radiation damage can take place, much of the stored energy is released, and catastrophic rise can be prevented. It may be mentioned as an example that, due to the release of stored energy, large-scale overheating occurred in a British plutonium-producing reactor at Windscale in 1957. It was partly for this reason that operation of the reactor was not resumed after the accident.

V. BERYLLIUM[12,23-26]

Beryllium, on account of its having the lowest absorption cross section for thermal neutrons of all metals (the value being one sixth of its nearest rival, magnesium), its large scattering cross section, and its high melting point, is an excellent structural, as well as moderator and reflector, material. Beryllium has the advantage that (n,2n) and (γ,n) reactions in beryllium under pile conditions contribute a few extra neutrons to the chain reaction. The high strength-to-weight ratio of the metal, particularly at high temperatures, combined with its good corrosion resistance, makes it an attractive proposition for other applications where these characteristics are important. However, the ductility of beryllium at room temperature is poor, and from an engineering standpoint it must be considered to be brittle metal when subjected to triaxial stressing at room temperature. Technological difficulties at extraction and fabrication have resulted in the price of the metal being high, and this has been an additional factor limiting its use.

Table 8
ADVANTAGES AND DISADVANTAGES OF BERYLLIUM AS STRUCTURAL MATERIAL

Advantages	Disadvantages
Very low thermal neutron absorption	Poor ductility
High thermal neutron scattering	Toxicity
High melting point	High cost, scarcity
High strength-to-weight ratio	Fracture rupture
Good thermal stability, low vapor pressure	
Good corrosion resistance	

Table 9
BERYLLIUM RESOURCE PROCESSING ROUTES

Silicofluoride sintering process
1. Grind the beryl ore finely
2. Mix with sodium silicofluoride or sodium ferric fluoride (serve as source for fluorine)
3. Heat to about 750°C, which results in the conversion of beryllium values into water-soluble sodium beryllium fluoride
4. Treat the solution with caustic soda, which yields insoluble beryllium hydroxide
5. Wash and ignite to give BeO

Sulfuric acid leaching process
1. Melt and quench beryl ore
2. Heat-treat the quenched ore. The operation releases beryllium locked up by the silica
3. Treat with sulfuric acid — the operation solubilizes beryllium and aluminum as sulfates
4. Add ammonium hydroxide, which enables the bulk of aluminum separation by crystallization as ammonium alum
5. Add sodium hydroxide to convert beryllium sulfate to beryllate
6. Heat beryllate to decompose it to beryllium hydroxide, which is fired at 750°C to convert to BeO

A summary of the various advantages and disadvantages of beryllium as a structural material, moderator, and reflector for nuclear reactors is given in Table 8. A minor application of the metal associated with its nuclear properties is related to the construction of neutron sources, where use is made of its capacity to produce neutrons when bombarded with alpha particles or gamma radiation. Examples of reactions, such as Pu-Be, Ra-Be, Po-Be, and Sb-Be, induce (α,n) or (γ,n) reactions for such neutron production (${}_4Be^9 + {}_2He^4 \rightarrow {}_6C^{12} + {}_0n^1$; ${}_4Be^9 + \gamma \rightarrow {}_4Be^8 + {}_0n^1$). In addition to this application, the metal also finds use in the construction of X-ray tube windows. A major application of the element is as a 2% alloy with copper, which accounts for over 80% of the present-day consumption of the metal.

A. Production[27-30]

Beryllium does not occur free. It is a constituent of about 30 known minerals, but the only commercial source of the metal is beryl, which is a beryllium-aluminum silicate containing approximately 4 wt% beryllium.

The first stage in the production of pure metal from beryl involves the production of pure beryllium hydroxide. Two processes, as given in Table 9, are in use for this operation. The end product from both these processes is BeO.

For metal extraction there are two major processes that are currently in use. Both use purified BeO as the raw material and convert it into a halide intermediate from which the metal is won. The first of the metal extraction processes involves magnesium reduction of the beryllium fluoride. The starting material, beryllium hydroxide, is converted to ammonium beryllium fluoride by treatment with ammonium bifluoride and/or hydrofluoric acid. Im-

Table 10
PROCESSING STEPS FOR BERYLLIUM CHLORIDE FOR ELECTROWINNING OF BERYLLIUM

1. Mix $Be(OH_2)$ with carbon and tar
2. Roast at 250°C for 4—5 h for moisture removal
3. Recrush, mix with tar, and extrude to cylindrical shapes
4. Heat to 750°C to sinter and remove volatiles
5. Chlorinate at 900°C in a carbon tube to give crude beryllium chloride intermediate
6. Reduce the chlorinated product to convert iron into nonvolatile ferrous state
7. Redistill for crude chloride purification

purities remaining in the beryllium hydroxide, which may include manganese, copper, lead, arsenic, and aluminum, are removed by a series of chemical precipitations, and the purified ammonium beryllium fluoride is subsequently recovered by crystallization. This salt is subsequently heated to about 950°C to give pure anhydrous beryllium fluoride, the ammonium fluoride being returned to the process. The beryllium fluoride is then reduced with magnesium. The reduction is carried out in a graphite-lined furnace, to which magnesium metal and beryllium fluoride are gradually added at a temperature of 900°C. An excess of beryllium fluoride is added for lowering the melting point of the slag and for giving improved metal separation. The beryllium is formed as a fine powder dispersed in a slag of magnesium fluoride-beryllium fluoride. The temperature is raised to coalesce these particles, and the charge is subsequently poured into molds. The beryllium is recovered as small pebbles from the slag matrix. The metal contains an appreciable amount of magnesium and is normally relatively high in metallic impurities.

The second process for beryllium metal extraction is the electrolytic reduction of chloride. The steps of the chloride production process are shown in Table 10. The pure chloride thus prepared is mixed with sodium chloride, or a mixture of sodium and potassium chloride and electrolyzed at 350°C. The electrowon metal is separated form the electrolytic bath and first leached with iced water to minimize reactions arising from the hydrolysis of the beryllium chloride contained in the adhering electrolyte. It is then washed with caustic soda and finally with nitric acid; then it is centrifuged and dried. Depending upon the current densities and voltages at which electrolysis is carried out, electrodeposit beryllium forms range from flaky to dendritic types.

Both methods yield the as-reduced metal with comparable levels of oxygen, being of the order of 0.05%. The powder from both sources is higher in oxygen and may be as high as 1.0%. Finer grades have even more than this. This has to do with the specific surface of the material. The beryllium pebbles from the magnesium reduction route contain appreciable quantities of magnesium and magnesium fluoride, and thus necessitate a vacuum melting operation for their removal. The electrolytic route seems to generate a metal with somewhat improved purity. There is, however, no evidence that electrowon beryllium in any way exhibits significant differences from that of the magnesiothermic beryllium with respect to the physical or mechanical properties.

The fabrication processes for beryllium fall entirely within the domain of powder metallurgy. Difficulties associated with the melting and casting route to produce fine-grained cast beryllium have led to the adoption of powder metallurgy techniques for the production of fabricated beryllium products. The main objective of using powder metallurgy techniques is to ensure the production of finished components having a fine-grained structure together with adequate mechanical strength and ductility. Cold-pressing and sintering, vacuum hot-pressing, and loose sintering processes have been used for consolidation of beryllium powder. Among these, the vacuum hot-pressing process is now widely being used for producing very large slabs of beryllium. The loose sintering technique, by virtue of its simplicity, has

Table 11
PHYSICAL PROPERTIES OF BERYLLIUM

Density (g/cm^3)	1.8477	
Phase transformation	Crystal lattice	Temp range (°C)
Alpha phase	hcp	20—1250
Beta phase	bcc	1250—1290
Melting point (°C)	1283 ± 3	
Boiling point (°C)	2970	
Specific heat (cal/g-°C)		
20°C	0.41	
200°C	0.50	
Thermal conductivity (cal/cm-s-°C)	Sintered powder	
100°C	0.395	
200°C	0.390	
Linear thermal expansion coefficient (10^{-6}/°C)	Transverse	Longitudinal
100°C	15	11
300°C	18	14

considerable advantages where a requirement for a multiplicity of small objects permits the required degree of standardization to be achieved. However, up to the present, this process has not been used as extensively as the vacuum hot-pressing route.

B. Properties

Some of the physical properties of beryllium are given in Table 11. Beryllium has a hcp structure from room temperature to 1250°C. It is on account of this structure that some of the physical, thermal, and mechanical properties of the metal show anisotropy. Beryllium products prepared by hot-pressing and sintering, however, display approximately isotropic properties because of the random orientation of the small grains.

The mechanical properties of beryllium show variation with powder technology, fabrication process, and treatment. The vacuum hot-pressing route for beryllium has proved to be a successful fabrication process for producing isotropic properties. Vacuum casting and extrusion processes, however, develop anisotropic properties. For purposes of comparison, the mechanical properties of vacuum hot-pressed, and cast and extruded beryllium are given in Table 12. Beryllium is not particularly corroded in air at temperatures below 670 K. At higher temperatures, a white, adherent layer of beryllium oxide appears on the surface of the metal. In terms of water corrosion resistance, beryllium rates are lower than zirconium. The corrosion rate decreases when pH is more than 6.5, and is influenced by the presence of such impurities as copper and chlorine in the aqueous medium.

C. Irradiation Effects

Beryllium shows good stability under irradiation. The exposure of beryllium to neutron fluences up to the order of 2.4×10^{21} n/cm^2 does not bring about any sizable dimensional changes. Also, no change occurs in electrical resistivity and corrosion resistance. Regarding the other properties, the thermal conductivity decreases slightly, the tensile strength increases, and the elongation decreases. The hardness value increases by about 30%. The generation of helium as a result of irradiation causes beryllium to become brittle.

D. Conclusion

The poor ductility, the high price of the metal, and handling difficulties associated with the toxicity of the metal have hitherto counterinfluenced the extremely attractive nuclear properties. The toxicity hazard and the associated necessity for protection will inevitably play a significantly important part in determining the detailed process used for both extraction

Table 12
MECHANICAL PROPERTIES OF HOT-PRESSED, CAST, AND EXTRUDED BERYLLIUM

	Strength and modulus	Kg/cm^2
Hot-pressed Be	Tensile strength	
	Room temp 20°C	3150
	Room temp 500°C	2240
	Tensile yield strength (0.2%) offset	
	20	2100
	500	1560
	Tensile modulus of elasticity at 20°C (10^3)	3080
	Compressive yield strength (0.2%) offset	
	20°C	1960
	500°C	1500
	Compressive modulus of elasticity at 20°C (10^3)	2940
	Poisson's ratio at 20°C	0.024
	Elongation at 20°C (%)	2.0—2.8
Cast Be as extruded at 1000°C	Tensile ultimate strength	
	Longitudinal	1130
	Transverse	580
Cast premium powder as extruded at 1000°C	Tensile yield strength	
	Longitudinal	565
	Transverse	475
Hot-pressed cast	Hardness (Rockwell B number)	75—85

and fabrication. From intensive research carried out in the field of beryllium metallurgy it is known that the metal is one of the most difficult of all metals to fabricate. This is in part attributable to its crystal structure. The associated impurities also contribute to this. The necessity of producing beryllium metal containing less metallic and nonmetallic impurities can hardly be questioned. However, it does not seem likely that metal of a higher degree of purity can be attained from the current production routes. The electrolytic route seems to be inherently more capable of improvement from the point of view of metallic impurities. This extraction route, in association with such operations as grinding and leaching, appears to be potentially capable of producing metal of lower oxygen content. This is because the fluoride processing is associated with a subsequent melting operation in a beryllia crucible. However, the full advantage of reducing oxygen content is not likely to be realized until the beryllium oxide can be brought down to a level below the solubility of oxygen in metal. As far as nuclear applications of beryllium metal are concerned, the emphasis that can be used most advantageously by the reactor design engineers is the limited ductility of the metal as it is known today.

REFERENCES

1. **Smith, C. O.,** The role of materials in reactors, in *Nuclear Reactor Materials,* Addison-Wesley, Reading, MA, 1967, 37.
2. **Glasstone, S. and Sesonske, A.,** Diffusion and slowing down of neutrons, in *Nuclear Reactor Engineering,* D Van Nostrand, Princeton, NJ, 1962.
3. **Gerasimov, V. and Monakhov, A.,** Water Radiolysis, in *Nuclear Engineering Materials,* Mir, Moscow, 1983, 152.
4. **Benedict, M.,** Review of heavy water production processes, in *Proc. Int. Conf. Peaceful Uses of Atomic Energy,* Vol. 8, United Nations, New York, 1956, 377.

5. **Benedict, M. and Pigford, T. H., Eds.,** Separation of isotopes of light elements, in *Nuclear Chemical Engineering,* McGraw-Hill, New York, 1957, chap. 11.
6. **Becker, E. W.,** Heavy Water Production, Rev. Ser. No. 21, International Atomic Energy Agency, Vienna, 1962.
7. **Levnis, D. M.,** Heavy Water Production, A Review of Processes, AAEC Rep. TM 562, 1970.
8. **Murphy, G. M., Ed.,** *Production of Heavy Water,* McGraw-Hill, New York, 1955, 4.
9. **Malkov, M. P.,** *Vydelenie Deyteria Iz Vodoroda Metodam Glubokogo Ohladenyia,* Gosatomizdat, Moscow, 1961.
10. **Haywood, L. R. and Lumb, P. B.,** *Chemistry in Canada,* Vol. 27, 1975, 15.
11. **Nightingale, R. E., Ed.,** *Nuclear Graphite,* Academic Press, New York, 1962.
12. **Gerasimov, V. and Monakhov, A.,** Structural materials of core, in *Nuclear Engineering Materials,* MIR, Moscow, 1983, 215.
13. **O'Driscoll, W. G. and Bell, J. C.,** Graphite, in *Materials for Nuclear Engineers,* McIntosh, A. B. and Heal, T. J., Eds., Temple Press, London, 1960, 171.
14. **Lockett, E. E.,** The use of graphite as a moderator in nuclear reactors, SCI Conf. Carbon and Graphite, 1957.
15. **Eatherly, W. P.,** Nuclear graphite — the first years, *J. Nucl. Mater.,* 100, 55, 1981.
16. **Prince, N.,** Technology of graphite moderator structures, *Nucl. Energy,* 18(4), 267, 1979.
17. **Davidson, H. W., Losty, H. H. W., and Rose, A. M.,** The mechanical properties of graphite at elevated temperatures, SCI Conf. Carbon and Graphite, 1957.
18. **McIntosh, A. B., Heal, T. J., and Cowan, A.,** Physical and mechanical properties of graphite moderators, SCI Conf. Carbon and Graphite, 1957.
19. **Easp, F. K. and Hill, M. W.,** Oxidation of carbon and graphite, SCI Conf. Carbon and Graphite, 1957.
20. **Tyzack, H. and Cowan, H. C.,** The corrosion of reactor materials in carbon dioxide, *At. Energy Rev.,* 14(2), 263, 1976.
21. **Wood, C. J. and Wickham, A. J.,** Graphite oxidation and coolant chemistry in CEGB AGRs, *Nucl. Energy,* 19(4), 277, 1980.
22. **Best, J. V., et al.,** Inhibition of moderator graphite corrosion in CEGB Magnox reactors, *J. Br. Nucl. Energy Soc.,* 4, 31, 1976.
23. **Kaufman, A. R., Gordon, P., and Lillie, D. W.,** The metallurgy of beryllium, *Trans. ASM,* 42, 785, 1950.
24. **Schenzfeier, C. W.,** *The Metal Beryllium,* White, D. W., Jr. and Burke, J. E., Eds., American Society for Metals, Cleveland, 1955.
25. **Williams, L. R. and Eyre, P. B.,** Beryllium, in *Materials for Nuclear Engineers,* McIntosh, A. B. and Heal, T. J., Eds., Temple Press, London, 1960, 269.
26. **Darwin, G. E. and Buddery, J. H.,** *Beryllium,* Academic Press, New York, 1960.
27. *Proc. Met. Soc. Conf. Beryllium Technology,* Gordon and Breach Science, 1966.
28. U.S. Atomic Energy Commission Rep. COO-312, 1956.
29. **Kjellgren, B. R. F.,** Production of beryllium, *Trans. AIME,* 93, 122, 1948.
30. Academy of Science of the U.S.S.R., Technology of manufacturing items of pure beryllium and beryllium oxide for use in nuclear reactors, in *Proc. Int. Conf. Peaceful Uses of Atomic Energy,* Vol. 8, United Nations, New York, 1956, 587.

Chapter 3

COOLANT MATERIALS

I. INTRODUCTION

The nuclear chain reaction in the reactor core produces heat energy. A major portion of this arises from the kinetic energy of the fission fragments, which usually manifests itself as heat released entirely within the fuel elements. In fact, one of the unusual features of a nuclear reactor is that there is no theoretical upper limit to the rate of energy release (of power production) due to fission. Unlike a conventional power plant, where the temperature is limited to that resulting from the combustion of coal, oil, or gas, the temperature in a nuclear reactor could increase continuously until the reactor is destroyed if the heat removal were less than the rate of heat generation. It is therefore essential that the heat generated in a nuclear reactor is removed efficiently. Evidently, heat removal has to be at the same rate at which it is being generated in order to prevent core overheating and to transfer the energy out of the core where it can be converted to a convenient form for further utilization. This job is performed by selected fluids known as coolants.

The energy transferred to the coolant, as it flows past the fuel elements, is stored in the form of temperature and pressure, and is called the enthalpy of the coolant. In pressurized (nonboiling) water reactors, the heat acquired by the coolant is in the form of sensible heat, whereas in a conventional boiler it is mainly latent heat (this implies that the rate of coolant circulation is relatively high). In an electric power plant the energy stored in the fluid is further converted into kinetic energy through a prime mover which, in the case of nuclear reactors, is predominantly a steam turbine. (In one of the proposed schemes in HTGR systems one notices, however, a difference. In this, the enthalpy of the coolant gas could be converted into kinetic energy through direct turbine in a gas turbine. (Reference to this was made in an earlier chapter dealing with the reactor system.) Another conversion takes place in the electric generator where kinetic energy is converted into electric power as the final form of energy for distribution. Often, the coolant in a reactor transfers its heat, through a heat exchanger, to a second coolant, which then carries the heat to the power generating point. In such a case, the reactor coolant usually returns to the reactor core for further heat pickup. When two coolants are thus used, the coolant which flows through the reactor core is known as the primary coolant, and the cycle through which it flows, from the reactor to heat exchanger back to reactor, is known as the primary coolant circuit. The coolant which picks up the heat from the primary coolant is called the secondary coolant, and the cycle through which it flows is known as the secondary coolant circuit. When some coolants flow through a reactor, they become somewhat radioactive, and a radioactive substance flowing directly to the power generator would cause undue inconvenience at the generators. Hence, an intermediate heat exchanger is usually provided to permit heat to be transferred from the primary coolant to the secondary coolant without transferring radioactivity.

In addition to the generation and disposal of heat in a reactor as generally observed above, there are other special problems that are unique to reactor design. It is worth making an introductory reference to some of them. In a conventional heat engine or power plant, materials are chosen on the basis of their mechanical performance. In a reactor system, on the other hand, the choice is dictated to a considerable extent by nuclear properties. This often results in the selection of uncommon and costly materials. The volumetric heat release rates in a reactor may be higher than in any other thermal systems designed for continuous operation, and this adds to the problem of heat removal from a reactor. The combination of high operating temperatures — for maximum thermal output of the reactor and maximum

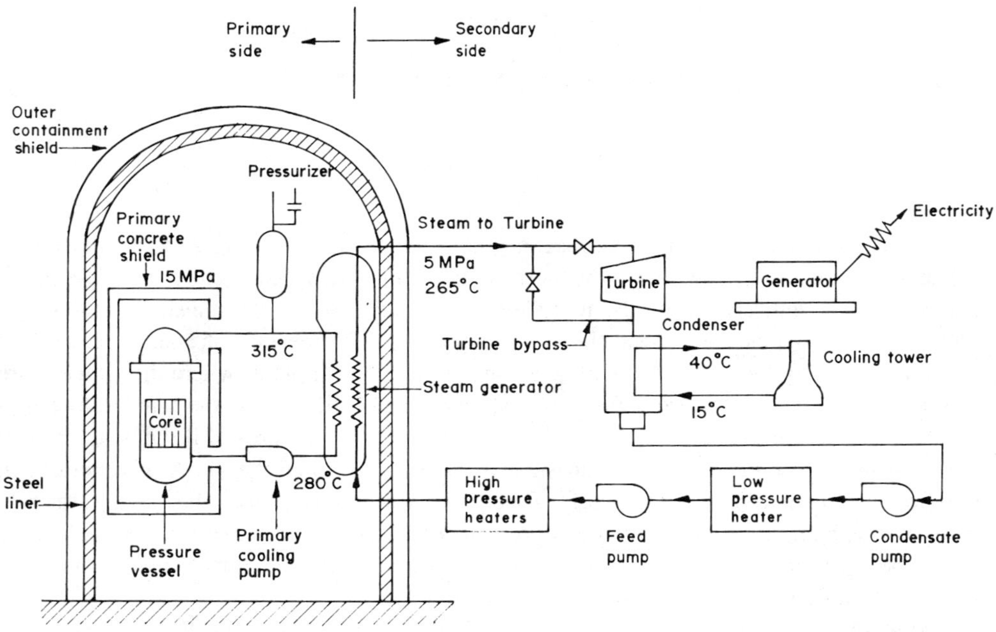

FIGURE 1. PWR system with its coolant circuit.

thermodynamic efficiency for ultimate conversion of heat into power — and large power densities impose special demands upon the reactor heat removal system and materials associated with it. This chapter concerns heat removal system materials: the coolant materials. The text is confined mainly to liquid and gaseous coolants used for solid-fueled reactors.[1]

II. COOLANT SYSTEMS

In nuclear reactors, there are essentially three types of coolant circuits or systems: (1) once-through coolant circuits, the simplest direct-cycle coolant circuits in which the coolant passes once through the reactor coolant channel and then to a heat sink; (2) double coolant circuits, consisting of a primary coolant circuit and a secondary coolant circuit, in which the closed primary circuit is in conjunction with the open secondary coolant circuit; and (3) triple coolant circuits, consisting of a primary coolant circuit, a secondary coolant circuit, and a third coolant circuit, in which the closed primary and secondary coolant circuits are in conjunction with the third coolant circuit, and parts of the primary and secondary coolant circuits constitute heat exchangers to produce steam in nuclear plants. Enumerated below are specific examples of some reactor systems with coolants and the flow patterns with them.[2-5]

A. Pressurized Water Reactor Coolant System

A simplified schematic of a PWR along with its coolant circuit is shown in Figure 1. The basic circuits and components consist mainly of (1) the reactor pressure vessel, which contains the nuclear core; (2) the primary coolant circuit; (3) a heat exchanger or steam generator; and (4) the energy conversion system where steam is used to produce electricity. The primary coolant circuit carries water which acts as both coolant and moderator. The primary system water, at about a pressure of 2200 psi (equivalent to about 15 MPa) and a temperature of about 572°F (300°C), enters the reactor vessel through inlet nozzles near the top. It then takes a downward flow in the annular space (called the downcomer) between the reactor pressure vessel and the core barrel, it picks up the fission-produced heat in the reactor core, and its temperature reaches about 629°F (332°C) as it exits into the upper plenum of the reactor vessel. In a typical PWR producing an electrical output of 1244 MWe (which requires production of about 3600 MW of thermal power), a water flow of the order of 150 million lb/h carries away the core heat and keeps the fuel cladding temperature below a comfortable 700°F (370°C). The hot primary water exchanges its heat in a heat exchanger or steam generator with water in a secondary loop where steam is generated and then returns to the core through the primary coolant pumps.

The diagram in Figure 1 shows, for simplicity, only one primary and one secondary loop. The secondary system is maintained at conditions of somewhat lower temperature and pressure. Typically, a temperature of about 510°F (265°C) and a pressure of about 950 psi (approximately equivalent to 5 MPa) prevail. In real situations, in order to increase the power output from a PWR plant while keeping the components down to a reasonable and available size, several loops are attached to the reactor pressure vessel. One-, two-, three-, and four-loop configurations have been utilized in various designs. The diagram also shows an additional vessel, called the pressurizer, connected to the primary system. This performs the function of maintaining control of primary system pressure.

The primary system, with all its piping and components, along with the steam generators, is enclosed in a specially designed structure called the containment. Any radioactivity that may escape from the fuel into the primary water via a potential leak remains contained and isolated from the environment by this provision. In the nuclear field, containment aspects are very important. The term containment refers to the technical means provided to contain radioactivity, primarily fission products, and prevent its escape into the environment. In an LWR there are successive barriers. The first is provided by the fuel matrix, the second is the fuel cladding, the third is the primary system (pressure vessel, pumps, steam generator primary side, pressurizer, etc.), and the fourth is provided by the reactor containment itself. The containment structures with their associated systems are designed to withstand a set of postulated external and internal loads and to effectively contain radioactivity that may be released from the primary system. They also stand to provide a shielding function: the protection of personnel from exposure to radioactivity contained inside containment. The name PWR stems from the fact that water in the primary circuit is kept at the very high pressure stated previously such that, at the maximum prevailing temperature in the system, no boiling can occur.

At the pressure employed in the PWR, the boiling temperature of water is about 650°F (343°C), which is well above the maximum temperature of 629°F (332°C) allowed in the primary water. Because the primary coolant must be kept below a certain temperature limit, under conditions dictated by primary systems pressure and materials behavior, the secondary fluid must necessarily be kept at an even lower temperature. Thus, the overall thermal efficiency of PWRs is a modest 32 to 33%, which is quite a bit lower than the thermal efficiency of about 40% achievable in modern fossil-burning power stations, where higher-temperature steam can be used.

B. Boiling Water Reactor Coolant System

A simplified schematic of the thermal cycle of a BWR is shown in Figure 2. The basic feature of the BWR is that water boils as it passes through the reactor core. This is a consequence of the use of a lower primary coolant pressure, typically about 1040 psi (equivalent to about 7 MPa). Feed water enters between the middle and top of the reactor vessel. Flow is induced downward in an annular space between the core shroud and the vessel wall by a set of jet pumps internal to the vessel. The coolant then takes a turn to an upward flow in the lower plenum and passes through orifices at the bottom of the fuel assemblies. As the water flows through the core, it is gradually heated to a boil, and bubbles

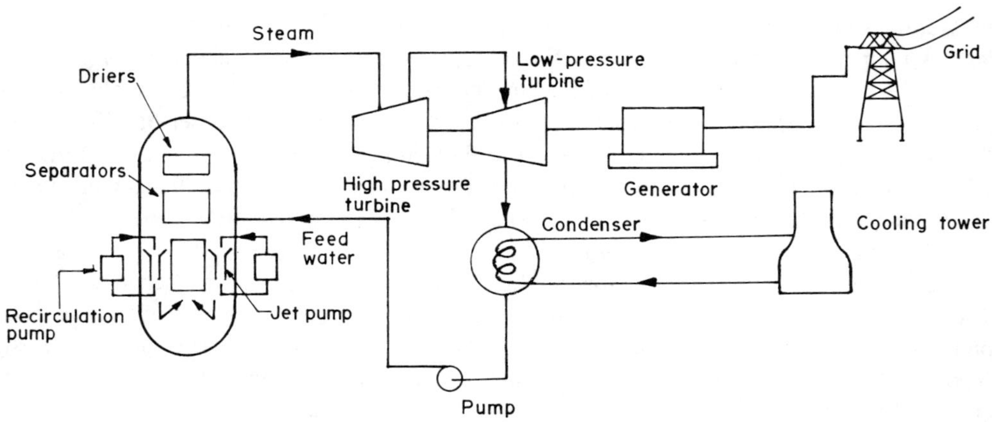

FIGURE 2. Thermal cycle of a BWR.

begin to form at an increasing rate as the coolant moves upward. It exits the core in a mixture of steam (13 to 10 wt%) and water (87 to 90 wt%). The water fraction is recirculated downward by the force of the jet pumps, while steam keeps its upward movement. Banks of steam separators and dryers are placed above the core. They provide a further separation of moisture. Steam is directly piped to the turbine for power production and is returned to the reactor vessel via feed pumps. A BWR is thus characterized by the occurrence of boiling in the core, the steam generated in the core passing directly to the turbine. There is no steam generator interposed between the reactor core and the turbine. This simplifies the energy conversion loop.

Because of these features, the BWR plants are classified as forced circulation, direct-cycle units. Steam coming out of the reactor vessel dome is typically at a pressure of 1040 psi, at which pressure water boils when heated to a temperature of 549°F (287°C). These conditions characterize saturated steam. This sets a limit to the enthalpy available for conversion with mechanical energy through expansion in the turbine. For this reason, the thermal efficiency of the BWR cycle is, as with PWRs, a modest 32 to 33%. Unlike PWRs and other plant types, the working fluid in BWRs, after removing heat from the core, enters the turbine directly, as stated. The radioactivity is no longer securely closed in a tightly sealed primary system. In fact, through the deaeration of the condenser, gases that may contain radioactivity are continuously removed from the system, and discharge from the stack becomes more critical. Also, shielding of turbine and associated piping becomes necessary for protection of plant personnel. Thus, in a BWR installation, the problem of handling radioactivity and protecting personnel from exposure to radiation is a more cumbersome and demanding activity than in other nuclear power plant types.

C. Pressurized Heavy Water Reactor Coolant System

A simplified schematic of a pressurized heavy water reactor (PHWR) cooling system is shown in Figure 3. It consists of two loops, 1 and 2. Each one ensures the cooling of half of the 380 pressure tubes in which fuel bundles (380 × 12) are introduced. The pressurized heavy water serving as coolant in loop 1 passes through the core in the direction indicated by the arrow (a) though 95 pressure tubes, and then through the connecting pipes (4) reaches the collector (5) and the steam generator (6). The heavy water is then sucked by a pump (7) and sent to the collector (8) from which, through the connecting pipes (9), it over passes the core in the direction of the arrow (b) through the second group of 95 pressure tubes (10) of the loop 1. Then, the heavy water reaches the collector (12) and the steam generator (13) through the connecting tubes (11). From here it is sucked by a pump (14) that brings it to

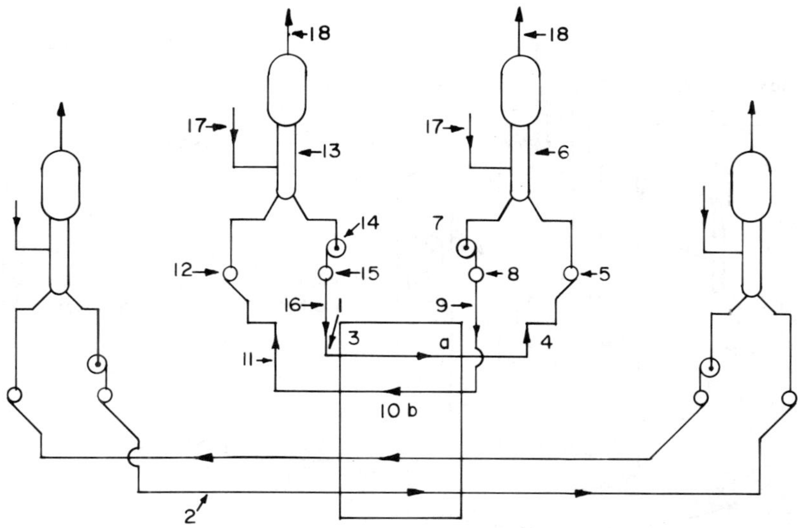

FIGURE 3. PHWR primary coolant circuit (1, 2, cooling loop; 3, 10, pressure tube; 4, 9, 11, 16, connecting pipe between pressure tube and collector; 5, 8, 12, 15, collector; 6, 13, steam generator; 7, 14, pump; 17, steam generator supply, secondary circuit; 18, steam outlet).

the collector (15); then, through the connecting pipes (16), the water reenters the first group of pressure tubes. The operation of loop 2 is identical to that of loop 1, as described. The temperature of the primary heavy water at the inlet of the pressure tube is 540 K, and the pressure is 11.05 MPa. At the outlet the temperature of heavy water is 585 K and the pressure 10.3 MPa. Generators (6 and 13) produce steam at 4.55 MPa and 531K.

D. Liquid Metal Fast Breeder Reactor Cooling Systems

A simplified schematic of an LMFBR cooling system is shown in Figure 4. It represents a departure from the PWRs, BWRs, and PHWRs described above. The LMFBRs operate at higher temperatures with liquid sodium as the heat transfer medium. This coolant gets intensively radioactive during performing its job of heat extraction from the reactor core, and therefore the designed coolant circuits for the coolant appear different from the others described. For LMFBRs involving cooling with sodium, two different types of heat exchangers are shown in the figure. In one, called the intermediate heat exchanger, heat is transferred from radioactive sodium flowing through the reactor to nonradioactive sodium. In the second, which is the steam generator, the nonradioactive sodium transfers heat to water and produces super saturated steam. The main requirement of the intermediate heat exchanger is that there should be no leakage from the radioactive primary to the nonradioactive secondary circuit. In the sodium-to-water steam generator, the main design problems arise from thermal stresses and the possibility of chemical interaction between sodium and water.

III. GENERAL CHARACTERISTICS OF REACTOR COOLANTS

The need for a nuclear reactor to have an adequate cooling system to prevent attainment of undesirable temperatures within the reactor warrants no further elaboration. It has sufficiently been emphasized in the preceding sections. The undesirable temperature regime within the reactor is specific situation dependent. It may first depend on the property of the fuel. For example, uranium metal, which undergoes a phase change at 665°C and accom-

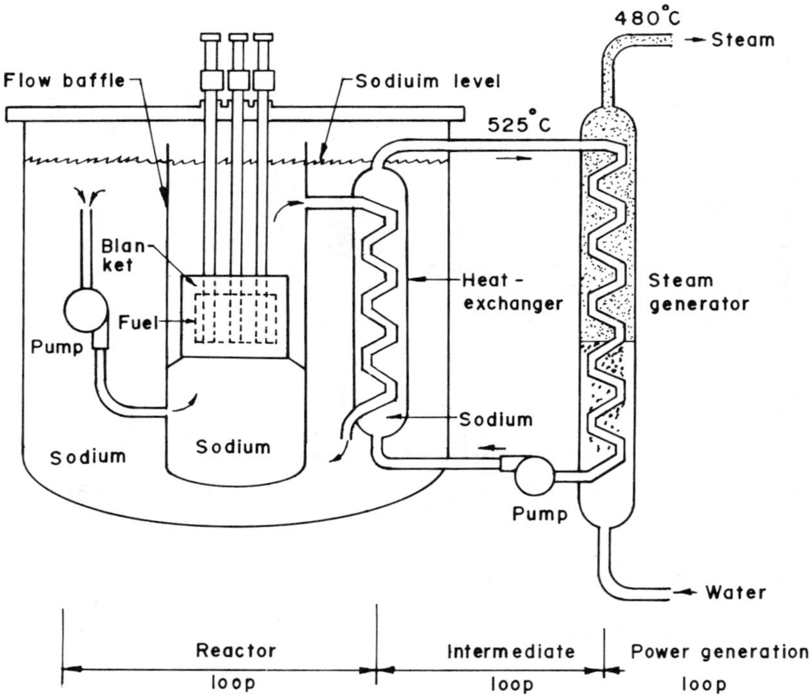

FIGURE 4. Cooling system for sodium-cooled fast reactor.

panying volume changes, sets a limit to permissible reactor temperature operation. It may also depend on the coolant. For example, it may be undesirable for the water to boil in a water-cooled reactor. The specific requirements for reactor coolants will largely depend on the rate of heat generation or, in other words, reactor power density, and the operating temperatures. In those cases where little or no use is made of the heat generated by fission, for example, for reactors of low power output, the cooling system employed may be relatively simple. On the other hand, for economic power production a reactor is required to be operated at high temperatures. At this time, one must give very careful consideration to making a choice of reactor coolants.

An ideal coolant should possess the following characteristics: (1) good thermal properties, (2) low viscosity, (3) high boiling point and low melting point, (4) thermal and irradiation stability, (5) small corrosive and erosive effect on both in-core and out-of-core reactor materials, (6) relative freedom from impurities having high neutron capture cross sections, (7) small capture cross section for neutrons, (8) nonhazardous nature including low induced radioactivity, (9) commercial availability in large quantities at an acceptable price, (10) minimum pumping power to remove heat at a given rate from core to boilers, (11) acceptable pressure employed in terms of reactor safety and structural material costs, and (12) easy and safe handling.

As the chief objective of coolant material is to remove and utilize heat or thermal energy generated in a nuclear power reactor, good heat transfer properties are the basic primary requirements of a reactor coolant. The heat transfer properties consist mainly of high specific heat, thermal conductivity, convective heat transfer coefficient, low melting point, high boiling point, low density, low viscosity, and least corrosiveness at elevated temperatures. The fraction of the power output required to pump the coolant through the reactor and heat exchanger should be small, and this requires a coolant with a high density and low viscosity. For application at high temperatures a liquid coolant should have a low vapor pressure. A strong pressurized system is not necessary in working with such liquid coolant. At the same time, the coolant should remain liquid down to low temperatures to avoid the possibility of solidification when the reactor is shut down. The former implies a high boiling point to minimize vapor pressure and the latter a low melting point to minimize coolant solidification. In the case of a gas as a coolant the problem considered here does not arise.

A coolant should show high thermal and irradiation stability. If the reactor has a circulating fuel, so that the secondary coolant is not exposed to a very high radiation field, the problem of radiation decomposition (and of induced radioactivity) is not as serious as when the coolant actually flows through the reactor.

It is also necessary that the coolant not attack the materials, either inside or outside the reactor, with which it comes into contact. Noncorrosiveness or chemical compatibility with the fuel, cladding, and other materials is desired for the coolants. Corrosion and erosion problems are often unusually serious in the coolant circuit because of the high temperatures (and sometimes pressures) involved. The corrosion rates, which may be negligible otherwise, can be significant in a reactor system. When there are temperature differences in a circulating system, as in the case of a liquid metal coolant, especially at high temperatures, corrosion of the mass transfer type can take place. Metal can be dissolved by the circulating liquid in hot regions identified by high temperature and metal solubility, and is deposited in low-temperature regions of the system, where the solubility of the metal in the coolant is low. With gaseous coolants, other than helium, of course, chemical reactions may occur, and, in addition, there may be erosion as a consequence of the high flow rates which are necessary for obtaining adequate heat removal.

The coolant, apart from being relatively free of neutron-absorbing impurities, should itself consist of element or elements having small cross sections for neutron capture. This restriction is less important for fast reactors because the capture cross sections for high-energy neutrons are invariably small. It is desirable, in any event, that the cooling medium be nontoxic and nonhazardous, and that it not become a hazard as a result of exposure to radiation. Finally, the coolant should be readily available at relatively low cost.

For a thermal reactor, it is both desirable and advantageous if the coolant also acts as a moderator and makes some contribution to the slowing down of the neutrons. As it is, there is no single substance (or mixture) which satisfies all the requirements laid down for a satisfactory coolant. In actual practice the coolant chosen for any particular application will represent a compromise among conflicting requirements.

A. Coolant Types and Their Characteristics

The fluid medium, or coolant, is basically of two types: liquid and gaseous. Considering the important part they play and the special working conditions to which they are subjected, the coolant materials must meet certain special requirements as generally outlined above. For the liquid coolant materials, the requirements are as follows: (1) a low melting temperature, a critical temperature as high as possible, and a critical pressure as low as possible; (2) low viscosity; and (3) small neutron capture cross section. This third general requirement is more important for liquids than for gases, where the absorption macroscopic cross section, $\Sigma_a = \sigma N$, is small, due to low values of N. For the gaseous coolants, the following further requirements are in order: (1) a good chemical and radiochemical stability; (2) low levels of induced radioactivity and short half-life of the nuclei generated by neutron-induced reactions; (3) high density, specific heat, and thermal conductivity; (4) compatibility with the structural and moderator materials with which they come into contact; and (5) plentiful availability and cheap price (except in the case of helium). The liquid used for cooling fast reactors must also have a moderating power as weak as possible, to avoid the growth of the fraction of low-energy neutrons, which would reduce the efficiency in the generation of fissionable nuclides.

B. Coolant Materials

The principal coolants which have been used for nuclear reactors are (1) liquids (e.g., water, heavy water, and organic fluid); (2) gases (e.g., carbon dioxide and helium); and (3) liquid metals (e.g., sodium). Before considering different coolants individually, the most important nuclear, physical, and thermal properties of selected potential coolants are given in Table 1. A list of commercially used reactor coolants is shown in Table 2.[6] A general discussion comparing the coolants appears in the text below.

The volumetric heat capacity of a gaseous coolant is normally less than that of a liquid. The total rate of heat production by a reactor q_{tot} is equal to the product of total mass flow rate w, of the coolant, its specific heat Cp, and the average temperature rise Δt_c in the bulk coolant temperature, i.e., $q_{tot} = wCp\Delta t_c$. Since w may be replaced by $v\rho A_f$, where A_f is total flow area, it follows[7] that $q_{tot} = (\rho Cp)vA_f\Delta t_c$. From this relationship it emerges that the removal of a certain quantity of heat will usually require a much higher coolant velocity for a gas than for a liquid. The assumption here is that $A_f\Delta t_c$ is not greatly different in the two cases. Hence, the pumping power [pumping power = pressure drop × volume flow rate = $(\Delta p)(A_f v)$, where Δp is the total pressure drop due to all causes, and A_f is the total cross-sectional or flow area of all coolant channels] requirements for a gas will be much larger than for a liquid coolant. The discrepancy can be partially overcome by increasing the gas density, which means operation at high pressures.

This general discussion of coolants is hardly complete without a comparison of their heat transfer characteristics. A coolant qualifies as a good one when it possesses a high heat transfer coefficient (and a large volumetric heat capacity) under the conditions that exist (or can exist) in a reactor. (The fundamental equation of convective heat transfer, for both free and forced motion of the fluid, is neutron's law of cooling and is written as $q = hA_h\Delta t$, where q is the rate of convective heat transfer to and from a surface of area A_h when the temperature difference is Δt. The quantity h defined by the given equation is commonly called the heat transfer coefficient. Some, however, prefer the term unit thermal conductance. It is usually expressed in units of Btu/(h)(ft^2)(°F) for engineering calculations.) Thus, if h is large, it is possible to secure a specified heat removal from the fuel elements with a small heat transfer area. It is also possible to reduce the coolant flow rate and, as a result, the pumping power required. The reduction in the flow rate will reduce the h. In a situation where h is high, especially for a liquid metal, the reduction in the flow rate may not be serious.

For a given flow rate and coolant channel diameter, large heat transfer coefficients are associated with high thermal conductivity, density, and heat capacity. Water and many liquid metals may be expected to possess good heat transfer characteristics, and such materials will therefore afford considerable flexibility in the heat removal system design of the reactor.

Gases have the disadvantages of low density and thermal conductivity. The low density can somewhat be compensated by high operating pressures. The high flow rate required with a gaseous coolant also helps to increase the heat transfer coefficient. This is attended by a very significant increase in pumping power. In reactors designed for useful power production, the flow rate of liquid coolants is generally about 15 to 20 ft/sec. In gas-cooled reactors (GCRs), in comparison, the flow rate is approximately ten times as great.

The volumetric heat capacity ρCp, given earlier in the relationship for q_{tot}, can be taken as the basis for comparison of coolants. It is useful in assessing the volume of coolant required to remove a given quantity of heat. The relationship indicates that for a given coolant flow rate v and total flow area A_f a large ρCp will make possible a specified rate of heat removal without an excessive rise in the coolant temperature. The liquid metals are somewhat superior to water with respect to heat transfer. The water has a better volumetric heat capacity. Thus, at 400°F, ρCp for water is 53.7, whereas that for liquid sodium is 17.1 Btu/(ft^3)(°F). At 10 atm pressure the volumetric heat capacity of carbon dioxide is only 0.174 Btu/(ft^3)(°F); thus, very high flow rates are necessary for gaseous coolants.

Table 1
SOME CHARACTERISTICS OF SOME POTENTIAL COOLANTS

Coolant	Mp (°C)	Bp (°C)	Temp (°C)	Density (g/cm^3)	Specific heat, Cp (cal/g-°C)	Viscocity (g/cm-s)	Thermal conductivity, K (cal/s-cm-°C)	Volumetric specific heat, ρCp (cal/cm^3-°C)	Thermal neutron absorption cross section (b/nucleus)	Induced radioactivity
Hydrogen (1 atm)	—	Gas	100	6.6×10^{-5}	3.43	10^{-4}	5.33×10^{-4}	2.26×10^{-4}	0.33	None
Helium (1 atm)	—	Gas	100	1.4×10^{-4}	1.25	2.2×10^{-4}	4.0×10^{-4}	1.75×10^{-4}	NO	^{41}Ar (1.8 h), 1.4 MeV gamma
Air (1 atm)	—	Gas	100	9.5×10^{-4}	0.240	2.1×10^{-4}	7.57×10^{-5}	2.28×10^{-4}	1.5	^{16}N (7.3 s), 6 MeV gamma, ^{41}Ar
Carbon dioxide (1 atm)	—	Gas	100	1.5×10^{-3}	0.218	1.8×10^{-4}	5.0×10^{-5}	3.27×10^{-4}	0.003	^{16}N
H_2O (liquid) (1 atm)	0	100	100	1	1	7×10^{-3}	1.7×10^{-3}	1	0.22	^{16}N
Sodium	98	883	100	0.928	0.33	6.9×10^{-3}	0.206	0.306	0.50	^{24}Na (15 h), 1.38 and 2.75 MeV gamma
NaK, 22 w/o (eutectic)	−11	784	100	0.843	0.227	4.68×10^{-3}	0.0583	0.192	1.5	^{24}Na, ^{42}K

From Smith, C. O., *Nuclear Reactor Materials*, Addison-Wesley, Reading, MA, 1967, 50. With permission.

Table 2
COMMERCIALLY USED NUCLEAR REACTOR COOLANTS

Reactor type	Coolant	Max temp (°C)	Pressure (MPa)[a]	Max efficiency, electricity/heat (%)
Magnox	CO_2	410	1.5	35
AGR	CO_2	650	4.0	44
PWR	H_2O	320	15.0	33
BWR	H_2O	280	7.0	33
SGHWR	H_2O	280	7.0	33
PHWR	D_2O	320	8.0	32
HTR	He	800	4.5	45
THTR	He	800	4.5	45
LMFBR	Na	580	0.1	42

[a] 1 Mpa ≈ 10 atm.

IV. WATER COOLANT

A. Specific Features

The overwhelming role of water in its various capacities in the nuclear energy field is well known. The use of such a familiar material as water certainly holds considerable attractions. Light water functions as coolant, moderator, reflector, and shielding and working substance to generate electricity all at once, making it one of the leading materials in LWRs. It is, however, predominantly recognized as acting simultaneously as a moderator and coolant. It has a high specific heat, although only a fair thermal conductivity compared to liquid metals; it also has a relatively low viscosity and is easily pumped. The pumping power for equivalent heat removal is roughly on the order of one tenth that required with a gaseous coolant at 10 atm pressure. The fact that more energy can be removed by the same quantity of water as latent heat than as sensible heat, combined with other desirable properties, makes water attractive as a boiling coolant. Drawbacks of water as a coolant include a fairly large neuton absorption cross section, decomposition by radiation, corrosive action on metals, and low boiling point at normal pressures. The high neutron capture cross section is compensated by the use of enriched fuel. As the boiling point is low and vapor pressure soars at high temperatures, the reactor must operate at high pressures. This is an obvious source of engineering problems.

Besides light water, heavy water is also attractive as a reactor coolant (a coolant-moderator). The only significant difference between light, or ordinary, water and heavy water lies in the appreciably small cross section of the latter for the capture of thermal neutrons. A good moderator, heavy water yields an inherent advantage in neutron economy. The MR is nearly 400 times greater for heavy water than it is for light water (the moderating ratio for heavy water is 21,000, and that for light water is 58). The improvement in neutron economy means that less fissile material is required for making a reactor critical. Use of natural uranium as fuel is entirely feasible. The additional cost of heavy water is not compensated by any gain in coolant properties. The heavy water is therefore, on the first count, normally specified as moderator and may incidently serve as a coolant.

B. Water-Cooled Reactor Systems[8,9]

LWRs are principally of two types: PWRs and BWRs, depending upon whether or not the design permits the water to boil on its passage through the core. About 70% of the total number of reactors operating worldwide belong to the LWR class. Present construction and future plans are based predominantly on this type of reactor. Historically, in fact, the first civilian applications of nuclear power began with the PWR, specifically the Shipping port reactor in Pennsylvania.[10] It attained criticality in December 1957, and full power operation (60 MWe) was realized 3 weeks later. This installation was followed by other reactors of the same type, including the Yankee reactor;[11] the Indian Point reactor;[12] and the reactor for the first merchant ship, the *N.S. Savannah,* to be propelled by nuclear power.[13] The development of the PWR was based on the early basic assumption that pressurized liquid water would keep the reactor stable and that boiling water in the core might cause the reactor to be unstable.

The second major type of LWR, the BWR, came into being after development originally from the BWR experiments (BORAX I, II, III, etc.) and experimental boiling water reactor (EBWR) experiments at Argonne National Laboratory in 1956. The first commercial version of the BWR was a dual-cycle reactor quite similar to a PWR, with the use of steam generators. Examples of such plants are the Dresden 1 plant (180 MW), which is still in operation in Illinois, and the Kraftwerk Union (KWU) plant (640 MWe) at Wuergassen, West Germany. It eventually became evident that full potential advantage of boiling in the core could be realized by the use of a single cycle. The sole vendor of BWRs in the U.S. is the General Electric Company. Manufacturers in other countries, including Japan, Germany, and Sweden, have also built BWRs.

Besides the dominant BWR, there are several varieties of reactor where the principle of boiling has been used. The Canadian reactor, using natural uranium and heavy water, can be designed to use boiling light water as its coolant. The Gentially 1 reactor, a 250-MWe reactor, called a boiling light water reactor (BLWR) but with heavy water moderator, owned by Hydro Quebec, has been in operation near Becancour, Canada since 1972. This particular type has not, however, found any imitators. The U.S.S.R. has built a large number of units using a boiling water pressure tube design with graphite as moderator, known as the light water graphite-moderated reactor (LWGR).[14] Future Russian plans for nuclear expansion are, however, based on PWRs and LMFBRs.

In making any reference to reactor systems which are both moderated and cooled by heavy water, the name that comes to mind first is Canada's CANDU (Canadian deuterium-moderated uranium-fueled) reactor. CANDU is almost synonymous with heavy water moderation and cooling. This is the one successful reactor type that has withstood the test of time. Not only has this reactor type been spectacularly successful in Canada, but it has proved attractive in the international marketplace in nuclear power generation. Apart from Canada, CANDU units have been installed in India, South Korea, Argentina, and Romania. In the list of examples of heavy-water-moderated reactor systems, the SGHWR which is the U.K. version of the BLWR, is another addition. The cooling in this system is, however, accomplished by light water. Although this variation causes some neutron penalty, natural uranium can still be used. Like the BWR, reactors of this type dispense with boilers by allowing the water to boil in the core and separating steam for turbines by means of steam drums.

C. Coolant Water Chemistry[15-17]

In a nuclear power plant there are many auxiliary systems which play a supporting role in the main functioning of the plant. Present reference is being made to the coolants in the primary and secondary circuits in LWRs. In all LWRs, the coolant chemistry, in both primary and secondary systems, must be closely controlled for good plant operations.

1. Primary Water Coolant Chemistry

The primary coolant gets fouled with impurities through wear and corrosion, and possibly through leaks in some fuel pins. On accumulation these impurities would cause progressive surface fouling and crud formation in certain parts of the primary system with resultant problems for plant operation and maintenance, and unfavorable impact on plant availability. In order to combat this problem, the chemistry and volume control system (CVCS) is provided to perform essentially three functions: (1) removal of fission and corrosion products; (2) provision of a constant supply of water for the primary system and compensation for losses; and (3) maintenance of the boron concentration at specified levels. The system is also known as the makeup and purification system.

As the primary coolant circulates, a certain portion of it is continuously removed from the primary loop through a letdown cooler; then is passed through a demineralizer, where impurities other than boron are removed. It is then introduced to a makeup surge tank, from which it is pumped back to the reactor coolant system as and when required. The demineralizers used in the CVCS and in several other applications in power plants are special components designed to remove dissolved materials from the reactor coolant water and various other liquid waste. They are essentially steel tanks filled with a bed of resins, and through a process of chemical ion exchange the resins absorb and retain radioactive and nonradioactive materials dissolved in water. The system is used to perform another important function at the plant: chemical sampling operations and chemical addition. The chemical sampling is essentially a means of monitoring the concentration of various chemicals in the reactor coolant and auxiliary systems. Two parameters are of particular importance: the pH level, which measures the degree of solution acidity or alkalinity; and the oxygen concentration and conductivity of the coolant (a high value means dissolved salts). These two parameters have the potential of promoting corrosion in the plant if not carefully controlled.

In order to reduce corrosion due to oxygen in the coolant, oxygen levels must be kept under control. Deaerators at several points of the system allow gases to escape from the coolant. The process does not, however, allow oxygen levels to drop to the required low levels. Chemical treatment is used to supplement the process of oxygen removal. Through the chemical addition system, which is part of the CVCS, chemical substances can be injected into the system for oxygen control. Hydrogen injection can be used. Hydrazine (N_2H_4) is extensively used for oxygen control via the reaction $N_2H_4 + O_2 \rightarrow 2H_2O + N_2$. Hydrazine solution, a clear, waterlike, alkaline liquid, introduces alkalinity to the water. It produces no solids. The nitrogen gas can be removed through the gaseous effluent control system. Dissolved hydrogen is employed to control and scavenge oxygen produced by water radiolysis in the core region. For this, sufficient partial pressure of hydrogen is maintained in the volume control tank to attain the specified equilibrium concentration of hydrogen in the coolant. A pressure control valve in the vapor space of the volume control tank can be adjusted to provide the correct hydrogen concentration needed. Apart from oxygen, the pH factor, as pointed out, is also a matter of great concern. An acidic water tends to be corrosive, since the corrosion process occurs as metal ions replace hydrogen ions. The pH control chemical employed is lithium hydroxide, and its concentration is maintained in the range specified for pH control (to keep water alkaline well above the 7.0 pH level). It may be pointed out that the degree of corrosion is also dependent upon the type of materials used. All the inner surfaces of the primary pressure boundary in a PWR, where high-pressure and high-temperature water circulates, are provided with a stainless steel surface layer.

To the list of chemical additives to the primary water coolant in PWRs one must add boric acid (H_3BO_3). (The solubility of boric acid is 63.5 g/l at room temperature and 267g/l at 100°C.) Since boron is a strong neutron absorber, the boric acid as a source of boron dissolved in the primary coolant is used in conjunction with the absorber rods to control the chain reaction. The maintenance of proper chemical control of the primary system is very important, and plant operation must maintain certain standards that have evolved. These are summarized in Table 3.

Table 3
PRIMARY WATER CHEMISTRY STANDARDS

Primary Water	
pH	4.5—10.5
Conductivity at 25°C (μmho cm)	1—30
Oxygen, max (ppm)	0.1
Chloride ion, max (ppm)	0.15
Fluoride ion, max (ppm)	0.1
Hydrogen (cm^3 (STP)/kg H_2O)	25—35
Total suspended solids, max (ppm)	1.0
Boric acid (ppm B)	0—4500
Makeup Water	
pH	6.0—8.0
Conductivity at 25°C (μmho)	2.0
Oxygen (ppm)	0.1
Chloride ion (ppm)	0.15
Fluoride ion (ppm)	0.1
Total dissolved solids (ppm)	0.5
CO_2 (ppm)	2.0
Solids (m)	25

Note: Variations exist in standards from vendor to vendor, but typical ones are represented here.

From Lish, K. C., *Nuclear Power Plant Systems and Equipment,* Industrial Press, New York, 1972. With permission.

2. *Secondary Water Coolant Chemistry*

The chemistry of the secondary-loop water strongly influences the performance of the steam generator and the steam turbine (particularly the discs and blades). All PWRs utilize chemical treatment of the secondary-side water to control pH. This necessity stems from the observation that an acidic environment tends to promote many forms of corrosion. With the chemistry of the secondary water in the alkaline region, however, most of the materials used in LWRs become passivated. This greatly reduces corrosion. The majority of nuclear plants control the pH of the water either by addition of phosphates or by the addition of volatile chemicals.

The phosphate control method involves adjustment of water alkalinity by the addition of phosphates. The phosphates can release sodium hydroxide through the reaction $Na_3PO_4 + H_2O \rightleftharpoons Na_2HPO_4 + NaOH$. In the conventional method, called the coordinated phosphate treatment, the formation of free NaOH is avoided by maintaining a balance of phosphate to sodium (3:1). The reaction proceeds both ways; excess NaOH in the solution drives the reaction to the left, while excess acidity neutralizes the NaOH and drives the reaction to the right. The main advantage of the phosphate treatment is its ability to buffer. It protects against imbalances both in the acidic side and the basic side, and maintains a pH value between 9 and 11.

The all-volatile treatment (AVT) is an alternative method for pH control. In this method, the introduction of solids into the system is avoided by using volatile alkaline chemicals to control the pH. This method is, therefore, particularly suitable for once-through steam generator plants. A main disadvantage of AVT compared to the phosphate treatment is that volatile chemicals do not buffer the solution, and the protection offered against acidic and alkaline excursions is not as strong as with phosphates. In the AVT treatment, the most widely used chemical is ammonia. Another chemical used with AVT in secondary water is hydrazine. It acts to remove oxygen, which is generally a harmful element, in the same way that it acts in the primary system. Hydrazine solution causes removal of oxygen through the reaction $N_2H_4 + O_2 \rightarrow N_2 + 2H_2O$. Any excess hydrazine not used up for oxygen adds alkalinity to the solution. This happens because of the decomposition reaction of N_2H_4 at the temperature prevailing in the steam generator ($3N_2H_4 \rightarrow 4NH_3 + N_2$ or $2N_2H_4 \rightarrow 2NH_3 + N_2 + H_2$).

In the case of BWRs, control of coolant chemistry using volatile chemicals such as hydrazine and hydrogen is more difficult, since, unlike PWRs, the BWR is an open system. BWR primary coolant contains 0.2 to 0.4 ppm oxygen during normal operation, and these levels are inherent in the operating characteristics of a BWR (oxygen content is the direct result of radiolysis). Hydrogen cannot be easily introduced into the system to scavenge oxygen. Unlike a PWR, boron and lithium are not added to the coolant. The chemicals can leak out through the turbine glands and condenser seals. This is one of the primary reasons why BWRs operate close to neutral pH. Thus, BWRs can be said to operate using high-purity water without additives. Impurity levels in the coolant are controlled by the reactor cleanup system and by using high-quality feed water. To reduce impurities, several procedures are employed, including oxygen control, full-flow condensate demineralization, and reactor shutdown when the level is prohibitively high. Materials used in the components are also important (stainless steel is used for feed water heater tubes; superior performance is being obtained with titanium alloys for condenser tubing).

V. GASEOUS COOLANT

A. General

The recommendation for the use of a gaseous substance as a reactor coolant can be made on the basis of general radiation stability, ease of handling, and absence of hazardous conditions. Although the heat transfer characteristics are not as good as those of water and liquid metals, it appears that, by suitable design, gas-cooled power reactors may be competitive with those using liquid coolants. Reference should be made to certain design developments. In the first place, improvements in thick-walled pressure vessels and the advent of the PCVRs have made it possible to use the coolant at high pressure, thereby enhancing its heat transfer properties. Second, the use of ceramic fuels permits operation at higher temperatures than have been achieved with metallic fuels. Finally, heat removal can be increased by means of fuel elements having large heat transfer areas. Simple examples of finned fuel elements are the longitudinal fin type used in the Brookhaven graphite reactor and the transverse fin type used in British Calder Hall power reactors.

As inherent virtues of gaseous coolant, a number of general points can be recorded. With gaseous coolant the operating temperature is independent of the pressure. Thus, a high gas temperature can be used, limited only by the core and circuit materials, to give good steam conditions from the boiler is essential for modern high-efficiency turbines; the optimum pressure can be selected separately from considerations of safety and the economics of pumping power and pressure circuit costs. Gas has certain safety advantages. The most fundamental property of any gas is that it cannot condense over the range of temperatures encountered and that it occupies the entire space in which it is confined. It obeys ideal gas laws, having a linear temperature-pressure relationship. There is no liquid-gas interface, and unambiguous signal (pressure) defines the properties of the coolant. The most important aspect of this is that rapid depressurization is possible without such matters as void formation, local core dry-out, and cavitation of pumps. Unlike water systems, gas systems can be cooled rapidly. Pressurization with gas cooling is much below that of cooling with water.

The dry nature of the core with gas cooling eliminates many of the corrosion, crust formation, and blockage dangers present with water-cooled reactors, and facilitates the maneuvering of fuel and control rods. Since the gas can undergo no phase change as a result of rising temperature or falling pressure, there can be no discontinuity in cooling under fault conditions, and flows and temperatures can be predicted with greater confidence. Continuity of fuel loading for on-load refueling is more easily achieved with a gas. In addition, with a gas there is no risk of fuel-coolant interaction of the kind that could result in certain circumstances from the dispersion of melted fuel in a liquid coolant. Finally, a gas carries a relatively low burden of activated corrosion products, gives low radiation levels for maintenance around the circuit, requires small effluent plants, and gives rise to only low radiation doses to the operators.

Offsetting these virtues is the combination of low density and low specific heat of gases. Even with fairly high pressures, this requires comparatively large temperature differences to transfer the heat between the fuel and the gas, and between the gas and the boiler surfaces. As a result, core ratings tend to be low, and core and boilers tend to be large. Gas also requires large volume flows to transport the heat, and therefore requires large circulator sizes and powers.

B. Gaseous Coolant Reactor Systems

Some examples can be given of the various reactor systems using various gaseous coolants. The concept of the GCR (graphite-moderated) originated from that of the Chicago Pile 1 (CP-1) and Hardford reactors designed for plutonium production as long ago as 1943. The coolant was air in these early reactor concepts involving gas cooling. In the U.K., the successful operation of Windscale reactors led to the development of dual-purpose Calder Hall and Chapel Cross nuclear stations for the production of plutonium and electricity generation. These served as prototypes for a series of GCR power plants. These reactor types (known as Magnox in Britain) were actually the first stages of the British and French programs.

The fuel is natural uranium metal clad, or canned, with a thin cylindrical tube, or can, made of a magnesium alloy in a graphite-moderated core. The coolant is carbon dioxide, circulated under pressure between the reactor core and the steam generator. The steam is led to pass through the turbogenerator system for power production. Because of the relatively low melting point of uranium metal and the chemical reaction between CO_2 and graphite, the maximum coolant temperature for this power reactor system is about 345°C. As a consequence, the operating temperature, pressure, overall (plant) thermal efficiency, and power capacity are low.

To increase the operating temperature, pressure, overall thermal efficiency, and power capacity, two major but different approaches have been proposed and advanced. The advancement and development of the two different approaches resulted in two major types of GCR: AGR and HTGR. The AGR was designed, tested, and developed early in the U.K. and some other countries. Here, the fuel elements are made of slightly enriched UO_2 canned in a stainless steel tube. The coolant is helium or carbon dioxide with the moderator remaining graphite. With this system, the operating temperatures, pressure, overall thermal efficiencies, and power capacities have appreciably increased. The HTGR was developed and tested by the heat transfer reactor experiment (HTRE) and the experimental gas-cooled reactor (EGCR), and subsequently demonstrated by the HTGR at the Peach Bottom, PA, and Fort St. Vrain power plants,[18] respectively. The fuel consists of about 14% ^{235}U, 1% ^{238}U, and 85% ^{232}Th in the form of their respective carbides dispersed in the graphite moderator. The fuel coat, or cladding, is made of dense, impervious pyrolytic graphite coated at the outer surface of each fuel zone. The reactor coolant is helium.

The foregoing coverage has briefly given an account of the gaseous coolants in the field of nuclear power generation. Table 4 gives examples of gas-cooled reactors around the world. As stated earlier, the first stage of the nuclear program in France was based on the use of gaseous coolants. The table lists these French reactors. In the U.S., a land dominated by the LWRs, gas cooling has been hard to champion. The HTGR was developed through the small experimental unit at Peach Bottom and the 330-MWe commercial demonstration plant at Fort St. Vrain. The development of GCRs in West Germany dates back to the late 1950s. From the beginning, the helium-cooled HTGR was chosen because of its higher temperature potential compared, for example, with CO_2-cooled graphite moderated reactors. In addition to electricity production, nuclear high-temperature process heat was recognized as a future application. The first HTGR in Germany was AVR, an experimental reactor with a 15-MWe capacity, located at Julich Nuclear Research Centre (KFK), pebble-bed core type. As the next step in the German HTGR development program, construction of a 300-MWe prototype power plant, THTR-300, was started in 1971 at Uentrop near the town Hamm. These various reactors based on gas cooling around the world have provided a wealth of experience. It is in the U.K., however, that the GCR has been taken to major commercialization. The development of nuclear power in Britain has been pioneered through the Magnox GCR stations.

A comparison of gaseous coolants can be made. From the characteristics of thermal stability, irradiation stability, ease of handling, abundance, and economics, air is apparently a natural choice as a reactor coolant. Air, however, is not a good transfer material. It requires high pumping power in a cooling system. At high temperatures, the oxygen and nitrogen constituents of air attack the graphite moderator and the structurals of the GCRs. Hydrogen, from the point of view of both heat transfer and pumping power, seems to be an ideal cooling medium. However, such advantages associated with hydrogen are greatly outweighed by the problems with handling this highly diffusive and hazardous gas under reactor operating conditions, and by the embrittling effect it has upon many materials with which it is bound to come into contact. Of the gases that have suitable thermodynamics, and nuclear and chemical properties, two stand out: carbon dioxide, which is dense, cheap, but not chemically inert; and helium, which is inert, has a high specific heat, but is costly.

C. Carbon Dioxide

From what has been said thus far, it can justifiably be stated that carbon dioxide is the most widespread among gas coolants. It is readily available and is used in other branches of industry in large quantities, and its transport and storage as a liquid with subsequent evaporation require only a standard and relatively simple plant. Its reputation as a reactor coolant is based on its low cost in addition to its other attributes. In number, CO_2-cooled reactors rank immediately after LWRs. Carbon dioxide is a colorless gas, readily liquefying on cooling or compression. The thermal and thermodynamic properties that are pertinent to the use of carbon dioxide as a cooling medium are relatively modest. As a result of the low specific heat of this gas, high flows are necessary to extract heat from the reactor core.

1. Corrosion under Irradiation

Corrosion phenomena[19] in a carbon dioxide environment are of particular importance from the standpoint of reactor operation. There are two types of corrosion that can occur: thermal and radiolytic.[20] Thermal corrosion is a purely chemical reaction between the graphite moderator and the carbon dioxide coolant according to the chemical equation $CO_2 + C \rightleftharpoons 2CO$. The forward reaction causes the CO content of CO_2 to increase as the coolant is recirculated through the reactor. The reverse reaction, causing the deposition of carbon, occurs at points of lower temperature. These phenomena, however, proceed at a very slow rate below 625°C and are not important even for the hottest inner fuel sleeves of AGRs operating at temperatures of up to 675°C. The corrosion that occurs radiolytically involves

Table 4
TYPICAL EXAMPLES OF GAS-COOLED REACTORS AROUND THE WORLD

Plant		Reactor type	Coolant	Moderator	Power, net (MWe)
U.K.	Wylfa[a]	Magnox (natural U)	CO_2	Graphite	Twin reactor, 1180
	Seven twin reactor stations	AGR (enriched U)	CO_2	Graphite	Total generating capacity, 8600
France	G 2, G 3	Natural U	CO_2	Graphite	2 × 28
	Chinon 1				68
	Chinon 2				198
	Chinon 3				476
	St. Laurent 1				487
	St. Laurent 2				516
	Bugey 1				547
West Germany	Julich AVR	HTGR	He	Graphite	15
	THTR Uentrop	HTGR	He		300
Italy	Latina	Magnox	CO_2		200
Japan	Tokaimura	Magnox	CO_2		157
Spain	Vandellos	Natural U	CO_2		480
U.S.	Peach Bottom 1	HTGR	He		40
	Fort St. Vrain	HTGR	He		330
Czechoslovakia	Bohurice A1	Pressure tube	CO_2	Heavy water	110
France	ELA				70
West Germany	KKN				100
Switzerland	Lucens				80

[a] One among the Magnox series of first generation of reactors in the U.K.

decomposition by fast neutron and gamma radiations giving rise to carbon monoxide and an active oxidizing species that reacts with the graphite. The site of corrosion is predominantly within the pores rather than on exposed surfaces of graphite. The corrosion proceeds at a rate proportional to the mass of gas contained in the pores and the radiation flux, but is substantially independent of temperature within the range of practical interest.

Radiolytic corrosion phenomenon is important to the designer. Moderator gets removed, and if the weight reduction amounts to 2%, an open porosity of graphite increases by approximately 40%. More open pores contribute to the expansion of the contact surface of graphite and coolant, which intensifies the oxidation process. The mechanical strength of graphite is reduced by corrosion, and if the graphite oxidation is excessive it would surely limit the operating life of the reactor. It may be added here that the rate of weight loss due to corrosion would be intolerably high even in the relatively low-rated Magnox reactors. However, the reaction is more than halved if the CO in the CO_2 coolant is allowed to build up naturally to about 1% by volume, and under that situation corrosion of the graphite moderator no longer limits the life of a Magnox reactor. The inhibiting action of CO on corrosion is the result of the scavenging or deactivation effect it has on the active oxidizing species (a product of CO_2 radiolysis process).

Compared with the Magnox, the AGRs present a more agressive environment. The coolant gas pressures and radiation flux levels are higher. A greater need arises here for more effective inhibition of radiolytic corrosion. Increasing the CO level does not by itself produce a very large reduction in corrosion rate. The rate decreases markedly at concentrations up to 0.4 % CO. It has been shown that an addition of a few hundred parts per million by volume of methane (CH_4) is very effective in bringing down the rate of attack. A further reduction in corrosion can be accomplished by using graphite with fewer pores. It is on account of this that the porosity of graphite for AGRs is specified to be lower by a factor of about 2 than that of the graphite for Magnox. The corrosion inhibition mechanism by CH_4 is different from that of CO. It has already been pointed out that CO inhibits by deactivating the oxidizing species in the gas phase. Compared with it, CH_4 is a much more effective inhibitor. It forms a protective deposit on the surface of the graphite pores, the sites of corrosion.

The mechanism of radiolytic corrosion and inhibition can thus be represented by the following. In the gas phase, radiolysis is through the reaction $CO_2 \rightarrow CO + O$, deactivation is through the reaction $CO + O \rightarrow CO_2$, and deposition is through the reaction $CH_4 + O \rightarrow P$, where O is an oxidizing species formed by radiolysis of CO_2, and P is a protective deposit formed from oxidation of CH_4. On the surface, the graphite oxidation is through the reaction $O + C \rightarrow CO$, and the protective deposit oxidation is through the reaction $O + P \rightarrow OP$, where OP is the deactivated gaseous product of methane destruction. This implies that if inhibition is to remain effective, the methane that is destroyed within the graphite pores must be replaced.

However, in the absence of radial pressure drops, which would induce a permeable flow of coolant gas through moderator bricks, methane transport can only occur by diffusion. The diffusivity of AGR graphite is quite low, coupled with high CH_4 concentrations at sites within the moderator bricks that are remote from external surfaces. The local oxidation rate is related to the local gas composition, and thus the rate of graphite loss within the bricks can be markedly higher than at the surface. There is, however, a limit to the freedom one can have in order to compensate for the effects of CH_4 depletion, because an excessively high level of CH_4 in the coolant can give rise to carbonaceous deposition on fuel pin surfaces. This may lead to impairment of the heat transfer performance of the fuel pins. More CH_4 is needed to cause fuel pin deposition when the CO concentration is low than when it is high, but the CH_4 destruction rate increases when the level of CO is reduced, thus enhancing methane depletion within moderator bricks. It is therefore difficult to achieve an optimum coolant gas composition in a diffusion-controlled situation.

Table 5
GAS COOLANT COMPOSITION AT HINKLEY POINT B SINCE POWER RAISING

Elapsed time at power	CO(%)	H_2O(vpm)	CH_4(vpm)
2500—5000 hr	0.5	200—300	—
0.3 year	1.0	200—300	—
1 year	1.0	200—300	130
After 1 year	1.0	300	165
From January 1981	Reactor 3, 1.2 Reactor 4, 1.0	300	165

In light of all this, it is clear that control of coolant composition has become a major operational requirement. For improving methane accessibility, two options exist: (1) inducing an impressed flow of coolant gas through the bricks by a radial pressure drop, which would provide an adequate supply of methane to replace that destroyed within the graphite pores and result in a substantially uniform concentration throughout the bricks; and (2) drilling a number of axial holes through the bricks, which would reduce the length of transverse diffusion plants within the bricks and raise the internal methane levels.

In summing up the corrosion inhibition process, it may be said that throughout its life the graphite core is subjected to attack by short-lived oxidizing species produced by irradiation of the CO_2 coolant. This oxidation process results in the gasification of a small portion of the graphite brick to CO, with a resulting weight loss of graphite from the bricks and an accompanying reduction in their strength. This process can be inhibited by CH_4, which, when introduced into the coolant, acts by forming a protective layer on the graphite surfaces within its pores. A second inhibitor is CO, which enhances the probability that any oxidizing species will recombine to form CO_2 rather than attack the graphite. Coolant compositions are, in fact, a compromise between the requirement to inhibit graphite corrosion and the prevention of deposition of carbon on the fuel, based upon the levels of CO, CH_4, and H_2O in the CO_2 coolant gas.[21-23] For the purpose of illustration, the coolant gas strategy employed at Hinkley Point B is shown in Table 5. It has been stated[22] that the future strategy at Hinkley Point is a gradual increase in CO and CH_4 levels to provide greater inhibition, but with close monitoring for any carbon deposition on the fuel.

2. *Induced Radioactivity*

Carbon dioxide, like helium, is used only in closed circuits, designed to be virtually leak tight not only to conserve the gas but also to prevent the short-lived activity arising from impurities (particularly nitrogen and argon) from contaminating work areas.[24] A typical analysis of CO_2-based coolant is as follows (in volume percent): CO_2, 99.5; CO, 0.38; H_2O, 0.001 (in weight percent); Ar, 0.0003; H_2, 0.0016; O_2, 0.0017; N_2, 0.0094; C_2H_6, 0.0001; and CH_4, 0.0013. Radioactivity in the gas is induced within the reactor owing to absorption of neutrons by the constituents ^{16}O, forming ^{16}N; ^{18}O, forming ^{19}O (as in water coolants); and ^{13}C, present in natural carbon to the extent of 1.1% and having a cross section for thermal neutrons of about 0.9 mb. The reaction is $^{13}C\ (n,\gamma)^{14}C$. Carbon-14 has a half-life of about 5700 years. Because of the long life of ^{14}C activity, the level continuously builds up as the coolant is recirculated through the reactor and power plant.

3. *Reactor Material Corrosion*

Corrosion of reactor materials in CO_2 is an important matter that deserves mention. To date, considerable experience has been gathered regarding the corrosion properties of a wide variety of materials in atmospheres containing hot CO_2.[20,25] As far as reactor applications are concerned, reference can be made to three materials: steels, magnesium, and zirconium.

a. Steels

In the case of Magnox reactors, at the design outlet temperature of 400°C, mild steel is found to have adequate corrosion resistance to CO_2. With coolant outlet temperatures up to about 670°C, it becomes necessary to use alloy steels. An 18% Cr-12% Ni-Mo steel has excellent corrosion resistance to CO_2. This steel, however, suffers stress corrosion with water, and therefore it is essential to have a cladding on the water side of the heat exchanger with 9% Cr ferritic stainless steel. A stainless steel containing 20% Cr and 25% Ni-Nb was chosen for fuel cladding in AGRs, as this ably resisted corrosion in CO_2 at 825°C. Other steels have been observed to have adequate corrosion resistance in hot CO_2 and titanium-stabilized austenitic 18% Cr steel (used for purposes such as cover plates and supporting members), and the nickel-based alloy PE 16 (used for tie bar construction). Mild steels, when close to the maximum temperature, tend to be covered with Fe_3O_4 (magnetic) film by CO_2. In the process, CO_2 is reduced to CO. A smooth protective film forms first. Corrosion changes to a very rapid 'breakaway' process after a certain time. Steels containing about 9% Cr and appreciable Si extend the time to the development of breakaway rates of oxidation.

b. Magnesium

For their uranium metal fuel rod cladding, the first generation of GCRs has made extensive use of magnesium alloys referred to by the generic term Magnox. The entire range of such nuclear reactors are designated Magnox reactors after the canning material used for the fuel elements. Magnox reacts with CO_2 at high temperatures, but a protective film is formed that prevents further corrosion. The film thickness increases markedly with temperature.

c. Zirconium

The best combination of mechanical and corrosion properties for service in CO_2 is provided by zirconium alloys containing 0.5 to 0.8% by weight of copper and 0.7 to 1.0% by weight of molybdenum. Between 400 and 700°C, such alloys require an activation energy of 108 kJ/mol to induce prebreakaway oxidation. A very negligible extent of diffusion of oxygen into zirconium occurs at high temperatures under CO_2 exposure of the metal, which can be the cause of cracking.

D. Helium

A noble gas, helium is the lightest of the inert gases and next to hydrogen in weight of all known gases. It is the preferred coolant for graphite-moderated reactors operating at temperatures above 600°C. Helium is the coolant of choice of HTGRs. Table 6 records certain examples of reactors that have used helium as a coolant. The paragraphs below highlight the advantages of helium cooling.

A principal advantage of helium is that it results in less neutron absorption and moderation than sodium or other coolants. The lower absorption increases the fraction of neutrons available for breeding, while the harder spectrum significantly increases the total number of neutrons produced per fission.

Helium has relatively good heat transfer properties in comparison with air and carbon dioxide, and requires less pumping power than air and carbon dioxide because of its low density. However, in comparison to liquid metals, particularly sodium, it is at a disadvantage with regard to these properties. This will be discussed later.

There are no problems with helium regarding instability under irradiation. Regarding induced radioactivity, only tritium (^3He), which is found in a concentration of 0.00013% in natural helium, is important. As a result of the (n,p) reaction, tritium generation results; tritium is beta active, with a half-life of 12.1 years. The only serious radioactivity in a helium coolant is due to other gaseous impurities in it. Those that may exist in high-grade helium are hydrogen, water vapor, and air with its various constituents, including argon;

Table 6
SOME EXAMPLES OF HELIUM-COOLED REACTORS

Reactor	Outlet temp of helium (°C)	Helium pressure (MPa)	Power (MWe)	Remarks
Peach Bottom	728	2.4	40	First HTGR plant in the U.S. built in a riverside in Pennsylvania; went into full power operation in 1967; shut down in 1974; provided a remarkably good availability over its 7-year test program; at the time of decommissioning this was the world's only commercial HTGR
Fort St. Vrain	780	—	330	Presently the only operating HTGR in the U.S.; first went critical in 1974
AVR	950	1	15	Plant at Julich, West Germany; reached full power operation in 1967
THTR	740	4	300	Construction started in 1972 at Uentrop near Hamm in West Germany; construction trial operations conducted to verify specified plant performance and availability successfully completed on May 25, 1987
HTGR	850	4.8	1100	In the design stage

and put together the total amount does not normally exceed 150 ppm. The problem of induced radioactivity in high-grade helium is thus not of a serious nature. Consequently, the intermediate coolant loop normally present in LMFBRs (to isolate radioactive primary coolant from the steam generator) can be omitted for cooling with helium. This simplification leads to lower capital cost.

As has been pointed out, helium is remarkably inert chemically. Incompatibility problems between coolant and structural or fuel materials are essentially absent. This permits the potential for higher exit coolant temperature. The transparent helium gas also allows visual observation of components, presumably resulting in more flexibility and ease in maintenance. A safety feature of the property of helium inertness is that helium does not react chemically with either water or air, a problem inherent in the LMFBR system.

A few other advantages deserve mention. Since helium is always employed as a single-phase gaseous coolant, there is no possibility of an abrupt change in cooling characteristics. This also projects the design potential of employing a direct cycle, i.e., ducting the primary helium to the turbine without resorting to the intermediate step of generating steam. Fission gas escaping through a cladding defect or rupture site cannot appreciably alter the heat transport capabilities in helium cooling. A similar circumstance in an LMFBR could result in displacement of liquid coolant, possible partially starving the fuel pins of adequate coolant in the immediate vicinity of the rupture site.

It should not be construed that cooling with helium is devoid of demerits. The heat capacity of helium is below that of carbon dioxide. This implies that for accumulation of a significant amount of heat in helium, the inlet to and the outlet from the reactor core must differ considerably in temperature. This interferes with the use of helium as the coolant. The heat conductivity of helium is, however, ten times that of carbon dioxide, and this facilitates heat transfer. Because of the inherently poorer heat transfer properties of helium relative to sodium, it is necessary to induce turbulence by roughening the surface on the fueled portion of the fuel pin cladding exterior walls. Another direct result of the poor heat transfer properties of helium is the high pumping power required for adequate circulation.

The cost of helium is high. It must be operated in a closed coolant system, or loop. Thus, it introduces another problem, that of a leakproof piping system and pressurized reactor vessel to reduce the leakage and loss of helium coolant in HTGRs or AGRs.

Another disadvantage of helium cooling is that helium must be pressurized up to the 7- to 10-MPa (1050- to 1500-psi) range in order to provide sufficient cooling for the high power density cores of interest for economic power generation. High system pressures are required, in part, to compensate for the low heat capacity of helium. Sodium systems, in contrast, it may be recalled, can operate at atmospheric conditions (the only pressurization necessary being that required in the inlet plenum to obtain adequate flow).

In discussing helium cooling, a brief discussion of the reactor systems is in order. Apart from HTGRs, helium cooling has attained the greatest importance for the gas-cooled fast reactor (GCFR),[26] which is under development in the U.S. and Europe as an alternative to the LMFBR. At present, perhaps the strongest barrier to widespread adoption of the GCFR concept is that there is very little experience with such a system. The heavy international committment to LMFBRs has uncovered several problems, although none appears to be without solution. If a similar committment were to be given to the GCFR, it is likewise expected that problems not yet evident would come to light. Nonetheless, there are enough attractive features inherent in the GCFR concept to warrant careful study of such systems.

VI. LIQUID METAL COOLANT[27-29]

A. General

The liquid metals are well known for their commendable role as reactor coolants. Their special place as coolants is for reactors with high thermal fluxes, operating at high temperatures. They have excellent thermal properties (high thermal conductivity and low vapor pressure). Some metals of low atomic weight, such as lithium and sodium, are characterized by their relatively high specific heats and volumetric heat capacities. The use of liquid metals as coolants in nuclear reactors in place of water has a number of advantages. Irradiation of liquid metals, as a rule, produces no such phenomena as the radiolytic decomposition of water. The heat transfer characteristics of several liquid metals are superior to those of water, so that it is possible to achieve the same rate of heat removal with smaller areas of contact between fuel and coolant. On the other hand, the most essential drawback of the majority of the liquid metal coolants is their low heat capacity per unit volume (volumetric heat capacities) as compared with water. For removal of the same amount of heat, the flow rate with liquid metal coolants is higher than that with water. The pumping power depends on several factors, and it may be greater or less than for water. For metals of low mass number it is generally somewhat less, while for metals of high mass number (bismuth and lead) it is larger. On the whole, however, pumping power is generally not taken as an important consideration in the choice of a coolant among various liquid metals and water.

The liquid metals, while recognized as being excellent coolants, present a whole range of novel handling problems on account of their chemical reactivity, burning readily in air and reacting violently with water. This, of course, means leakproof structural requirements, which may be difficult to maintain with relatively corrosive liquid metals. In general, liquid metals beome intensely radioactive. As mentioned earlier, this necessitates in power plants the use of an intermediate coolant loop, also of liquid metal for good heat transfer purposes. The intermediate loop separates the reactor primary loop, where induced radioactivity exists, and the working fluid (steam).

Consequently, liquid metals were not considered as coolants for commercial reactors until fast reactors added extra criteria to the coolant selection considerations. In a fast reactor where there is a requisite of high breeding ratio, the neutron energy is necessarily to be maintained as high as possible. This in turn means the absence of moderating materials from

Table 7
PROPERTIES OF SOME LIQUID METAL COOLANTS

Metal	Mp (°C)	Bp (°C)	Density at 400°C (kg/m^3)	Specific heat at 400°C (kJ/kg)	Thermal conductivity at 400°C (W/m K)	Neutron capture cross section (10^{-28} m^2)
Sodium	98	883	850	1.281	71.162	0.56
Potassium	64	757	750	0.766	40.186	2.5
Lithium	179	1327	490	4.186	37.67	65
Mercury (300°C)	−39	357	12,900	0.134	13.81	430
Lead	325	1740	10,510	0.155	15.91	0.17
Bismuth	271	1477	9,910	0.147	15.49	0.033

the core. This eliminates water as a contender and leaves the choice among liquid metals, gases, and steam, the latter two being acceptable on account of their low density of moderation atoms. Besides these, there are several advantages in using a gas or steam as coolant for a fast reactor. The main ones are (1) high specific power; (2) compatibility with water (unlike sodium); (3) possible use in closed cycle and in direct cycle power plants (in the case of steam), reducing capital costs; (4) absence of thermal shock problems; and (5) inertness and low induced activities. Helium, carbon dioxide, and steam have been considered as fast reactor coolants. Other gases are not suitable because of chemical activity or metallurgical considerations. Of the three gases considered, helium has the lowest pressure drop-to-heat transfer ratio, thus reducing flow-induced vibrations in the normally high-speed coolant flows necessary to obtain adequate heat transfer in fast reactors. Steam is less costly and offers the possibility of direct cycle application, but suffers from lower breeding ratios than gases because of the softer spectrum associated with it. Steam has lost much favor because of the inability of the known cladding materials to withstand the corrosive atmosphere of steam at high temperatures.

The thermal and hydrodynamic properties of liquid metals allow efficient cooling of the moderator-free, compact cores that are typical of the current fast reactors. A drawback of the liquid metal coolants (except mercury) is that they solidify at room temperature. This requires a heating system to melt the metals, somewhat complicating the construction.

B. Liquid Metal Coolant Materials

In discussing liquid metals as coolants, a general reference can be drawn to a list of some potential metals consisting of bismuth, lead, lithium, potassium, and sodium. Table 7 gives the main properties of these metals.

Bismuth and lead have satisfactory low cross sections, but their melting points are fairly high. A significant drawback of the use of bismuth in a reactor is the fact that it undergoes the (n,γ) reaction to form ^{210}Bi, a beta particle emitter with a 5.0-d half-life, decaying to yield ^{210}Po, which constitutes a hazard. It is very toxic and difficult to contain. The thermophysical properties of lead are not good enough. Its heat capacity is nearly one tenth that of sodium, thermal conductivity is low, and specific gravity is relatively high.

Lithium is the lightest metal and in terms of chemical properties approximates the alkali earth metals. It excels sodium in heat transfer and heat capacity per unit volume at 773 K. Natural lithium, however, contains 7.4% stable nuclide ^6Li, which possesses a large thermal neutron absorption cross section. Reducing the content of this isotope is an expensive process.

The advantage of potassium as a coolant over sodium is in its lower melting point. The other properties are less attractive. The physical and thermal properties of potassium are nearly the same. As to the chemical properties, potassium is more reactive than sodium. Naturally occurring potassium consists of two isotopes, containing about 93.10% ^{39}K and about 6.88% ^{41}K. The remainder, ^{40}K, is a beta and gamma emitter with a long half-life and thus has a very low level of activity. Of the three isotopes, ^{41}K is converted to a radioactive isotope upon neutron absorption. It converts to ^{42}K, a beta and gamma emitter with a 12.40- half-life, which in turn converts to stable ^{42}Ca. While ^{40}K converts to ^{41}K, which is susceptible as mentioned above, its natural abundance is so low that its contribution to activity is not important. Sodium-potassium alloys containing 40 to 90% of potassium are in the liquid state at room temperature. This essentially eliminates the necessity of having the heating systems to melt a liquid metal coolant prior to starting the reactor. An austenitic alloy containing 77.2% potassium has a minimum melting point of 261.5 K.

Liquid sodium has attracted much attention as a reactor coolant, and the general opinion is that it is, on the whole, the most suitable liquid metal generally available as the coolant for a reactor operating at high temperatures, particularly for fast reactors. From our brief consideration of the properties of liquid metal coolants, it can be generally summarized that mercury, bismuth, and lead have all been discarded because their high densities (and consequently high mass flow rates) require excessive pumping power. The sodium-potassium solution (NaK) was used in an early design because of its low melting point (liquid at room temperature). Potassium, however, is a fairly strong neutronic absorber. Taking all this into account, it is easy to understand why the mantle has finally fallen to sodium as the most suitable coolant for fast breeder reactor systems. Sodium satisfactorily meets the primary breeder reactor coolant requirements, which are as follows: (1) it must minimize neutron moderation, (2) it must remove heat adequately from a high power density system (nearly a factor of four higher than for an LWR), and (3) it must minimize parasitic neutron absorption.

The general account of the liquid metals must make reference to the pumping systems. Two different types of pumps are used for liquid metals: centrifugal pumps and electromagnetic pumps. Centrifugal pumps have been built with capacities up to 13,000 gal/min and operating temperatures up to 540°C. A primary consideration in the design of a centrifugal pump is the method of sealing the shaft. The free-surface vertical shaft pump, in which a liquid-gas interface is maintained in the pump case, is the most widely used. The electromagnetic pumps take advantage of the electrical conductivity of a liquid metal to force it to flow in a pipe under the influence of a magnetic field. In the pump the liquid metal is the conductor which passes through a duct located between the poles of an electromagnet. A current from an external force is passed across the liquid metal in a direction perpendicular to both the magnetic field and the desired direction of flow of the liquid metal. This represents the principle that, if a conductor in a magnetic field carries a current flowing at right angles to the direction of the field, a force is exerted on the conductor in a direction perpendicular to both the field and the conductor. Thus, in the electromagnetic pumps, the force exerted on the conductor by the magnetic field causes the flow to take place. An advantage of the pump is that it has no moving parts, so that with proper design, maintenance is greatly simplified. This is important not only on account of the nature of the liquid being pumped but also on account of possible contamination of the pump with radioactivity (the situation with ^{24}Na in the case of liquid sodium).

C. Sodium

The liquid metal sodium is the coolant of choice for all the major fast breeder power reactor projects underway around the world. The outstanding property of sodium is its high thermal conductivity. It melts at 98°C and has a high boiling point (880°C at 1 atm pressure). The hottest sodium in a typical LMFBR system is around 530°C, which is approximately 350°C below the boiling point of sodium at atmospheric pressure. Indeed, this represents a good margin against boiling if there should be a power transient. The high boiling point of sodium allows it to be used over any likely range of reactor operating temperatures. In the case of sodium, therefore, high reactor exit temperatures are possible with low system pressures. This results in both higher power cycle efficiencies and lower system pressure in the reactor vessel than in water-cooled reactors. The low specific heat of sodium, however, reduces its capacity as a heat sink in the case of coolant loss. In the fast breeder reactor this low specific heat is counteracted by the use of very large volumes of sodium in the primary circuit. In the liquid state, both the density and viscosity of sodium are of the same order of magnitude as those of water.

Apart from the thermal considerations as outlined above, neutronics, hydraulics, and compatibility considerations are of considerable importance in addressing liquid sodium in the eminent role it plays as a reactor coolant.

In terms of neutronics, reference may first be made to a moderating effect of the coolant on the neutron spectrum. The degree of moderation is proportional to the atomic mass and the density. Sodium, with a mass of 23, is clearly heavier than either helium or steam. With density taken into account, the helium-cooled reactor yields the hardest spectrum, steam the softest spectrum (because of the hydrogen content), and sodium an intermediate spectrum. Because of this spectrum difference, helium can be regarded as having a slight intrinsic edge over achievable breeding ratios. A drawback associated with the use of sodium as a reactor coolant, either alone or as an alloy, is regarding its neutronic activities. The activation attains a high degree of short-life activation. All naturally occurring sodium is made up of ^{23}Na (thermal neutron absorption cross section 0.53 b; fast neutron cross section about 1 mb at 0.25 MeV). When subjected to neutrons it undergoes the reaction ^{23}Na (n,γ) ^{24}Na. ^{24}Na is a radioisotope of about a 15-h half-life which emits gamma radiation mainly of 2.76- and 1.38-MeV energy. It beta-decays into stable ^{24}Mg, which has a low neutron cross section. The level of activity of sodium, of course, depends upon the time spent in and out of core. Therefore, radiation shielding is necessary for the sodium coolant and cooling system, e.g., coolant tanks, piping system pumps, heat exchangers, etc. It has also led to inclusion of an intermediate coolant loop in order to ensure that all radiation is confined to the primary loop. This eliminates the possibility of radioactive contamination in the steam generator. This loop further protects the core from possible pressure surges or positive reactivity effects of hydrogen moderation should a steam generator leak occur and result in a sodium-water reaction.

The chemical reactivity of sodium is so important that it warrants elaboration. Sodium reacts vigorously with most noninert gases and liquids. In the solid state, it tarnishes readily when exposed to air. This is because of the formation of an oxide film on its surface. In the liquid state, when exposed to air it burns with a low flame and evolves a thick smoke of Na_2O. Its reactions with water are vigorous, and the reactions become explosive if they take place in a confined volume. If H_2 and H_2O enter a sodium system in nonexplosive quantities (via impurities or leakage), they form sodium hydride and sodium hydroxide. This appreciably increases the moderating ratio of the coolant, an effect that may not be desirable in fast reactors. Alkalis other than sodium exhibit similar reactions as indicated for sodium. Potassium, however, is more reactive than sodium. Sodium-potassium (NaK), with many of its mixtures liquid at room temperature, is more reactive than either of its components.

The coolants must be evaluated from the point of view of hydraulic considerations. In order for a reactor coolant to perform its heat removal function, a fraction of the total power output must be used to pump it through the core. It is therefore of interest to compare the pumping requirements for the various coolant candidates. Sodium coolant requires the least pumping power, and helium requires the most. Other liquid metals are ruled out largely because of their much larger pumping power requirements.

1. Corrosion Aspects

Compatibility must be considered in the assessment of the capability of a coolant to

Table 8
PERMISSIBLE IMPURITY CONCENTRATION IN MOLTEN SODIUM

Element	Conc (ppm in weight)	Element	Conc (ppm in weight)
A1	10	K	200
B	2	Li	5
Ba	10	Mg	10
C	15	Mn	5
Ca	10	N	5
Cd	5	Ni	10
Co	5	0	10 ± 5
Cr	5	Pb	10
Fe	25	Rb	50
H	10	Si	10
Halogens	20	Sn	10

perform the function of heat removal from a high power density core. Sodium is quite compatible with stainless steels, which constitute the preferred cladding material for fast reactor systems. However, because sodium has been chosen as the base coolant, much work has been directed toward every facet of its in-core environment. Some of the categories in which effects have been focused on in order to isolate problem areas are discussed

General corrosion[30,31] — The main metallic elements in the cladding are iron, chromium, and nickel. They are slowly dissolved from the hot core region and deposited in the cooler areas. This long-term uniform attack is called cladding thinning or wastage. Although this amounts to only tens of micrometers per year at 700°C, it must be considered in the overall load-bearing capacity of the cladding, which is stressed by internal fission gas pressure. Trace impurities, especially O_2, in liquid sodium induce corrosion attacks on the structural material at high temperatures near 650°C. At the same time, the reaction of sodium with oxygen forms the sodium monoxide, Na_2O, which is extremely corrosive when pure. Because of the low solubility of Na_2O in Na, it may plug partially the liquid metal cooling system. Sodium free from O_2 does not attack such materials as stainless steels, nickel alloys, etc. at temperatures below 650°C. Corrosion attack on structural materials takes place seriously above 650°C. It is therefore necessary to keep the oxygen content low. In stainless steel systems the method of cold-trapping is generally adequate. The cold trap is a region, usually in a bypass of the main coolant circuit, which is maintained at a lower temperature of about 150°C. Because the solubility of sodium oxide in sodium is very low in the cold region, it deposits as a solid and is removed from time to time. For operation at higher temperatures, the metals niobium, titanium, and vanadium may be considered as alternatives to stainless steels. In some situations, such as for the niobium, titanium, or vanadium circuits, the cold trap does not reduce the oxygen content of the sodium to a sufficient extent to inhibit corrosion. The use of a hot trap is then advisable. This contains a metal, such as zirconium, which reacts chemically with sodium oxide at a temperature of 600 to 700°C and reduces it to sodium.[32] Permissible impurity concentrations in molten sodium are shown in Table 8.

Selective leaching — Each of the principal metallic elements mentioned has a different dissolution rate in sodium. Hence, selective leaching takes place, and the composition of the steel at the outer surface is eventually altered. The influence, though small, is normally deleterious, since Cr and Ni are removed faster than Fe. This causes the surface layer of austenitic steels to be transformed to ferritic steels.

Thermal gradient transfer — In this, materials are dissolved by the liquid metal in a high-temperature region, where the solubility is high, and precipitated in a cooler region, where the solubility is low. While the rates of such mass transport are low, extended operation and coolant circulation may result in noticeable corrosion of the hotter regions and plugging in the cooler regions of the systems. The deposits can eventually become so thick that coolant driving pressure must be increased in order to maintain the desired flow.

Radioactivity transport — Some of the major long-lived radioactive species produced in the cladding are ^{54}Mn, ^{58}Co, and ^{60}Co. Though small in quantity, sodium transport and deposition of these radionuclides in the intermediate heat exchanger and other primary system compounds can cause high enough local activity to effect routine maintenance.

Carbon transport — Carbon in the cladding can migrate fairly readily to the outer surface. Sodium can therefore transport carbon, as well as all the metallic elements, to colder surfaces.

Sodium-fuel interaction — In the event of a small cladding breach, a direct interaction of sodium and fuel could lead to the production of Na_3UO_4 or Na_3PuO_4. The concern is whether this substance could be swept into the coolant stream as a contaminant.

It is obvious that absorption of oxygen by the coolant should be avoided. An inert gas blanket should be provided over all free sodium surfaces. The inert gas may be helium, which has the advantage of no induced radioactivity, or argon, which has the advantage of being heavier than air, thus facilitating the blanketing effect. Nitrogen is less expensive but is also soluble in sodium. This solubility, though low, results in the mass transport of N_2 through the system. This in turn results at high temperatures in nitriding and damage to thin-walled components such as cladding, valve bellows, etc.

2. Summary

To summarize, the advantages of liquid metal sodium are excellent heat transport properties, low pressure system, low pumping power requirements, lowest fuel cladding temperature, potentially high breeding ratio, inherent emergency cooling of fuel, extensive sodium reactor experience, and potential for vented fuel. As disadvantages, the following can be mentioned: radioactivity (intermediate sodium loop incorporated), unfavorable coolant reactivity void coefficient, chemical reactions with air and water, nonvisible refueling procedure, solidity at room temperature, and impedance of primary system maintenance by radioactivity.

REFERENCES

1. **Dietrich, J. R. and Zinn, W. H.,** *Solid Fuel Reactors,* Addison-Wesley, Reading, MA, 1958.
2. **El-Wakil, M. M.,** *Nuclear Energy Conversion,* Intext Educational, London, 1971.
3. **Holman, J. P.,** *Heat Transfer,* 3rd ed., McGraw-Hill, New York, 1972.
4. **Lish, K. C.,** *Nuclear Power Plant Systems and Equipment,* Industrial Press, New York, 1972.
5. **Gerasimov, V. and Monakhov, A.,** *Nuclear Engineering Materials,* Mir, Moscow, 1983, 69.
6. **Smith, C. O.,** *Nuclear Reactor Materials,* Addison-Wesley, Reading, MA, 1967, 50.
7. **Glasstone, S. and Sasonke, A.,** *Nuclear Reactor Engineering,* D Van Nostrand, New York, 1963, 395.
8. **Roddis, L. H., Jr. and Ward, J. H.,** Operating experience with commercial central-station light-water reactors in the USA, in Proc. 4th U.N. Conf., 2, 1970, 3.
9. **Davis, W. K. et al.,** U.S. Light water reactors: present status and future prospects, in Proc. 4th U.N. Conf., 2, 1970, 21.
10. *The Shipping Port Pressurized Water Reactor,* Addison-Wesley, Reading, MA., 1958.
11. **Shoup, W. E. et al.,** Proc. 2nd U.N. Conf. Geneva, 8, 1958, 492; *Power React. Technol.* 4(3), 47, 1961; Nucleonics, 19(3), 53, 1961.
12. **Milne, G. R.,** Proc. 2nd U.N. Conf. Geneva, 8, 1958, 483.
13. **Kramer, A. W.,** Nuclear Propulsion for Merchant Ships, U.S. Government Printing Office, Washington, D.C., 1962.

14. **Vasiliev, V. A.,** Nuclear power in the USSR, *Nucl. Energy,* 21(2), 113, 1982.
15. **De Paul, D. J., Ed.,** *Corrosion and Wear Handbook for Water Cooled Reactors,* McGraw-Hill, New York, 1957.
16. **Balakrishnan, P. V. and Allison, G. M.,** Some in-reactor loop experiments on corrosion product transport and water chemistry, *Nucl. Technol.,* 29, 105, 1978.
17. U.S. AEC, Conf. Aqueous Corrosion of Reactor Materials, AEC Rep. TID-7587, 1960.
18. **Brey, H. L. and Olson, H. G.,** Fort St. Vrain experience, *Nucl. Energy,* 22(2), 117, 1983.
19. **Tyzack, H. and Cowan, H. C.,** The corrosion of reactor materials in carbon dioxide, *At. Energy Rev.,* 14(2), 263, 1976.
20. **Diamant, R. M. E.,** *Atomic Energy,* Ann Arbor Science, Ann Arbor, MI, 1982.
21. **Shennan, J. V., Wood, C. J., and Barr, W.,** Controlling corrosion of AGR graphite moderators, *Nucl. Energy,* 22(2), 113, 1983.
22. **Hall, R. W. and Chaffey, C. A.,** Review of the operational experience of the Hinkley Point 'B' AGR over the past two years, *Nucl. Energy,* 21(1), 41, 1982.
23. **Best, J. V. et al.,** Inhibition of moderator graphite corrosion in CEGB Magnox reactors, *J. Br. Nucl. Energy Soc.,* 4, 319, 1976.
24. **Manly, W. D. and Bush, S. H.,** Nuclear Metallurgy, Vol. 3, Symp. Materials for Gas- and Water-Cooled Reactors, 1962, Fall Meeting, Statler-Hilton Hotel, New York, IMD Special Rep. Ser. No. 11, Met. Soc. of AIME.
25. **Gibbs, G. B. and Popple, L. A.,** Oxidation of structural steels in CO_2 cooled reactors, *Nucl. Energy,* 21(1), 51, 1982.
26. **Kemmish, W. B.,** Gas-cooled fast reactors, *Nucl. Energy,* 21(1), 77, 1982.
27. **Voice, E.,** Sodium as a Fast Reactor Coolant, U.K. Atomic Energy Agency, Dounreay.
28. **Lyon, R. N.,** Liquid-Metals Handbook, U.S. Atomic Energy Commission, 1952.
29. **Dwyer, O. E.,** Liquid Metal Handbook, Sodium and NaK Supplement, U.S. Atomic Energy Commission, 1970.
30. NASA, Liquid-Metals Corrosion Meeting, NASA Rep. TID-769, 1961.
31. **Fidler, R. S. and Collins, M. J.,** A review of corrosion and mass transport in liquid sodium and the effects on the mechanical properties, *At. Energy Rev.,* 13(3), 3, 1955.
32. **Starr, C. and Dickinson, R. W.,** *Sodium Graphite Reactors,* Addison-Wesley, Reading, MA, 1958.

Chapter 4

CONTROL AND SHIELDING MATERIALS

I. INTRODUCTION

In a nuclear reactor operation there is a need that an excess of reactivity be built into the core, and this is to compensate for the gradual depletion of fissile material with time, as burnup accumulates, and to compensate for other phenomena such as the buildup of fission products. This excess reactivity must be compensated as needed at any given time to keep the reactor critical for steady-state operation. Methods which can be employed to vary the reactivity include addition or removal of (1) nuclear fuel in the core, (2) moderator in the core for thermal reactors, (3) reflector surrounding the core, or (4) neutron absorbers in the core. Each one of these methods, or a combination of them has been used or proposed to accomplish control of nuclear reactors in existence or under design and development.

The task of controlling in thermal reactors has hitherto been widely accomplished by materials that are strong neutron absorbers. In general, the control elements perform the following tasks: power generation regulation according to demand, provision of quick shutdown and accounting for short- and long-term reactivity changes that ensue from temperature changes, fission product accumulation, and depletion of fissile material. In a fast reactor, obtaining control by means of neutron absorbers is not generally satisfactory because of their low capture cross sections for neutrons of high energy. With a strong absorber, it may be mentioned, there is a loss of neutrons, a situation which is unwelcome in fast breeder reactor systems. The recourse to accomplish fast reactor reactivity control is by removal of fuel material from (or by addition to) the core or by movement of part of the reflector. It may be pointed out that, when fuel or reflector is employed for control purposes, its direction of movement will be opposite to that of an absorber. For decreasing the reactivity, fuel or reflector must be removed, whereas an absorber, on the other hand, must be inserted into the core. This opens up the possibility of accomplishing control by combining absorber insertion with removal of part of the reflector.

II. METHODS OF CONTROL

There are principally three primary control methods currently used in power reactors: movable control absorbers, burnable neutron poisons, and chemical shim. Most power reactors employ control rods plus one or more of the other methods. The foremost characteristic of a control material used in these methods is its neutron absorption properties. In addition to neutron absorption, other properties are of importance in the reactor environment, and the form of control material must provide them. In order to perform their functions, control materials have to meet the following requirements:[1,2] (1) to exhibit fair mechanical, structural, and dimensional stability, since in control rods there is a need that materials free of dimensional changes (swelling) and changes in mechanical properties be used; (2) to be available at a reasonable cost; (3) to be stable under steep and cyclic heating; (4) to be of relatively low mass to allow rapid movement; (5) to be corrosion- and wear-resistant; (6) to exhibit a high thermal conductivity, for dissipation of heat generated because of absorption of neutrons; and (7) to exhibit high melting points (for solid materials) or high boiling points (for liquids), thus enabling use at high temperatures.

A. Movable Control Rods

It may be pointed out at the very outset that only movable control rods are capable of meeting the time requirements for all reactivity effects. Complete control by movable rods has a number of disadvantages, however, not the least of which is the mechanical difficulties associated with placing enough movable neutron absorbing material in the core to absorb 13 to 20% of the neutrons produced. The worth of control rods in a thermal reactor is roughly proportional to the ratio of control rod surface area to reactor volume. High values of control rod surface area to core volume result in closer spacing of control rods, smaller fuel assemblies, and more control rod drive mechanisms. All of these factors increase mechanical complexity and core cost. Movable control rods also give rise to large local power density perturbations in fuel elements adjacent to control locations following control rod movement, which can yield excessive fuel-clad mechanical interactions under some conditions. Changes with regard to the gross reactor power shape also take place following insertion or withdrawal of control rods, and these changes may result in an increase in peak-to-average power density and a reduction in operating margin. On the other hand, insertion of solid absorbers into the core usually results in a more negative moderator temperature coefficient of reactivity. In most systems of reactors, however, augmentation of movable rods with either burnable poison or chemical shim to satisfy portions of reactivity requirement is desirable.

1. Control Element Functional Classification

The term control rods is frequently applied in a very general way to any mechanical device which permits a change of the reactor reactivity. A complete reactor control system specification is generally discussed in terms of three main requirements: range, accuracy, and efficiency.[3] Range is concerned with the total amount of reactivity to be dealt with, accuracy refers to the closeness with which the actual behavior of the reactor responds to the control system formula, and efficiency is an overall, combined measure of neutron, power, and operation economics. With a single type of control rod it may prove difficult to achieve both range and accuracy. Consequently, most reactors have rods with different functions or serving different purposes. Most power reactors now differentiate between minute-to-minute type of control and the necessary long-range adjustments required because of constant buildup of neutron-absorbing fission products in a reactor. In such cases, the control elements, which provide the rapid minor adjustments in reactor operation, are still known as control rods. Those control elements, however, which are used only to adjust for the long-term buildup of high-absorbing fission products are called shim rods. Shim rods vary in position over long periods of time, except, for example, when new loadings of fuel are placed in a reactor. At such times, the buildup of fission products associated with the old fuel is removed, and hence shim rods must be inserted further into the reactor to accommodate this change.

Almost all reactors have a third type of control element. The first type, as described, is used to control the minute-to-minute reactor operation. The second type, again as described, the shim rods, is used to adjust for the long-range changes in the overall neutron absorption level of the materials in the reactor. The third type is used primarily as safety elements (or safety rods). The safety elements are kept completely out of the reactor, but are made to enter the reactor at great speed should an emergency condition arise which necessitates the immediate shutdown of the reactor. These safety elements are commonly referred to as scram rods. These rods contain large quantities of a high neutron absorbing material. The output of an operation of scram rod insertion into the reactor puts out the chain reaction in the core.

2. Control Rod Geometries

Volume within the active core region of a nuclear reactor is premium space. Volume that is unoccupied by energy-generating fuel must be held to a minimum. This is of special significance in LWRs, in which one is concerned not only with neutron absorption in control

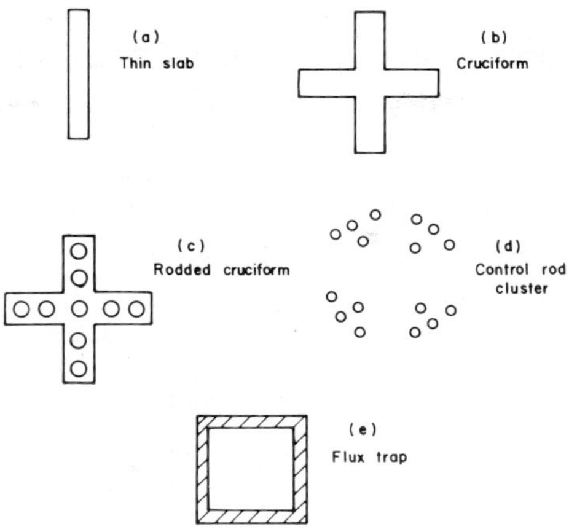

FIGURE 1. Typical control rod geometries.

rods while they are inserted, but also with local thermal neutron flux peaking and creation of coolant-flow short circuits (coolant flow bypass) when control rods are removed. The most efficient space utilization in a thermal spectrum reactor controlled with movable poison rods is accomplished by using high absorption neutron cross section materials in a control rod geometry that maximizes the ratio of surface to volume.

Typical control rod cross section geometries which achieve this high surface-to-volume effect are shown in Figure 1. The simplest is a thin slab of absorbing material (Figure 1A). Long lengths of such a slab are not very stable from a mechanical standpoint, however, and a preferred structural arrangement is obtained with a control rod which has a cruciform cross section (Figure 1B). This provides a surface-to-volume ratio quite comparable to that for the slab, but with considerably more stiffness. Figure 1C is essentially a variant of the cruciform control rod for a special type of neutron-absorbing material. It is for those cases where the neutron absorber is a ceramic material that possesses high neutron absorbing density but poor structural properties. In this, a high control surface-to-reactor volume can be achieved by filling steel tubes with the neutron absorber material and joining them in the form of a cruciform, as shown in the figure. The BWRs are users of this control rod shape. A high surface-to-volume ratio can also be achieved with absorber-filled tubes that are distributed within a fuel assembly volume, as shown in Figure 1D. The number of control rod drive mechanisms is reduced by joining nearby elements to a common control rod drive shaft outside the active core volume. This type of arrangement is called rod cluster control. A final configuration that has a high surface-to-volume ratio is the flux trap control rod. This concept involves the control region containing moderator, which is usually water in LWRs, and neutrons thermalized in this region are absorbed on the inside surface of the control rod. This type of geometry is, however, not used in currently operating power reactors.

3. Control Rod Design and Operating Criteria

There exist a number of design and operating criteria that enter into the final specification of a control rod design.[4,5] The following sections summarize the most important ones.

a. Thermal Worth

In thermal spectrum reactors, 60 to 80% of the total worth is due to absorption of thermal neutrons, where self-shielding of the absorber material is very significant. The primary reason for employing control rod shapes with very high surface-to-volume ratio is to obtain maximum worth per unit volume as well as maximum worth per unit weight of absorber.

Table 1
NEUTRON ABSORPTION PROPERTIES OF SOME TYPICAL CONTROL MATERIALS[6]

Materials	σ_a (0.025 eV) (b)	Σ_a^T (cm^{-1})	Thickness for $\Sigma_a^T t = 2.0$ (cm)	Resonance integral (b)	Absorber density (g mol/ℓ)
Cadmium	2,450	114	0.018	21	9.4
Gadolinium in gadolinium oxide (Gd_2O_3)	49,000	1,244	0.0016	381	13
Boron in boron carbide (B_4C)	759	83	0.024	341	36
Hafnium	102	4.7	0.428	2,008	92[a]
Silver-Indium-Cadmium	198	11.1	0.180	1,060	89

[a] Effect of absorbing chain of isotopes included.

A term usually used for the absorptivity expression of a slab for neutrons is $\Sigma_a t$, where Σ_a is the macroscopic absorption cross section of the absorber material, and t is the thickness. If this product is 2.0 or greater, the slab is taken to be a black absorber, which implies that it is opaque to neutrons. For the purpose of comparison, reference is made to Table 1, which presents absorption properties of a few control materials. It may be mentioned that consideration of both epithermal absorption and poison depletion results in thickness requirements being increased from those given in the table.

b. Epithermal Worth

The thermal neutron absorption no doubt accounts for a greater fraction of the worth of control rods in thermal reactors. Epithermal absorption is nevertheless an important aspect which should not be forgotten. The infinite dilution resonance integral, $(RI)_\infty = \int \sigma_a(E)dE/E$, for each of the control absorbers is given in Table 1, column 5. The $(RI)_\infty$ is the value of the energy-integrated microscopic cross section (σ_a) in a flux that has an I/E energy variation and is not perturbed by the presence of the absorber. These values indicate the relative effectiveness of the listed materials to absorb epithermal neutrons. It may be noted that hafnium, which as the lowest macroscopic thermal absorption cross section among the materials, has the highest resonance integral. Since the epithermal absorption is in high cross-section resonances, self-shielding and the maintenance of a high control rod surface-to-core volume ratio are also important in the epithermal energy region.

c. Depletion

In general, neutron absorption results in the loss of an absorber atom. A control rod must have high absorber density in addition to high thermal and epithermal macroscopic cross sections if it is to have a long lifetime. A control material with a low abundance of a very high cross-section isotope is quite inferior to one in which all atoms have intermediate to moderately high neutron absorption cross sections and higher absorber densities. From the density values in gram moles per liter (number of absorber atoms per liter divided by Avogadro's number) of the tabulated materials given in Table 1, it may be seen that the blackest thermal absorbers possess the poorest characteristics from the point of view of absorber density.

In the case of some elements, a chain of absorbing isotopes exists so that neutron absorption by one yields another with a high absorption cross section. It occurs, for example, in europium and hafnium. Figure 2 shows the chain for hafnium along with the natural abundance and absorption cross sections of the important component isotopes. This effect has been taken into account for evaluating the absorber densities given in Table 1.

Isotope	Hf^{177} + n →	Hf^{178} + n →	Hf^{179} + n →	Hf^{180}
Microscopic cross section, σ_a (barns)	365	86	45	13
Resonance Integral, $(RI)_\infty$ (barns)	7260	1950	600	43
Natural abundance, (%)	18.5	27.2	13.8	35.1

FIGURE 2. Absorption chain for hafnium[6].

4. Other Factors

In the evaluation of control rod performance, a number of other factors, such as mechanical integrity, heating effects, and cost aspects, deserve consideration in addition to reactivity worth and nuclear lifetime.

Among the factors as cited which demand consideration for control rod performance evaluation, perhaps the most important is mechanical integrity. In some absorber materials, such as cadmium and hafnium, neutron capture produces higher isotopes of the same element. Under these conditions, physical and chemical changes are minimized. If boron is the absorber material, however, a neutron absorption reaction involving it results in the production of different elements (helium and lithium). The control rod cladding with boron as the absorber material must then be so designed as to accommodate the pressure buildup due to helium under operating conditions. The aspect of potential corrosion by lithium oxides needs also to be taken into account.

Under some circumstances, control rod heating may pose a problem. Heating arises as a result of two sources. First, the energy of alpha or beta particles generated as the result of neutron absorption must be transferred to the primary coolant. Second, roughly 20% of the fission energy is in the form of gamma rays, which are absorbed in the reactor approximately in proportion to the mass of the constituents. In view of the fact that many control materials have a high mass density, heating due to gamma radiations in the control rod material is not insignificant. In this respect, if one goes by the criteria of high-temperature mechanical strength as the acceptance standard, both Cd and AgInCd present themselves as unsatisfactory. The duo do not dependably display as good a high temperature mechanical property as others in the tabulated list of materials.

A final important consideration in selection of control material and associated geometry is cost. Control rod cost depends on costs of raw material and fabrication. Even though a mechanical design for a particular control rod has been established, substitutes are often entirely feasible and possible in obtaining replacement control rods. For example, mention may be made of the fact that both Hf and B_4C may be considered substitute materials for the Ag-In-Cd alloy in rod cluster control absorbers.

B. Burnable Neutron Poisons

In most reactor cycles the largest single component of the reactivity control requirement is the depletion component. Using a high cross-section material that will increase core reactivity as it depletes to compensate for the loss in reactivity due to depletion of fuel and buildup of fission product is an ideal way for this control requirement.[7] Present LWR designs utilize burnable poisons. These are solid poisons placed in selected rods in the reactor. As they are subject to neutron irradiation, the absorber material is gradually depleted, thus roughly matching the depletion of fissile material. Burnable poisons are employed to counterbalance excess reactivity at the beginning of the fuel cycle and to provide a means for power shaping and optimum core burnup. Burnable poisons must have the following characteristics:

1. The capture cross section of the poison material must be much higher than that of the fissile material, so that there is very little of the poison left after the depletion of fissile

material. It must also not burn out before a loss of reactivity has taken place due to burnup of the nuclear fuel.

2. The reaction products of the burnable poison must have sufficiently low return capture cross sections, so that there is a significant reduction in their poison characteristics as they are eliminated.
3. The burnable poisons or their reaction products should display compatibility with the entire mechanical integrity of the reactor system. If a design calls for the use of separate burnable poison elements, it is necessary to verify the radiation and corrosion resistance of those elements. If the poison material is mixed with the fuel, its irradiation behavior should be evaluated and assessed so that fuel failure mechanisms are not enhanced.

1. Modes of Burnable Poison Utilization

There are two modes of burnable poison utilization: homogeneous and heterogeneous. If the poison concentration is dilute enough so that it is exposed to approximately the same average flux as the fuel, it is a homogeneous burnable poison. If the poison material is lumped or concentrated in selected elements so that significant spatial self-shielding of the poison exists, it is called a heterogeneous or lumped burnable poison. The difference between the two modes is the existence of significant self-shielding. In either case, core locations for burnable poison can be selected that will yield a more uniform gross power distribution than might have been obtained without the burnable poison.

In homogeneous burnable poison, the consequences of high and low burnable poison cross sections are important considerations. With a burnable poison having too high a cross section, the poison atoms are depleted so rapidly that most of the initial control requirement remains. There is a mismatch between the depletion of the burnable poison and depletion of fuel, which creates the reactivity rise. With use of material of lower absorption cross section, the reactivity mismatch decreases, but so does the achievable lifetime of fuel. When the burnable poison absorption cross section is decreased, the reactivity mismatch is lower, but the burnable poison residue at the end of the cycle is greater, causing a lifetime deficit. A trade-off exists, then, between control requirements and neutron economy. There exists an additional practical problem. The number of potential burnable poison isotopes with cross sections in the range of interest is quite limited, and the reactivity behavior is quite sensitive to the value of microscopic cross section.

From the reflected problems arising from the use of homogeneous burnable poison, conclusions can be drawn regarding an ideal isotope. The ideal isotope should possess an absorption cross section comparable to, but greater than, the fuel for the first two thirds of an operating cycle for matching the fuel depletion characteristic, and very high near the end of the cycle to rapidly burn out the remaining poison atoms and minimize the lifetime deficit. The application of self-shield burnable poisons brings us much closer to that ideal. The mismatch in reactivity that occurs with homogeneous burnable poisons can be compensated for by locating the poison in discrete elements in sufficient concentration to achieve reduction of the average flux to which the poison is exposed.

2. Examples of Reactor Systems

Present LWR design utilizes burnable poison as mentioned earlier. In PWRs, boron-based materials are used in the form of either aluminum oxide-boron carbide pellets or borosilicate glass. They are kept, in varying numbers, in thimble locations (otherwise reserved for control rods) of specified assemblies, or in fuel rod locations. In BWRs, pellets of a homogeneous mixture of uranium oxide with 1 to 5% of gadolinia are used to entirely fill selected rods in each fuel assembly.

The uranium oxide and gadolinia homogeneous dispersion have been used in BWRs for some time because BWRs did not use soluble poison to control reactivity over the fuel cycle life. The introduction of burnable poisons in PWRs was primarily to avoid the need for a high boron concentration at the beginning of life, which could cause a safety concern for the following reasons: with a sudden increase in power, and consequently temperature, the water expansion causes the loss of some boron poison from the core; the negative moderator temperature coefficient is thus not as negative as it would be if the water was free of boron. This concern is particularly acute at the beginning of life when boron concentration is highest to compensate for the high reactivity of fuel. The incorporation of burnable poison reduces the reactivity of the fuel at the beginning of life, and hence the required boron concentration is lower.

Two materials have found use in PWRs as burnable poisons. The first is aluminum oxide contained with 3.5 to 6.9 vol% of boron carbide as prepared by hot-pressing or cold-pressing and sintering powders. The produced pellets are cladded with zircaloy tubes and put in selected assemblies. They occupy either thimbles reserved for control rods or certain fuel rod locations. The second is borosilicate glass containing about 12.5 wt% B_2O_3. Use of burnable poison in this form minimizes any possibility of leaching. All of the helium gas is released in this core, and therefore an adequate free volume must be available inside the stainless steel cladding to avoid excessive stresses.

There exists considerable interest in increasing use of gadolinium as burnable poison in PWRs. One good reason is that it provides good reactivity hold-down at the beginning of fuel life. Another reason for the trend is that gadolinia, unlike boron compounds, can be readily mixed with uranium oxide and does not have to be lumped in separate rods. Its odd-mass number (A) isotopes, ^{155}Gd and ^{157}Gd, account for its high thermal cross section, which results in a more complete burnout toward the end of the cycle, which yields better neutron economy and hence a higher fuel utilization. As additional advantages of increased usage of gadolinium in PWRs, mention may be made of (1) no displacement of water, and practically no displacement of fuel, resulting in an easier lattice utilization; and (2) elimination of the problems of handling and disposal of the extra burnable poison rods at the end of cycle. Potential disadvantages include (1) a degraded thermal conductivity and lower melting point of the mixed oxide, which leads to a lower power generation; (2) a more complex configuration which is harder to analyze; and (3) an increased cost of fuel assembly fabrication. As it stands, the advantages outweigh the disadvantages, and the fuel manufacturers are moving toward enhanced use of gadolinia.

C. Chemical Shim

The term chemical shim refers to the use of a soluble neutron absorber which is dissolved in the primary reactor coolant. The absorbing compound must meet the following requisites: (1) it should be soluble in the coolant over the full range of desired concentrations, (2) it must be chemically stable, (3) it must not cause corrosion to primary system components, and (4) the neutron capture products should have the same characteristics. Reactor control is accomplished by varying the concentration of the absorber in the coolant. This, of course, is a slow process and is employed only to control slowly varying reactivity effects, in addition to conventional control rods. Since chemical shim control involves flexibility for concentration and dilution, the number and/or size of rods may be reduced, resulting in simplified design and reduced costs. A chemical absorber does not by itself materially affect the spatial power distribution, since it is uniformly distributed throughout the core. The use of chemical shim results, however, in improvements in spatial power distribution and therefore increased average to maximum power density in the core.

The principal disadvantages of chemical shim system are its effect on the moderator temperature coefficient of reactivity and the limited rate at which control material can either be introduced or removed from the fluid system. The moderator temperature coefficient in a chemically shimmed reactor is less negative than in a fully rodded one and may even be positive. This is because the rods contribute to the negative coefficient, and in a chemically shimmed core they are fewer, less black, or partially inserted; and because the boron

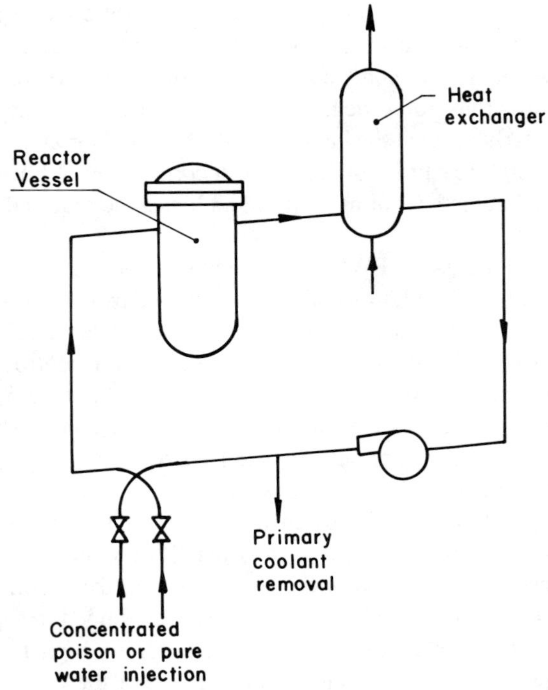

FIGURE 3. Fluid systems for chemical shim.

concentration decreases, together with the water density, temperature of the latter increases. This results in a positive contribution to the moderator temperature coefficient. It may be noted, however, that the moderator temperature coefficient becomes more negative with time due to the decreased absorber concentration with burnup.

Inserting and removing control material requires a fluid system that can inject either pure water (to increase reactivity) or concentrated neutron absorber solution (to reduce activity) at some point in the primary system. In order to maintain the system volume, primary coolant must be removed at an equal rate from another point in the system. This is called a feed and bleed process. The coolant that is removed is either processed to reconcentrate the poison solution for reinjection or sent to a separate waste disposal system for processing and removal. The system is shown in Figure 3.

Among the possible chemical absorbers, boric acid (H_3BO_3) has obtained much wider acceptance. It has good water solubility and has been used experimentally in both PWRs and BWRs.[8-10] The boric acid used in this functional role is often referred to as soluble boron or simply solbor. Its use in commercial power reactors, however, is now restricted to PWRs. Since boron has no radioactive isotopes, no coolant radioactivity problems arise from it. The concentration of boric acid in the coolant is changed at start-up and during the lifetime of a core to compensate for (1) changes in reactivity resulting from changes in moderator temperature, (2) changes in reactivity due to the buildup of xenon 135 and samarium 149 concentrations in the core, and (3) reactivity losses due to fuel depletion and buildup of long-lived fission products other than xenon and samarium. Rapid reactor transients are handled by usual control rods. It has been observed that boron does not increase corrosion rates and that most materials suitable in neutral or high-pH water at high temperature are also suitable in a boric acid solution.

Two problems with chemical shim are referred to as hideout and plate-out. Hideout is understood as the boron precipitation from solution onto solid surfaces or in deposits of corrosion products adhering to surfaces of the core and coolant system. It may later make an entry to the system as a result of changes in operation or water conditions. This is plate-out. Hideout and plate-out are of course not desirable features, since they would result in positive and negative reactivity drifts. In fact, they stand as the only significant safety problems resulting from the use of chemical shim.

In the consideration given to overall power plant control problems, it should be noted that there is an important difference between an all-rodded PWR and one using chemical shim. It is in the reduction in magnitude of the negative moderator temperature coefficient of reactivity when chemical shim is present. In the design of automatic load following between reactor and turbine generator into the overall control system, a large negative coefficient is relied upon. It may be noted that an increase in load increases steam flow and therefore reduces reactor inlet temperature, which in turn increases reactivity, and a new equilibrium is reached. The extent or degree of reduction in the temperature coefficient with chemical shim is not constant with time but decreases with burnup as the amount of dissolved absorber is reduced. It is therefore necessary to take into account these considerations in designing control systems of a chemically shimmed PWR plant.

III. CONTROL ROD, BURNABLE ABSORBERS, AND CHEMICAL SHIM

In order to control a nuclear fission reactor, particularly an LWR, three ways are most practical: (1) control rods with B, B_4C (natural or enriched ^{10}B), Cd, Ag-In-Cd, Hf, Ag-Hf, Ag-Ir-Hf, or rare earth elements as the control material; (2) burnable absorber with the oxide of the rare earth elements. Eu_2O_3, Er_2O_3, or Gd_2O_3, as a control material; and (3) chemical shim with boric acid, H_3BO_3, as an absorber or control material. The control rods perform the function of primary control, while the chemical shim and burnable absorber perform the function of secondary, or supplementary, control of LWR. Also, the chemical shim has an interesting feature in which solubility of H_3BO_3 increases with temperature distributed in the reactor core. The PWRs employ the combination of control rods and chemical shim. The BWRs, on the other hand, use the combination of the control rods and burnable absorber for reactor control and safety.

From the control rod, burnable absorbers, and chemical shim, the main nuclear properties of thermal and epithermal (resonance) absorption (capture) cross sections of neutrons for control materials are given in Table 2. An account of these control materials is given in following sections.

A. Boron-Based Control Materials

Boron is one of the best neutron absorbers (Table 2). The estimated concentration of boron in the earth's crust is about 0.001%. Natural boron is made up of two isotopes, 19% boron-10 and 81% boron-11. The former isotope has a very large absorption cross section for thermal neutrons. Natural boron has a reduced absorption ability because of the dilution effect of boron-11.

1. Chemical Metallurgy of Boron

Boron does not occur in nature as the free element. Its oxygenated compounds (borates) have been known and used since ancient times. Borax (a sodium borate) is the most common compound. It occurs in lake brines and as the mineral tincal. Principal uses of boron minerals and chemicals in the U.S., shown in Figure 4, typically illustrates the wide and varied uses of boron.[11] Figure 5 shows the refining and production of boron compounds.

As far as nuclear applications of boron-based materials are concerned, the use of boric acid in chemical shim control has already been pointed out. Apart from this, elemental boron and boron carbide, natural or enriched in ^{10}B, are the most popular forms of boron for nuclear use. The physical properties of boron are given in Table 3, and those of boron

Table 2
NUCLEAR PROPERTIES OF CONTROL MATERIALS

Material	Isotopic abundance (%)	Thermal absorption σ_a (b)	Macroscopic thermal absorption Σ_a (cm^{-1})	Principal resonances: Neutron energy (eV)	Principal resonances: Epithermal absorption σ_a (b)
Boron (natural)		759	97.23		
Boron-10	20	3,810		None	
Cadmium (natural)		2,450	113.60		
Cadmium-113	12.3	20,000		0.18	7,200
Silver (natural)		63	3.69		
Silver-107	51.3	31		16.6	630
Silver-109	48.7	87		5.1	12,500
Indium		194	7.44		
Indium-113	4.2	58			
Indium-115	95.8	197		1.46	30,000
Samarium		5,800	163.0		
Samarium-149	13.8			0.096	16,000
Samarium-152	26.6			8.2	15,000
Sm_2O_3		11,200	143.0		
Hafnium		105	473		
Hafnium-117	18.4	380		2.38	6,000
Hafnium-178	27.1	75		7.80	10,000
Hafnium-179	13.8	65		5.69	1,100
Hafnium-180	35.4	14		74.0	130
Europium		4,300	89.0		
Europium-151	47.8	7,700		0.46	11,000
Europium-153	52.2	450		2.46	3,000
Gadolinium		46,000	1,401		
Gadolinium-155	14.7	61,000		2.6	1,400
Gadolinium-157	15.7	240,000		17.0	1,000

carbide in Table 4. Elemental boron is a dark brown powder in the amorphous form and a dark gray, hard, brittle solid in the crystalline form. It melts at 2300°C and has an atomic weight of 10.81. It is marketed in several grades ranging from 90 to 99+% purity. The isotope ^{10}B is used as a control for nuclear reactors, a shield for nuclear radiation, and in neutron detection instruments. Boron carbide has a rhombohedral crystalline structure. It is a highly refractory material and one of the hardest substances known. Most commonly used for both abrasive and abrasion-resistant applications, as well as nuclear control and shielding, boron carbide is marketed in technical (67 to 75%) and high-purity (75 to 80%) grades.

Preparation of boron is accomplished by fused-salt electrolysis.[12] The electrolytic process involves use of potassium fluoride, potassium chloride, and boron trioxide (KF-KCl-B_2O_3) molten bath operated in an inert-atmosphere electrolytic cell. In the process, a low-carbon iron has been observed to serve as the most satisfactory cathode material. The electrolytic process has been proposed to proceed analogously to aluminum oxide-cryolite baths, with boron depositing at the cathode and oxygen evolving at the anode.

Preparation of boron carbide is accomplished by carbothermic reduction of a boron source, either B_2O_3 or H_3BO_3, in an electrical resistance furnace with a central heating core.[13,14] The carbide formation takes place according to the reaction, $2B_2O_3 + 7C \rightarrow B_4C + 6CO$. With H_3BO_3 as a source for boron, secondary reactions which influence the product include the water vapor equilibria $CO + H_2O = CO_2 + H_2$ and/or $C + H_2O = CO + H_2O$. The volatility of boric acid in the presence of water vapor is also important, and these reactions need to be carefully taken into account in formulating the furnace burden. The principal reaction is highly endothermic. The free energy decreases to a value of zero or about 1000°C.

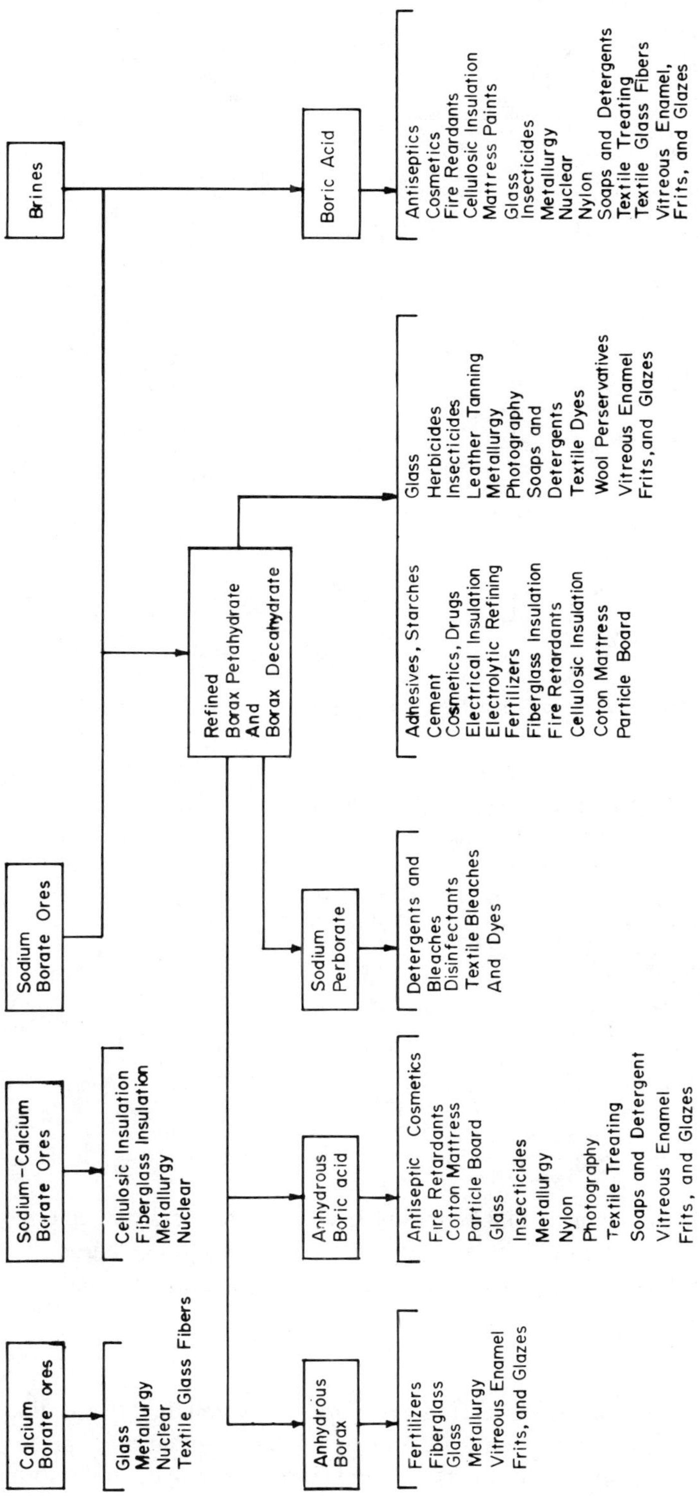

FIGURE 4. Principal uses of boron minerals and chemical compounds in the U.S.

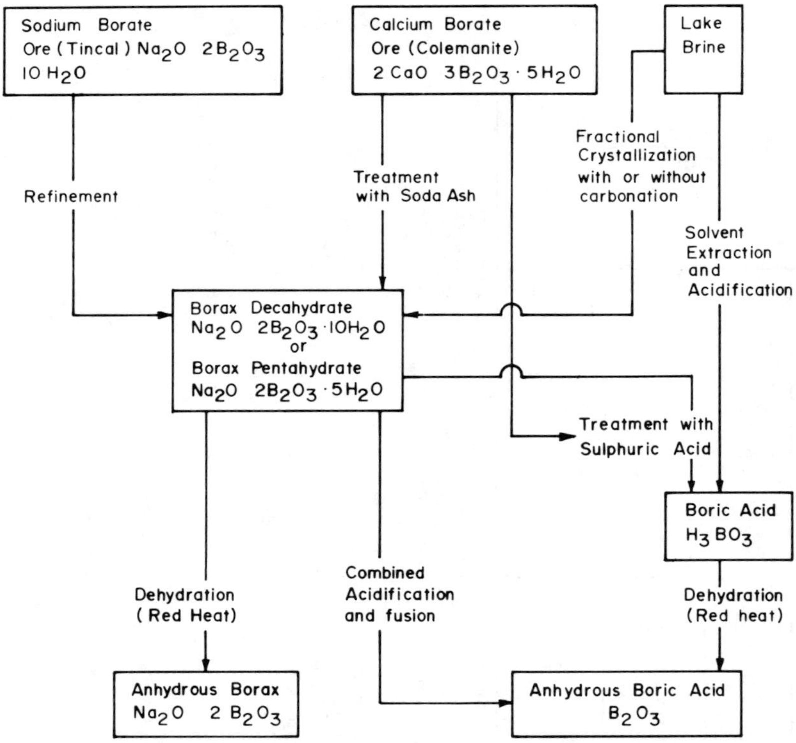

FIGURE 5. Refining and production of boron compounds.

Table 3
PHYSICAL PROPERTIES OF BORON

Density, (kg/m^3)	
Amorphous	2.35×10^3
Crystalline	2.48×10^3
Melting point (K)	2273—2348
Boiling point (K)	2823
Knoop hardness	3300
Specific heat (J/kg · K)	
At 273 K	600
At 573 K	651
At 1073 K	1760
Thermal expansion coefficient at 293—1023 K (10^{-6}/K)	8.3

It implies that boron carbide formation should be feasible above this temperature, and this is what is actually observed. Boron carbide is not only an economical form of boron, but also possesses superior high-temperature properties, extreme abrasion resistance, and considerable chemical inertness.

2. Applications

Boron carbide is principally used in nuclear reactors to control the neutron flux. Its function as a shielding material is referred to in Section V.C. Structural integrity of absorber pins is provided by cladding material, thereby allowing boron carbide to be employed in the form of either pellets or powder. For use in control rods, sintered boron carbide is occasionally employed (the process employs hot-pressing in graphite molds at 2200°C). In general,

Table 4
PHYSICAL PROPERTIES OF BORON CARBIDE

Formula weight	55.29
Boron content, B_4C (%)	78.28
C content, B_4C (%)	21.72
Crystal structure	Rhombohedral
Lattice constant (Å)	a_o = 5.60; C_o = 2.10
Color	Black, shiny
Specific weight ($g \cdot cm^{-3}$)	2.51
Melting point (°C)	2450
Boiling point (°C)	3500
Thermal expansion ($°C^{-1}$)	5×10^{-6}
Thermal conductivity at 25°C ($cal \cdot cm^{-1}\ sec^{-1} \cdot °C^{-1}$)	0.07—0.016
Microscopic capture cross section (b)	600
Knoop hardness at 100 g loading ($kg \cdot mm^{-2}$)	2900—3100
Mohs hardness	9.5+
Bending strength ($kg \cdot mm^{-2}$)	30—50
Compressive strength ($kg \cdot mm^{-2}$)	280
Elastic modulus ($kg \cdot mm^{-2}$)	45,000

however, control rods are made by filling boron carbide powder, of special particle size distributions, into stainless steel tubes, which are then vibrated to increase the bulk density of the powder. In addition to the density of boron, the chemical purity of the boron carbide is also important. This is because, during neutron capture, many elements form radioactive isotopes which emit, to some extent, high-energy gamma radiation in addition to alpha particles, etc. Boron carbide is the predominant control material for BWRs. It is also used in PWRs.[15] The carbide in which boron is enriched up to 90% ^{10}B is used in fast neutron reactors.

The principal effects of irradiation on boron carbide are helium generation and matrix swelling. The reactions with neutrons are as follows: $^{10}B + n \rightarrow {}^{3}H + 2[{}^{4}He]$ and $^{10}B + n \rightarrow {}^{7}Li + {}^{4}He$. Both the He and Li atoms resulting from neutron absorption process are larger than the original boron atoms. Hence, swelling occurs due to both Li and the portion of He atoms remaining in the matrix. The amount of gas released during burnup is sensitive to both radiation temperature and exposure level. The percentage of gas release builds up with irradiation temperature, with a maximum around 830°C. Either this gas release must be vented to the coolant, or sufficient plenum volume must be provided in order to prevent excessive loading on control cladding material. As far as tritium is concerned, about 80% of it is retained in B_4C at below about 800°C. Unfortunately, cladding will not contain tritium gas which escapes the carbide matrix. The released helium gas and the swelling can stress the stainless steel cladding tube and lead to failure. If this happens, the boron can be subjected to leach-out by the coolant water, which accelerates boron depletion. Despite these problems, because of other desirable properties control rods containing boron have probably been used more extensively than any other type.

B. Cadmium-Based Alloys

The most interesting neutron absorber after boron is cadmium. Cadmium is a constituent of zinc ores because of its close similarity to zinc in properties. Cadmium is a silvery metal that is softer and more malleable than zinc. It is extremely ductile and can be fabricated by rolling, forging, drawing, etc., with ease. Its melting point is 321°C, and its boiling point is 767°C; thus, the metal is useful as a control material for low-temperature applications only. Table 5 presents certain physical and mechanical properties of cadmium. Cadmium does not have good corrosion resistance, and so it is clad with either aluminum or stainless steel. Cadmium exhibits a high cross section for thermal neutrons (2550 b). At neutron

Table 5
PHYSICAL AND MECHANICAL PROPERTIES OF CADMIUM

Density ($kg \cdot m^{-3}$)	8.64×10^3
Melting point (K)	594
Boiling point (K)	1038
Rupture strength (MPa)	72
Elongation (%)	50
Elasticity modulus (MPa)	0.5×10^5
Poison ratio	0.30
Brinell hardness	21—23
Specific heat at 293 K (J/kg · K)	230
Thermal conductivity (W/m · K)	
At 293 K	96
At 373 K	94
At 473 K	92
Thermal expansion coefficient at 293—373 K (10^{-6}/K)	31.8

energy higher than 0.178 eV, the cross section starts to decrease rapidly. Therefore, the element has a low absorption capacity for epithermal neutrons. This feature of the element presents itself as a drawback to its use in water-moderated power reactors, where a considerable portions of the neutrons are in the epithermal (or resonance) region.

Cadmium- and silver-based alloys are used as control materials only for research reactors, where temperatures reached by the control rods are relatively low. The Ag-Cd alloys combine silver's large absorption cross section for resonance neutrons with cadmium's large absorption cross section for thermal neutrons. Indium is another absorber material that remains effective for resonance neutrons. Its presence improves absorption properties, mechanical strength, and corrosion resistance at high temperatures of Ag-Cd alloys. An alloy containing 80% Ag, 15% In, and 5% Cd (Ag-In-Cd) obtained by molding and hot extrusion and encased in stainless steel, with which the alloy makes a metallurgical bond, is the dominant material used in control rods in PWRs. Some PWRs (those made by combustion engineering) use boron carbide, but the other vendors are also moving toward use of this material for economic reasons. The Ag-In-Cd alloy was originally developed based on the neutronic considerations and as a less expensive substitute to hafnium, but has now greatly surpassed it in cost. The alloy has a reactivity worth less than that of boron carbide and poor corrosion resistance in high-temperature oxygenated water. Consequently, corrosion would be a problem in the case of cladding developing a leak. Another problem with the alloy rods is the production of long-lived activation products that emit gamma rays, and this complicates the disposal of control rod assemblies at the end of their useful life. The main disadvantage of Ag-In-Cd alloy is, however, high cost, primarily from the high cost of silver and indium.

C. Hafnium and its Alloys

The high cost of Ag-In-Cd and the technical performance problems (gas-producing) of B_4C have accounted for a search for alternative materials for reactor control. Hafnium is a control material that is attracting increasing attention. As cadmium is associated with zinc, hafnium is closely associated with zirconium in all its minerals. Hafnium is obtained by separating it from zirconium, and the metal is produced in the same way as zirconium by the well-known Kroll process, involving reduction of a chloride intermediate with magnesium.[16] Purified hafnium is produced by iodide refining.[17] A molten salt electrorefining process has also been reported to be effective in producing purified hafnium from metallothermically reduced hafnium metal.[18]

Naturally occurring hafnium has six stable isotopes, a microscopic thermal neutron cross section of 105 ± 5 b, and a scattering cross section of ±2 b. Of the isotopes ^{177}Hf, ^{178}Hf,

Table 6
PHYSICAL AND MECHANICAL PROPERTIES OF HAFNIUM

Physical Properties

Density (g cm^{-3})	13.64 13.0 (commercial nuclear grade with 4.5 wt% Zr)
Crystal structure	
Alpha phase to 1760°C	hcp
Beta phase at 1760—2230°C	bcc
Transformation temp, alpha to beta (°C)	1760
Melting point (°C)	2222
Boiling point (°C)	3100
Coefficient of linear thermal expansion at 0—1000°C (°C $^{-1}$)	5.9×10^{-6}
Thermal conductivity [cal · cm^{-1} s^{-1} $°C^{-1}$]	
At 50°C	0.533
At 400°C	0.494
Specific heat (cal/g − K)	
At 298.15 K	6.15
At 1300 K	7.98

Mechanical Properties[a]

	Rod (swaged)	Plate (hot-rolled)	Strip (cold-rolled)
Tensile (min psi)	65,000	55,000	60,000
Yield (min psi)	15,000	25,000	24,000
Elongation (min %)	22	22	19

[a] Fully annealed; test direction, in longitudinal; room temperature.

and ^{179}Hf account for approximately 92% of the effective cross section of natural hafnium. Hafnium absorbs both thermal and epithermal neutrons, and its worth does not appreciably diminish by irradiation. The unique ability of hafnium to absorb neutrons above thermal energies makes it doubly effective as an absorbing material. Table 6 presents some of the properties of hafnium. It is a hard, shiny, ductile metal with a color similar to stainless steel. It has a hexagonal close packed crystal (HCP), structure below 1760°C and a body centered cubic structure above this temperature. Due to the nature of deformation systems in hafnium, the wrought product is very anisotropic. Hafnium is superior in corrosion resistance to zirconium in water and steam, molten alkali metals, and air. Like other refractory metals, hafnium is resistant to chemical attack due to the thin tenacious layer of metal oxide on the surface of the metal. The corrosion properties are affected as this thin layer is penetrated by reactants at increasing temperatures. However, only at temperatures of a few hundred degrees centigrade does hafnium begin to react appreciably with water, oxygen, nitrogen, and hydrogen. In aqueous solutions, hafnium is soluble in HF and concentrated H_2SO_4. It is resistant to dilute solutions of HCl and H_2SO_4, but is unaffected by nitric acid in all concentrations. Hafnium is very resistant to alkalis and boiling or concentrated sodium hydroxide.

The combination of unique neutron absorption, good mechanical strength, and corrosion-resistant properties make hafnium ideal as a control rod material or as a poison material for increasing the stacking density of spent fuel storage or a container for corrosive media in spent fuel reprocessing plants. Hafnium, hafnium alloys (Ti, 12 wt% Hf; Zr, 15 wt% Hf), and sheathed hafnium oxide are prime candidates for these specific fields of application.

Many U.S. naval reactors have made extensive use of hafnium. The Shipping Port PWR has also employed this material. The metal has been established as qualified for BWR control application where unclad solid hafnium rods alloyed with a small amount of zirconium have been projected. It may be added that alternative alloyed materials with hafnium, such as pyrohafnates (alloyed with one or more rare earth), Hf-In-Ag, and Ag-Hf, have received attention for application as control rods. The binaries or the ternaries are basically intended to provide the best attributes of the constituents with which they are formed.

D. Rare Earth-Based Control Materials

The rare earths are a group of 16 elements: yttrium (at. no. Z = 39) and lanthanum to lutetium (Z = 57 to 71). However, promethium (Z = 61), a product of fission reaction, is not found in nature. The remaining 15 of this group occur together in nature. The rare earths can be grouped into (1) light (or the cerium group) and (2) heavy (or the yttrium group). The cerium group consists of La, Ce, Pr, Nd, Sm, Eu, and Gd; and the yttrium group consist of Y, Tb, Dy, Ho, Er, Tm, Yb, and Lu. Among the large number of minerals containing rare earths, only a few are rich and are of commercial importance. They are (1) monazite, an orthophosphate essentially of the cerium group; (2) bastnasite, a fluorocarbonate of the same group; and (3) xenotime, an orthophosphate of the heavy group of rare earths, but less abundant than monazite. The extractive metallurgy aspects, basic properties, and industrial applications of rare earths have been excellently discussed in the literature referenced.[19-21]

Some physical properties of the rare earth elements are given in Table 7. Among the rare earth elements, europium, samarium, gadolinium, and dysprosium can be used as control materials because of their outstanding neutron absorption cross section.[22] The rare earths are weakly resistant to corrosion, being easily oxidized; and to overcome this shortcoming, in the manufacture of control rods, rare earth oxides and dispersions in metallic matrices are employed instead.

Europium in the sesquioxide form Eu_2O_3 can be dispersed in stainless steel. These metalloceramic dispersions may contain Eu_2O_3 up to 50 wt%. Control rods are obtained by cladding the dispersion in stainless steel claddings, sometimes followed by rolling or extrusion. The stainless steel cladding is meant to avoid the contact of the dispersion with water, with which Eu_2O_3 might react. Control rods like the one described are used in special reactors that require a small core. Recent designs of fast reactors also contemplate the use of Eu_2O_3-based dispersion as control materials. This particular attention toward europium is because of its high fast-spectrum absorption cross section. Unfortunately, a full assembly with europia has a reactivity worth equivalent only to that for an unenriched B_4C assembly in a fast breeder reactor. The other advantages include (1) the high level of radioactivity induced into the europium decay chain; (2) the limited supply (as a rare earth); and (3) low thermal conductivity (which requires fairly small diameter pins). An interesting attempt is to combine the attributes of both Eu and B to form the compound EuB_6. The europium boride compound has a reactivity worth equivalent to approximately 25% ^{10}B-enriched B_4C, and it is worth about 10% more than europia. Furthermore, the loss of reactivity is lower than boron carbide. It is, however, possible to enrich the boron in the europium boride compound to further enhance its absorption capability. The expense, however, goes up accordingly. The reactor experience with this is limited. The dimensional stability properties appear attractive. The only major problem is that helium gas release is much larger than for boron carbide, so that venting is probably required.

Regarding samarium and gadolinium as control materials, their properties differ from those of europium. Thus, for samarium, the natural nuclide ^{149}Sm provides the greatest contribution to the capture cross section. When this isotope becomes ^{150}Sm, the cross section drops by a factor of 10^2 to 10^3, which implies that the element is a consumable absorbent.

Table 7
SOME PHYSICAL PROPERTIES OF RARE EARTH ELEMENTS

Element	Chemical symbol	At. no.	At. wt.	Abundance (ppm)	Melting point (°C)	Boiling point (°C)	Density (g/cm^3)	Crystal structure	Absorption cross section for thermal neutrons (b, 10^{-24} cm^2)
Yttrium	Y	39	88.90	28	1522	3338	4.469	hcp	1.27
Lanthanum	La	57	138.91	18	918	3464	6.145	dhcp	8.9
Cerium	Ce	58	140.12	46	798	3433	6.770	fcc	0.7
Praseodymium	Pr	59	140.90	5.5	931	3520	6.773	dhcp	11.2
Neodymium	Nd	60	144.24	24	1021	3074	7.007	dhcp	46
Promethium	Pm	61	147.00	—	1042	3000[a]	7.260	dhcp	—
Samarium	Sm	62	150.35	6.5	1074	1794	7.520	rhomb	5500
Europium	Eu	63	151.96	1.0	822	1529	5.243	bcc	4600
Gadolinium	Gd	64	157.25	6.4	1313	3273	7.900	hcp	4600
Terbium	Tb	65	158.92	0.9	1356	3230	8.229	hcp	44
Dysprosium	Dy	66	162.50	4.5	1412	2567	8.550	hcp	1100
Holmium	Ho	67	164.93	1.2	1474	2700	8.795	hcp	64
Erbium	Er	68	167.26	2.5	1529	2868	9.066	hcp	166
Thulium	Tm	69	168.93	0.2	1545	1950	9.321	hcp	118
Ytterbium	Yb	70	173.04	2.7	819	1196	9.965	fcc	36
Lutetium	Lu	71	174.97	0.8	1663	3402	9.840	hcp	108

Note: hcp, hexagonal close-packed; dhcp, double hexagonal close-packed; rhomb, rhombohedral; fcc, face-centered cubic; bcc, body-centered cubic.

[a] Estimated.

Like samarium, gadolinium is a consumable absorbent. Its use as burnable poison has been described earlier.

E. Summary and Comparisons of Control Materials

The rates of neutron generation and loss are directly related to the effects of reactivity and control of a nuclear fission reactor. The most effective control materials for the fission reactor are strong neutron absorbers. The disadvantage of reactor control by a strong neutron absorber, however, is the loss of useful neutrons and impairment of the neutron economy.

Certain rare earth elements, cadmium, boron, and hafnium, on the basis of their high neutron absorption cross section, have received attention as control materials. The very high thermal neutron absorption cross section, the notable feature of rare earths, make them admirably suitable as control materials. The familiar form is the rare earth oxide dispersion in stainless steel. This has, however, not been taken to any large-scale use. Gadolinia has found established use as burnable poison. In the rare earth control material family, europium has a special place, in that it remains effective in epithermal neutron energy range. This rare earth element also has another attribute, in that its higher isotopes have a reasonably high neutron absorption cross section; and thus europium does not deplete in its performance as a control material, unlike others in the list of the rare earths, such as gadolinium and samarium.

The next important material is cadmium; its characteristic of being extremely ductile is a boon as far as shape fabrication is concerned. Its melting point being low, the application is limited to low temperatures. Its poor corrosion resistance mandates its use cladded with aluminum or stainless steel. Features of cadmium in terms of neutronic considerations include (1) it is totally ineffective for the epithermal neutron range at which it has almost negligible neutron absorption cross section, and (2) control rod worth life is limited since a higher isotope of cadmium is ineffective as a neutron absorber.

Turning to boron, the most widely used neutron absorbing material in the nuclear industry, several interesting features can be mentioned. Boron, either natural or enriched in ^{10}B, in its most economic form as boron carbide, is a control material of choice in the nuclear field. Neutron reactions with boron are different than those with the rare earths and cadmium. The n-alpha reaction associated with $^{10}B \rightarrow {}^{7}Li$ is particularly suited for neutron absorption, due to the lack of any appreciable gamma radiation. The weak radiation resulting, for example, from a 50% burnup of ^{10}B is easily absorbed because of the small energy associated with it: about 0.4 to 0.5 MeV. In contrast, cadmium emits gamma radiation of 6 MeV (rare earths emit even more). Like cadmium, the control rod worth life with boron is limited. The products of reaction with boron, lithium, and helium, accumulate in the structure containing boron carbide. It may cause straining and swelling, and eventual failure of the structure. In the cases of cadmium and the rare earths, neutronic reactions yield their respective radioactive elements an effect which is absent in the case of boron. When is taken as a basis for comparison between boron/boron carbide and camium, it will be noted that due to low atomic weight and small density, boron/boron carbide is as effective as cadmium.

In reference to hafnium, it will be noted that from a neutronic point of view there are similarities between it and the rare earth element europium. Like europium, it does not deplete, since the higher isotopes of hafnium produced as a result of interaction with neutrons possess appreciable neutron-absorbing properties. Since the daughter product is hafnium, there does not arise any lattice straining or any hostile situation in the parent hafnium structure. The element, again like europium, remains effective in the epithermal neutron energy range. The metal has a high melting point, good high temperature strength properties, and above all good corrosion resistance properties. All these contribute to qualifying the metal, in spite of its high cost and relative rarity, as a very attractive control element.

IV. SHIELDING MATERIAL

A. General

Shielding materials are meant to reduce the radiation and heat flows generated in the core of a nuclear reactor low enough to protect personnel and equipment from the damaging effects of radiation. In principle, the radiations which might escape from the reactor system include alpha and beta particles, gamma rays, neutrons of various energies (a nuclear reactor is by far the most prolific source of neutrons), and fission fragments. In a perfect vacuum, all forms of radiation will continue to move indefinitely. In solid, liquid, or gaseous media, however, all forms of radiation lose energy or are absorbed through a variety of mechanisms. Therefore, the source of interaction of radiation with matter has always been of considerable importance in the understanding of the materials used for shielding purposes.

B. Interaction of Radiation with Matter[22,23]

1. Alpha Particles

These are sometimes called rays, but addressing them thus is a misnomer. They are not rays at all, but consist of streams of positively charged particles called alpha particles, shown to be doubly charged atoms which are four times as heavy as hydrogen, identical with the helium ion He^{2+}, traveling with an average velocity about one tenth that of light. The alpha particles are emitted in groups with discrete energies that vary between 3 and 9 MeV, depending upon the emitting nuclide. By virtue of their electric charge they interact with electrons of the medium through which they are passing. The result of interaction with an electron is that part of the kinetic energy of the moving particle is transferred to the electron, and the velocity of the moving particles is slightly decreased. Eventually the particle will be stopped, but a very large number of collisions will be required. Through this interaction the atom of the concerned medium can be excited, i.e., an electron is brought to a higher energy level. The struck electron will sometimes be knocked out of the atom to which it belonged, so that a free electron and a positive ion (ion pair) are produced. This ionization is frequently used as a means of detection of charged particles, for example, in ion chambers and proportional counters.

Because the heavy charged particles such as alpha particles move in almost straight lines and lose energy in a large number of small dips, they have definite ranges. The alpha particles, being doubly charged and moving relatively slowly, are densely ionizing, and hence, in spite of their high energies, their penetrating power or range is extremely short. In fact, alpha-particles of energies up to 7.5 MeV are not capable of penetrating the protective layer of the skin in most parts of the body (e.g., the hand). Their range in air is only a few centimeters. Thus, shielding against this type of radiation poses no problem at all.

2. Beta Particles

The beta particles, either positive or negative, are emitted with a continuous speed of energies. They induce excitation, ionization, or both in the stopping media. They lose their energy essentially by the same mechanism as heavy charged particles like alpha particles. However, there are several important additional complications. Since the masses of beta particles and of the orbital electrons are equivalent, the former can lose a large fraction of its energy in a single collision. The beta particle undergoes a wide-angle deflection in such a collision; consequently, beta particles are scattered out of the beam path all along its length. The fact that scattering is more important in the case of beta particles than with the heavy charged particles means that the path of an electron through an absorber is tortuous and not well defined. For the same initial energy, beta particles have a greater velocity than alpha particles, since their electronic mass is approximately 1/7300 of the helium mass. This greater velocity results in a correspondingly lower specific ionization and consequently in a much greater range.

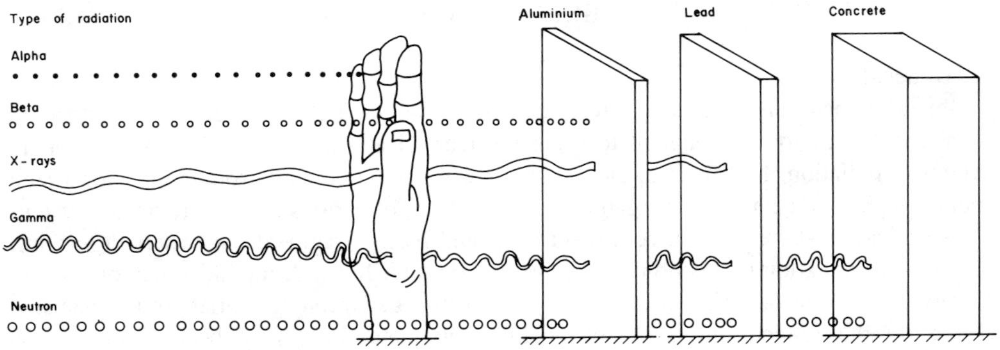

Radiation	Relative range	Relative Linear Energy Transfer (LET)
Alpha	1	10,000
Beta	100	100
Gamma	10,000	1

FIGURE 6. Schematic of ranges of radiations.

An additional situation is that when an electron passes near a nucleus, it may be deflected by the positively charged nuclear field. In such a fast-moving, very light particle, this deflection causes a loss of energy by emission of bremsstrahlung (decelerated radiation) or low-energy X-rays. This source of energy loss is not present with heavy charged particles. The loss of energy increases with beta energy and with the atomic number of the absorbed material. In aluminum, approximately 1% of the energy of a 1-MeV electron is lost by radiation and 99% by ionization, whereas in lead, loss by radiation is approximately 10%. In tungsten, a 10-MeV electron loses about 50% of its energy by radiation, whereas a 100-MeV electron loses over 90% of its energy by that mechanism.

3. Gamma Rays

Gamma rays are electromagnetic radiations like X-rays, UV radiation, visible light, IR radiation, and radio waves. Having no charge or rest (zero-energy) mass, gamma rays do not produce a change in Z or A in the emitting nucleus, but they do cause a decrease in mass corresponding to the energy carried away by the gamma rays. The gamma rays are emitted with discrete energy groups, like alpha particles, rather than a continuous spectrum of energies as positive or negative beta particles are. Since they have no mass or charge, gamma rays do not interact readily with matter and therefore possess great penetrating power. Whereas a 4-MeV alpha particle is stopped by a film of water 1 mm thick, a 4-MeV electron requires approximately 2 cm of water to stop it. By contrast, a beam of 4-MeV gamma rays passing through 20 cm of water would retain half the original number of rays, and 10% would survive passage through 70 cm of water. Figure 6 shows qualitatively the penetrating characteristics of the various radiations.

The absorption of gamma rays by matter takes place by a different mechanism compared to charged particles. The interaction is mainly with electrons, but the gamma ray can lose a large fraction of its energy or all of it in a single encounter. There is therefore no quantity corresponding to a range. The rays are absorbed according to an exponential law, $I = I_o e^{-\mu x}$, where I_o is the original energy intensity, I is the final energy intensity, x is the absorber thickness, and μ is the absorption coefficient. This is the basic equation which applies to shielding against gamma radiation. The interactions of gamma rays are categorized into three groups or processes: (1) the photoelectric effect, (2) the Compton effect, and (3) the pair production effect.

The photoelectric effect occurs when an incoming gamma-ray photon causes the ejection of an atomic electron from its orbit while itself being absorbed. The absorption of low energy gamma rays is mainly due to this effect. The kinetic energy of the electron is equal to the energy of the incident photon minus the being in energy of the electron in the atom from which it was ejected. The electron most probably comes from the K shell, provided that the gamma ray has sufficient energy to supply the K shell binding energy. The absorption due to the photoelectric effect decreases with increasing gamma-ray energy, but increases very rapidly with increasing absorber atomic number. Hence it is not important for absorption of low energy gamma rays in heavy elements.

The Compton effect involves gamma-ray photon interaction with (or scattering) a free or loosely bound electron. Compton scattering is the most important energy loss mechanism for medium-energy photons. The photon transfers part of its energy to an electron, but still remains a photon of lower energy which may then undergo photoelectric or Compton interaction. Although the fractional energy loss is much greater for high-energy photons, the probability that a photon will be Compton-scattered at all decreases about linearly with increasing energy. Therefore, Compton scattering makes only a small contribution to the absorption of high-energy photons.

Pair production is the third mechanism by which photons of energy greater than 2×0.51 MeV are absorbed. The photon vanishes, to be replaced by a positive-negative electron pair. The absorption coefficient μ in the equation given earlier is used as the sum of the three effects. Using the symbols μ_{pe}, μ_c, and μ_{pp} to refer to absorption coefficient for photoelectric effect, the Compton effect, and pair production respectively, the total coefficient is given by $\mu_{tot} = \mu_{pe} + \mu_c + \mu_{pp}$. The μ_{pe} falls rapidly within increasing photon energy; the Compton effect is a much more gradual effect which contributes mostly in the region above 1 MeV but below 10 MeV. Starting with the threshold energy of 1.022 MeV, pair production gradually increases to become dominant at high energies.

Nothing intercepts the short-wave, high-frequency electromagnetic radiation, the gamma rays, except matter itself. From the mechanisms of interaction of matter with gamma rays as outlined above, it is seen that interactions are mainly with electrons. Obviously, the more electrons there are for the gamma ray-photons to collide with in a given volume, the better it is. High atomic number and high density contribute large numbers of electrons per unit volume, and for this reason heavy elements, such as lead, are particularly good as shielding materials. It should be emphasized that gamma rays present the most difficult shielding problems: (1) they are very penetrating, since their interaction rate in material is rather low; (2) the materials which in general are the most efficient gamma stoppers have high atomic numbers and consequently high densities and large mass, and are cumbersome to handle; (3) gamma rays do not generally disappear in a given collision, but instead lose only part of their energy. The low-energy or scattered photon continues through the material and needs stoppage elsewhere. The cited literature deals further with this.[24]

Since any kind of matter will intercept and, in sufficient quantity, attenuate gamma radiation, the choice of shielding material is based on economics, costwise and spacewise. For example, uranium is a better shield than lead, but its relative cost would make its use prohibitive in many cases Even for lead, however, due to its relatively high cost, the preference is for cheaper materials such as high-density reinforced concrete with appreciable proportions of strontium and barium compounds added. These additions raise the average atomic number and thus present to gamma rays a large number of electrons with which to interact. If space is available and it is convenient to use, water is a very inexpensive and adequate shield. Although the ability of different elements to absorb gamma radiation varies considerably, one can state as a rule of thumb that equal masses of shielding materials have the same shielding effect. For example, if concrete, with a density of 2500 kg/m^3 is substituted for lead, which has a density of 11,344 kg/m^3, 11,344/3500 = 4.54 times the thickness will be needed. To shield to a factor of 10^{10}, which has been calculated as needing 465 mm of lead, one now needs $465 \times 4.54 = 2110$ mm of concrete.

Table 8
PROCESSES INVOLVING INTERACTION OF NEUTRONS WITH MATTER

Process	Symbolism[a]
Elastic scattering	X (n,n) X
Inelastic scattering	X (n,n, and γ) X
Radiative capture	X (n,γ) Y
Particle emission	
Charge particle emission	X (n,p) Y; X (n,α) Y′; etc.
Kn emission	X (n,Kn) Y
Fission	X (n,fission)

[a] $X + x \rightarrow Y + y$, where X is the target nucleus, x is the projectile particle, Y is the product value, and y is the product particle.

4. Neutrons

Being uncharged, neutrons behave in a manner quite different from the charged particles (alpha and beta particles). The behavior is also different from gamma rays. They interact only very weakly with atomic electrons. The slowing down of fast neutrons as they travel through matter is due almost entirely to collision of the neutrons with the atomic nuclei. Depending on their energy, neutrons interact with matter by various processes. These include

1. Elastic scattering. The neutron shares its initial kinetic energy with the nucleus, which suffers a recoil only and is not left in an excited state. The smaller the mass of the nucleus, the greater the fraction of the kinetic energy taken by it.
2. Inelastic scattering. Inelastic scattering is possible only with fast neutrons. The scattered neutron carries less energy than the neutron, and the nucleus goes into an excited state. The excited nucleus either emits a gamma ray or stays in a metastable state.
3. Radiative capture. In addition to being slowed down by elastic and inelastic collisions, neutrons may be captured by nuclei in the stopping medium. The compound nucleus formed in the process is excited and emits gamma radiation. This will be particularly probable if there exists some neutron energy for which the neutron absorption capture cross section of the medium is very large. For example, fast neutrons are rapidly slowed down by collision with hydrogen nuclei in passing through an aqueous solution of cadmium chloride. When the neutron energy is in the neighborhood of 0.178 MeV, resonance capture of ^{113}CdF becomes very probable because ^{113}Cd (n,r) ^{114}Cd cross section for neutrons of this energy is about 8000 b.
4. Particle emission. This type of interaction between incident neutrons with the target nucleus may lead to the emission of particles such as protons, neutrons, and alphas.
5. Fission. In this process the compound nucleus splits into two fission fragments with the emission of one or more neutrons.

The neutron-matter interaction processes are listed in Table 8. The absorption of neutrons occurs in two distinct stages. First, the fast neutrons are slowed down by elastic and inelastic scattering processes with nuclei, particularly light nuclei like carbon and hydrogen. Second, the neutrons thus slowed down are then captured, as the capture cross section for low-energy neutrons is high for most elements.

C. Reactor Shielding[25-28]

The design of a nuclear reactor and its associated equipment must take into consideration the provision for the attenuation of escaping nuclear radiations by some form of shielding.

The character of the shield suitable for a particular reactor is greatly dependent on the purposes of the reactor, so that shield design is to be regarded as an integral aspect of reactor design. A low or zero power reactor would require little or no shielding for normal operation. The attenuation effect of distance may suffice in this case. For a reactor which normally operates at high power, a shield of considerable thickness will be required. For a stationary system, a partial, although expensive, solution may be found in building the reactor underground. The degree of shielding necessary is influenced by the basic purpose of the reactor. In a power reactor, for example, only occasional access may be required outside the shield for inspection and maintenance. On the other hand, where the reactor is used for research and testing, it may require continuous attention, especially when measurements are being made. The permissible radiation level in the latter case will obviously be less than in the former. For specific cases, such as for mobile reactors, e.g., for propulsion of a submarine, the weight of the shield is of paramount importance and the cost of the shield may be of secondary significance. For a stationary reactor, however, cost becomes the primary factor, and weight becomes secondary.

The outline of interactions of radiations with matter, as given before, points out that as far as shield design is concerned, only gamma rays and neutrons need to be taken care of, since these are by far the most penetrating among the radiations from a nuclear reactor source. In principle, the problem of shielding the reactor itself involves three aspects: (1) slowing down of fast neutrons; (2) capture of the slowed-down neutrons; and (3) attenuation of all forms of gamma radiation, including primary radiations from the reactor core and secondary radiations formed as a result of various interactions between neutrons and nuclei in the shield. Any material which attenuates these radiations to a sufficient extent will automatically reduce all the others to negligible proportions.

Radiation shielding serves three major functions or purposes: (1) thermal shielding to protect the pressure vessel, coolant loop, and inner shield from intense heat liberated upon absorption of the nuclear radiation;[29] (2) biological shielding to offer protection to personnel; and (3) apparatus-instrument shielding that serves to protect electromagnetic apparatus and electronic instruments, which otherwise would not function properly in an intense radiation field.[30,31]

The attenuation of gamma radiation and neutron leaks is achieved by shielding materials that surround the core. A distinction can be made between thermal shielding and biological shielding. Thermal shielding functions to avoid exposure of the entire shielding to the heat generated in the reactor and to attenuate radiations nearest to the core. These are made of materials of high density, good thermal conductivity, and high melting temperatures. Thermal shielding absorbs the high-energy gamma radiations and reduces the energy of the fast neutrons by inelastic collisions. These two types of radiation carry most of the energy leaking from the reactor core. The absorption of the radiations accompanying the capture of slowed-down neutrons produces a considerable amount of heat in the thermal shield. In power reactors the heat is removed by the coolant and so contributes to the available energy. In research reactors, on the other hand, the heat is not utilized. In a materials testing reactor, for example, the thermal shield consists of two thick layers of steel, separated by a space through which air is passed to serve as the coolant. The air is then discharged into the atmosphere. For low-power reactors the heat liberated in the shield may not be high and in such cases the introduction of a thermal shield may be unnecessary.

The thermal shield is placed between the reactor core and the main shield, referred to as the biological shield. This is made of water and/or concrete, and its main task is to attenuate the secondary gamma radiations, as well as to thermalize and absorb the neutrons. As an example of the thermal and biological shields, reference may be made to the Shipping Port PWR. In this reactor the thermal shield consists of two concentric stainless steel cylinders which surround the reactor core. The inner cylinder is separated by an annular space from

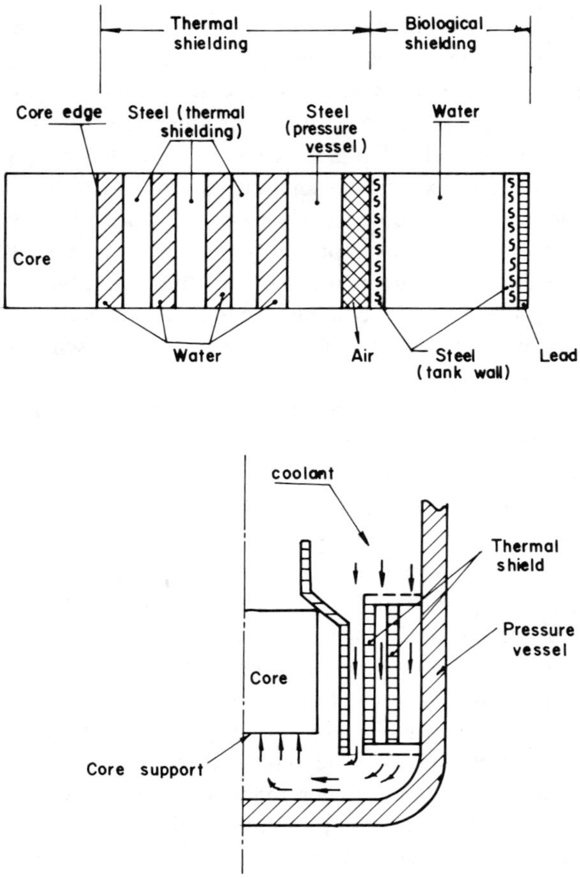

FIGURE 7. (a) Shielding made up of alternating layers of various materials; (b) configuration of a PWR thermal shielding.

the outer cylinder. Part of the cooling water entering the reactor flows in the annular spaces between the core and the inner thermal shield and between the outer shield and the pressure vessel, as well as between the two thermal shields.

As an illustration of the type of shielding, Figure 7 is presented for a PWR. It shows a series of radial water and steel layers which represent the thermal shield and part of the biological shield. In the system the fast neutron flux decreases regularly. The fast flux decreases regularly, although somewhat more rapidly in the steel layers because the macroscopic removal cross section is slightly larger than for water. The thermal neutron flux increases at first upon entering a water layer, as a result of moderation, and then decreases due to neutron capture. In the steel layers, however, the thermal flux first decreases because capture is significant, and then increases as the fast neutrons are slowed down. The primary gamma dose rate decreases steadily, the attenuation being more rapid in steels than in the water layers. The variation in the secondary gamma radiation is much less simple. It depends upon a host of factors such as the thermal neutron flux distribution, the capture cross sections in the water and steel, the energy spectrum of the capture gamma rays, the spectrum of the decay gamma rays resulting from neutron capture, and the attenuation of the various radiations in water and steel layers. These constitute a very important consideration in shield design.

Figure 7b shows the configuration of thermal shielding.[32] The system is made up of concentrically placed metallic cylinders around the reactor core. The primary coolant is circulated between them. In this way, direct thermal loading of the pressure vessel is prevented. One achieves also in the process a reduction of fast neutron flux down to 10^8 to 10^9 n/cm^2·s at the level of vessel walls, thus eliminating their radiative embrittlement. The thermal shield materials can be iron, carbon steel, or stainless steels. Apart from PWR shielding, other reactor examples can be taken for the purpose of elaboration. Thermal shielding of fast reactors is mainly obtained by means of a complex structure made of stainless steel, sodium, and a small-density structure made of steel alloyed with chromium. The thermal shielding of the Enrico Fermi fast reactor[33] consists of stainless steel plates with liquid sodium circulating in between to prevent the heat generated by irradiation. The materials of shielding of fast reactors can be either iron and boral or lead and boral screens arranged in an alternating manner.[34] In research reactors, shielding is generally provided by the water in which the core is immersed.

V. SHIELDING MATERIAL CATEGORIZATION

Materials employed in shielding broadly fall into three categories: (1) heavy and moderately heavy elements, (2) hydrogenous materials, and (3) boron-containing light substances.

A. Heavy and Moderately Heavy Elements

The heavy and moderately heavy elements serve to attenuate the gamma rays and to slow down fast neutrons by inelastic collisions. A commonly used material for thermal shields is iron as carbon steel or stainless steel. Such shields can absorb a considerable proportion of the gamma rays and fast neutrons escaping from the core. Layers of steel with water between them represent a very effective shield for both neutrons and gamma rays. The most common shielding materials for gamma rays are probably lead and lead alloys. These are abundant and economical. However, because of its low melting point, lead is limited in its application to situations where temperatures are not too high. Because of their high density and high melting points, the elements tantalum and tungsten may be of considerable interest in shield design. Their high cost prohibits them from general application and limits them to application in special cases only.

B. Hydrogenous Materials

Hydrogenous materials serve functionally to moderate fast neutrons, absorb slow or thermal neutrons, and attenuate gamma rays. The characteristics or properties on the basis of which hydrogenous material is evaluated for neutron shielding are (1) hydrogen content and neutron moderating power, (2) absorption cross sections for neutrons and gamma rays, (3) abundance, and (4) economics. The material which satisfies all of these characteristics is light water. In fact, water is probably the best neutron shield material. It is, however, poorly disposed toward absorption of gamma radiations. As other drawbacks to the use of water in reactor shields, mention may be made of its low boiling point at ordinary pressure and its susceptibility to decomposition by radiation. Water, of course, provides a ready means for removing the heat generated by radiation absorption. The pool type of research reactor is a well-known example of use of water as a shielding material.

As a general hydrogenous shield material, a special reference can be drawn to concrete. It is relatively inexpensive to build, easy to form into any desired shape, and quite stable against radiation damage and other environmental factors. It has high mechanical strength, especially prestressed concrete. These reasons account for its use more than any other nuclear reactor shielding. Ordinary concrete of density 2.3 g/cm^3 generally contains somewhat less than 10 wt% of water when cured. It has about 1.4×10^{22} hydrogen atoms per cubic centimeter, which is less than the hydrogen concentration in water. However, the larger proportion of oxygen present in the concrete presents itself to compensate for the difference to some extent by acting as an additional moderator (the macroscopic cross section by fast

Table 9
COMPOSITION OF HEAVY CONCRETES FOR REACTOR SHIELDS (wt%)

Barytes concrete (density 3.5 g/cm^3)	
Barytes	60
Limonite	22
Portland cement	11
Water	7
Iron concrete (density 4.5 g/cm^3)	
Steel punchings	57
Limonite	26
Portland cement	13
Water	4

From Glasstone, S., *Source Book on Atomic Energy,* 2nd ed., D Van Nostrand, Princeton, NJ, 1968, 279, 567. With permission.

neutron removal by concrete is 0.085 cm^{-1}, compared with 0.1 cm^{-1} for water). It must be remembered that concrete stands at a much superior position compared with water in regard to attenuation of gamma radiation. This is because of the presence in the concrete some 50 wt% of elements of moderately high mass number, such as calcium and silicon.

Concretes used for radiation shielding can be classified as ordinary concrete and special concrete.[35,36] As drawbacks to the use of concrete as a reactor shield, several points can be raised. There is variability in comparison and water content of the commercially available material, and this adds to uncertainties in shield designs to predict radiation distribution and attenuation in the shield. The water content of the concrete after its use is important. Its value is not well known. It appears that 7 w/o is adequate for neutron attenuation, but a higher amount is desirable. An increase in water content decreases both the density and structural strength of the ordinary concrete. Loss of water from concrete as it becomes hot with absorption of radiations is another problem. Extraneous materials added to concrete can bring improvements with respect to retention of water content in concrete up to high temperatures. One such use is the mineral serpentine, which is a hydrous magnesium silicate.

Special concrete refers essentially to those with higher than normal density by incorporating elements of high mass number for reactor shielding in special cases. With these heavy concretes a given amount of attenuation of both neutrons and gamma rays can be achieved by means of a thinner shield than is possible with ordinary concrete. This is certainly an advantage in particular situations; for instance, in research reactors, as an experimental convenience, it is desirable to have shield penetrations as short as possible. In applications where space is not an important consideration, ordinary concrete, on account of its lower cost, is preferred over the heavy ones. The heavy concretes which have been mainly used are barytes concrete and iron concrete. In barytes concrete the mineral barytes, consisting mainly of barium sulfate, largely replaces the sand and gravel of ordinary concrete. The density of barytes concrete is about 3.5 g/cm^3, and the macroscopic cross sections for the removal of fast neutrons and the attenuation of 4-MeV gamma radiations are 0.105 and 0.10 cm^{-1}, respectively. The corresponding values for ordinary concrete are 0.085 and 0.066 cm^{-1} The barytes aggregate may be partly or totally replaced by limonite, $2Fe_2O_3{\cdot}3H_2O$, which contributes both iron and bound water to the concrete. The compositions of two fairly typical heavy concretes for use in reactor shielding are given in Table 9.

Various other cements and concretes have been devised. One, consisting of iron, Portland cement, and water, has a density as high as 6 g/cm^3. In order to increase the probability of neutron capture without high energy production, boron compounds, as for example, colemanite ($2CaO{\cdot}3B_2O_3{\cdot}5H_2O$), have also been incorporated into concretes. As alternatives to water as the hydrogenous component of a reactor shield, there has been some interest in metal hydrides. They possess features suitable for high-temperature shields. Such materials would combine both inelastic and elastic neutron scattering with gamma radiation absorption properties. Among metal hydrides, titanium hydride and zirconium hydride have received attention.

C. Boron-Containing Substances

There are a number of common boron compounds, such as boric acid, borax, and various other borates, which are sufficiently soluble to be added to a water shield to enhance neutron capture rate. Borated water has found many applications as a reactor shield.

In addition to the above, boron carbide has received wide acclaim as a shielding material. A disadvantage of this material is its poor formability. Hot-pressed or sintered bodies are limited in their possible dimensions and are also relatively expensive, while loose powder does not necessarily meet all application requirements. As an alternative to the expensive hot-pressed boron carbide sintered pellets, methods have been developed to use binders such as copper[37] and aluminum[38] for boron carbide shapes for control rod application. Use of such binders allows the reduction of boron carbide shaping temperatures considerably. Reactor experience with these shapes of bonded carbide is not extensive.

For shielding application of boron carbide, a material called boral, which is essentially a dispersion of boron carbide in aluminum, has found extensive application. The exclusive choice of aluminum as a binding material for boron carbide in shielding applications is because it is inexpensive, light, easily workable, and ductile, and has good thermal conductivity. In addition, it suffers no damage due to radiation and does not emit very hard gamma rays. There are two ways in which boral can be fabricated. One is by a powder metallurgical route, which involves mixing of boron carbide (with a small amount B_2O_3 added to it for improving wetability of aluminum with boron carbide) and aluminum, hot-pressing the powder mix, jacketing and picture-framing the hot-pressed boral with aluminum, and finally hot-rolling to sheet forms. The other route for making boral involves mixing boron carbide with a melt of aluminum at around 666°C. For improvement of wetability some additions of B_2O_3 are made in this preparative process also. The solidified mix is then hot-pressed, and following the same procedure as described for the powder metallurgy route, final shapes of boral are prepared. The flowcharts of the two routes for making boral sheets are shown in Figure 8. Apart from boral, various other materials with boron carbide incorporated in epoxy resins, silicon rubber, and graphite have been developed for neutron shielding application. Among these, the material produced by incorporating boron carbide in silicon rubber (Bocarsil) is a flexible material and hence may be used conveniently for shielding with tubular or round geometries.

An aspect of shielding against neutrons is worth mentioning. In the neutron absorption process by the shield material, some special considerations need to be given to the emission of other particles. No doubt isotopes such as ^{10}B and ^{113}Cd, which have high neutron absorption cross sections, are the first choice. When ^{113}Cd is used to absorb thermal neutrons, the process involved is radiative capture, with subsequent gamma emission. Since these gamma radiations are very penetrating, another shielding problem is created as the neutrons are absorbed. If ^{10}B is used, however, the resulting process is charged particle emission involving alpha particles. Since alpha particles are very easily shielded, no secondary shielding problem is encountered. On this basis boron-based materials can be said to score over cadmium-based matereials in neutron shielding applications.

VI. SUMMARY

The three primary control methods currently used in power reactors are (1) movable

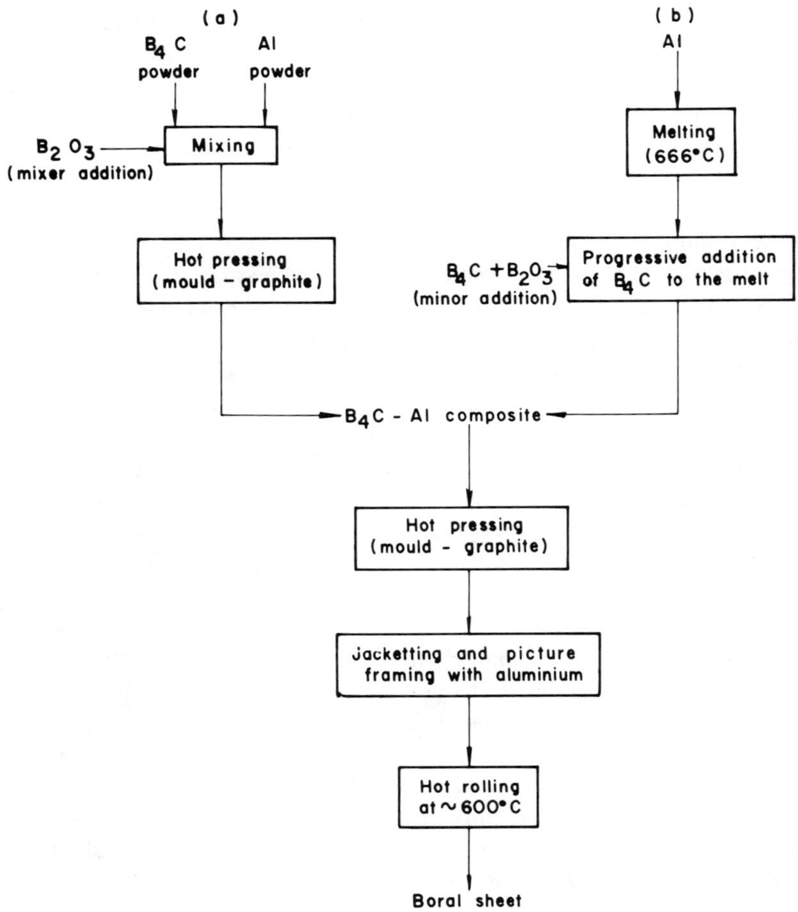

FIGURE 8. Processes for the production of boral sheet. (a) Powder metallurgy route; (b) melt route.

control absorbers, (2) burnable neutron poisons, and (3) chemical shim. In most power reactors, more than one method of control is used to take advantage of the unique characteristics of each in matching control requirements.

The most effective control materials for fission reactors are strong neutron absorbers. The results of use of such materials, however, are the loss of useful neutrons and impairment of neutron economy. Control materials for fission reactors consist mainly of (2) natural boron or enriched ^{10}B in the form of its compounds or cermets, such as B_4C, H_3BO_3, boral, or B_4C-stainless steel; (2) cadmium and its alloys, Cd-Ag, Cd-Ag-Hf, or Cd-Ag-Ir; (3) hafnium and its alloys, such as Hf-Ag or Hf-Ag-In; and (4) rare earth elements and their oxides, such as Eu_2O_3, Er_2O_3, or Gd_2O_3, dispersed either in UO_2 (as burnable absorber) or in stainless steel (as control elements). Some of these, such as B_4C and H_3BO_3, are also utilized as shielding.

Radiation shielding serves three major functions or purposes: (1) thermal shielding to protect the pressure vessel, primary coolant loop, and inner shield from intense heat liberated upon absorption of the nuclear radiation; (2) biological shielding to protect operating personnel and the general public; and (3) apparatus-instrument shielding to protect delicate electromagnetic apparatus and electronic instruments. The most important kinds of nuclear radiation, which require heavy shields, are (1) primary (fission) neutrons; (2) primary gamma rays (due to nuclear fission reaction); (3) slow and thermal neutrons; and (4) secondary

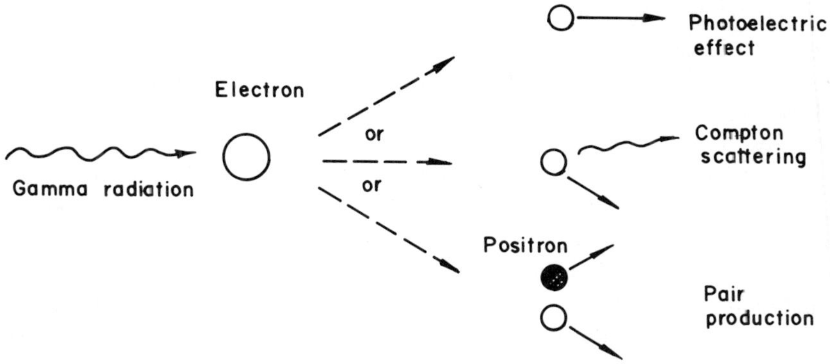

FIGURE 9. Photoelectric, Compton scattering, and pair production mechanisms of interaction of gamma radiation with an electron.

gamma rays (emitted by the capture of slow and thermal neutrons in the radiation shields).

A classification scheme for the radiations is based on the basic mechanisms by which they interact with various atoms and nuclei. The charged particles (alpha, beta, and fission fragments), electromagnetic radiation (X-rays, gamma-rays, etc.), and neutrons each behave in fundamentally different ways.

The net macroscopic effect of charged particle interactions is characterized in terms of range and linear energy transfer (LET). The range is the average distance traversed by a charged particle before stoppage, and the LET is the energy deposition per unit distance of travel [mathematically, LET = dE/dx for particle energy E and distance x; since the range R is the total distance of travel for initial particle energy Eo to be reduced to zero, R = $\int_0^{Eo}$dE/LET = $\int_0^{Eo}$dE/(dE/dx)]. The range determines the penetration distance, while the LET determines the energy distribution along the path. The fission fragments, on account of their large mass and high charge, result in a range of only a few centimeters in air and a fraction of a millimeter in solid material. The fission fragments generally stop very near their point of origin and deposit all their energy within this short distance of travel. They have, in other words, a high LET. Compared with fission fragments which have masses between 80 and 150 amu, alpha particles have a mass of 4 amu. They are doubly charged. At typical energies they have ranges only three to six times greater than fission fragments and LETs about an order of magnitude lower. This sheet of paper of the book is thick enough to stop alphas or nuclear reactor-produced fission fragments. The combination of low mass and single charge impart to electrons relative ranges about 100 times greater than those for alphas and LETs correspondingly reduced. The paths traced by electrons deviate greatly from the roughly straight paths of the heavy charged particles.

Photons of electromagnetic radiation interact directly with electrons and more rarely with nuclei. The three important mechanisms, as shown in Figure 9, are

1. The photoelectric effect, in which the photon energy is converted completely to kinetic energy of an orbital electron. Only this effect results in complete loss of an X-ray or gamma-ray photon.
2. Compton scattering, in which a photon transfers a portion of its energy to an electron and leaves the reaction at a correspondingly lower energy. The process reduces the energy and changes the direction of the incident photon. The energy imparted to the electron is dissipated as heat, as for any charged particle.
3. Pair production, in which photon energy is converted to mass and kinetic energy of an electron-position pair. This interaction can occur only for a photon whose energy exceeds the mass of two particles, i.e., twice the electron mass of 5.5×10^{-4} amu

= 0.511 MeV. When the positron ultimately stops and contacts another electron, the combined mass is converted to two 0.511-MeV photons called annihilation gammas. Thus, the net effect of pair production is the conversion of one high-energy photon into two photons of 0.511 MeV.

Figure 6, described earlier in this chapter, shows a very rough comparison of relative range and LET for alpha, beta, and gamma radiations. Since each of the important mechanisms depends more on the density of electrons than on the specific atomic composition, the relationships are relatively material dependent. These rule-of-thumb values demonstrate that gamma radiation is about 100 times as penetrating as beta particles. The betas, in turn, are more penetrating than alpha particles by about the same factor. The LET values are inversely related to the ranges.

The text has described a wide range of neutron interaction mechanisms. Of these, the absorption and scattering reactions are of the most interest in the context of radiation effects. Most absorption reactors result in the loss of a neutron coupled with production of a charged particle or gamma ray. When the product nucleus is radioactive, additional radiation comes out at some later time. Scattering reactions result in the transfer of energy from a neutron to a nucleus. The latter then interacts in the system as a charged particle. Neutron interactions may be characterized by mean free path as an average range. The mean free path values are extremely sensitive to neutron energy and material composition, and it is for this reason that generalized comparisons of neutron behavior with that of other radiations are not readily made.

The study of the mechanisms of interaction of radiation with matter has been of practical importance in the selection and development of shielding materials. The shielding materials used for nuclear radiations consist of (1) the heavy and moderately heavy elements, such as Pb, Bi, Ta, W, and Fe, to attenuate gammas and slow down fast neutrons; (2) hydrogenous materials, such as polyethylene, light water, ordinary concrete, and special concretes, which moderate and thermalize neutrons and absorb gammas; and (3) light substances containing boron, such as boric acid (H_3BO_3), borax ($Na_2B_4O_7{\cdot}3H_2O$), colemanite ($Ca_2B_6O_{11}{\cdot}5H_2O$), boron oxide ($B_2O_3$), and boron carbide, incorporated into such materials as aluminum, graphite, and silicon rubber to capture slow and thermal neutrons.

VII. NUCLEAR POWER PLANT OPERATION[39-42]

Nuclear reactor operation usually involves three discernable stages: start-up, operation at a desired power level, and shutdown.

A. Plant Start-Up

There are essentially three different types of start-up procedures:

1. Initial start-up, which applies to the starting of a new reactor or one which has been refueled. The initial start-up procedure may also be applicable to a reactor which has been shut down for several days. This may also be called start-up of a reactor from cold condition.
2. Start-up of a reactor from hot standby condition, such as after accidental or deliberate shutdown.
3. Start-up of a reactor after a severe reduction in power.

1. Initial Start-Up

The procedure of starting up a new reactor is somewhat different from that of starting up a reactor which is being operated after an extended outage. Initial start-ups are accomplished with very slow rates of changes in reactivity; control rods are moved slowly and very carefully. To cite an example, a PWR at its cold condition is started by using heat generated by the reactor coolant pumps. These pumps require about 5% of the normal electrical output of a plant. This energy is supplied from an outside source in the shutdown condition of the reactor. The system is brought to a hot condition by the pumps within 6 h by friction energy dissipated in the coolant. Borated water is then charged to the primary coolant system. It controls the neutron reactivity in the core prior to start-up and allows the reactor to operate with control rods almost completely withdrawn. The methods of controlling the reactivity in a PWR are (1) control rods, (2) boron addition to the coolant, (3) burnable poisons incorporated with the fuel, and (4) changing the coolant (i.e., moderator) temperature. Each of these four methods is used, and the choice of technique is governed by the extent to which a power change (i.e., change in reactivity) is needed and how quickly the change has to be brought about.

The next operation involves making all safety systems, including the ECCS, operational. This is followed by turning on the pressurizer heaters to form steam bubbles in the pressurizer. The bubbles greatly increase the pressure stability of the primary coolant system and allow the pressure to be regulated. During the start-up, the pressure of the primary coolant system must remain below a certain specified value for any given temperature. The pressure eventually reaches about 27 bars and is maintained until the pressurizer level reaches the no-load set point (25%). Up to this point, the reactor core is still shut down, with the control rods completely inserted.

After the preheating of the system using the main coolant pumps is complete, the reactor core is made critical. The system is brought up to low power as follows:

1. The shutdown group of control rods is taken out slowly in a prescribed sequence.
2. As criticality is approached, the count rate of the neutron monitors no longer reaches equilibrium between control rod withdrawals. The reactor coolant temperature increases, while the pressurizer spray maintains the system pressure required by the temperature-pressure specification.
3. The reactor power is then increased by further removal of control rods from the initial critical position. The control rod groups still remaining in the core are used for power range control. They are subjected to manual operation up to low power levels, and only beyond about 15 to 25% power are they put on automatic control.

As the reactor is brought to power and the system temperature reaches 205°C, steam is drawn from the steam generators. This steam is deployed for heating the secondary systems, to establish a vacuum in the condenser, and to start the turbine moving. The latter is not connected to the electrical grid at this point, and will not be until the plant start-up is completed and it is rotating at synchronous speed with the main grid. The heat-up and pressurization of the primary system continues until zero-level conditions are established (290°C), at 150 bar. Much of the description given for the PWR also pertains to a BWR plant start-up operations. The major exception is the actual sequence of operations needed to bring the plant on line. This is because of the differences in system design between a PWR and a BWR. A BWR is a direct-cycle plant, and boiling occurs directly in the core. This leads to a greater stability for BWRs and considerably different transient response. The initial approach to criticality entails the usual preoperational system status checking. The reactors are now stated to be in a hot condition, and from now on the procedures for bringing the plant to power are as described in the next section.

2. Start-Up after Hot Standby Conditions

The reactor in hot standby condition has coolant at almost the operating temperature and pressure. In addition, it also has a high neutron flux and accumulated fission products. The start-up period from this condition is considerably shortened. After verifying the status of the system, the coolant system is made ready by placing the main coolant pumps and the feed water pumps in gear, followed by starting up the condensate and condensate feedwater system. The condenser vacuum is next established. At this point, the reactor is ready to commence withdrawal of control rods (the control rods are withdrawn sequentially in groups). Unlike the PWR, a BWR is heated up and pressurized by nuclear heat via rod withdrawal from the core. The heat-up rate is usually 25 to 50°C/h. All control rods in a BWR are hydraulically actuated and are gradually removed. If the plant is on hot standby, the start-up begins at this point. As the plant reaches key pressure levels, various additional checks are made and various systems brought on line. When the plant reaches its operational set point, the turbine/generator is started up, synchronized with the grid, and loaded. The pressurizer spray and heaters are put on automatic mode. The feedwater and temperature controls remain on manual control. At this point, if the entire system is correctly functioning, the plant goes on automatic control. The power can then be increased to full power (the rate of increase is such that it does not exceed 5%/min). The time taken to attain full power is about 13 h during a normal start-up. The plants are now stated to be producing net power.

3. Start-Up after Severe Power Reduction

The start-up procedure in this condition is relatively less complicated than the two conditions outlined above. If the interval is not too long, the neutron level will be sufficiently high to start up readily.

B. Shutdown

There are essentially two ways by which a reactor is shut down: controlled and enforced (scram). Shutdowns due to malfunctioning of some instrument or component in the plant may require special rod motions. Whenever there exists a possibility, it is desirable to resort to slower means of shutdown rather than scramming.

C. System Chemistry

In all LWRs, the coolant chemistry, in both primary and secondary systems, must be controlled for good plant operation.

In BWRs, the additions of chemicals, as those for PWRs, are not made. High-purity water is used. Full-flow condensate demineralization, shutting down reactor operations when the impurity level is too high, and use of suitable materials in the circuit, are among the important procedural measures that keep the water quality in BWRs in the right condition.

D. System Transient Control

Nuclear reactors are designed for stable, steady-state operation. To acquire this, the key system parameters must be carefully controlled. The power level of the reactor is governed by a complex set of differential equations: (1) $dP/dt = (\rho - \overline{\beta}/A)P + \Sigma_i C_i$; and ($_i$2) $dC_i/dt = (\beta_i/A)P - \lambda_i C_i$, where C_i is the precursor concentration, P is the power, β_i is the delayed neutron fraction, β is the average delayed neutron fraction, A is the prompt neutron generation time, and λ_i is the parameter related to delayed neutron half-life. The key parameter in these equations for the present discussion of system transient control is the reactivity ρ. Its value at each instant determines system behavior. The net reactivity is calculated from control rod positions, thermal hydraulic feedback, and all other power feedback mechanisms of the system. It is perturbations of those reactivity mechanisms that give rise to transients, which go by such names as control rod withdrawal accident, loss-of-flow condition, turbine trip, cold water injection accident, and so on.

During the first minute of the transient, the power and moderator in the core slowly increase, until the reactor scrams; then both fall rapidly to low values. The response to transients varies considerably between reactor types, as does the reactivity in the core between hot and cold conditions, and so on.

When a transient occurs, the reactor control system automatically tries to counteract the transient and to restore the system to its original operating state. In mathematical terms, this requires that the reactivity change be brought back to zero. This can be achieved either by reversing the initiator of the transient or by modifying some of the other system parameters. An example of the former is restoring feedwater flow in a loss-of-feedwater transient. An example of the latter is partly inserting control rods to accommodate a step decrease in turbine load.

E. Power Distribution Transients

Reactors commonly undergo gross power transients during several normal operating procedures, such as reactor start-up, control rod adjustments, or load following. Three transients are accompanied by significant changes in the core ^{135}Xe inventory, which in turn causes a reactivity transient. ^{135}Xe is a fission product which accumulates in the reactor during operation. It has a very large neutron capture cross section of 3×10^6 b, many times greater than all other long-lived fission products together. The majority of ^{135}Xe comes from radioactive decay of another fission product, tellurium ($^{135}\text{Te} \xrightarrow{1\ \text{min}} {}^{135}\text{I} \xrightarrow{6.7\ \text{h}} {}^{135}\text{Xe} \xrightarrow{9.2\ \text{h}} {}^{135}\text{Cs}$). The buildup rate (6.7 h) and decay rate (9.2 h half-life) of ^{135}Xe are important in the concentration of this reactivity poison. When a reactor operates at constant power, ^{135}Xe equilibrium is reached in about 1 day. Because the amount of ^{135}Xe produced in a given part of the core is proportional to the local power and delayed in time by the 6.7-h ^{135}I decay rate (i.e., ^{135}Xe build-up rate), different parts of the reactor core respond differently as the control rods are introduced and removed at various times to adjust the power.

In order to maintain criticality during periods of changing xenon concentration, compensating reactivity must be added. In a PWR, the control rods are used for this purpose, while a BWR can adjust the moderator void fraction. Automatic control systems are available to execute control rod and void adjustment strategies. These systems considerably improve the operator's ability to respond to xenon transients. In a BWR, however, axial variations are important because of the reactivity feedback (the reactivity changes are such that they tend to destabilize the reactor, producing slow power oscillations, a mechanism analogous to the output-feeding-input mechanisms of some electronic devices) due to the voids in the core. When coolant flow decreases, the void fraction increases (i.e., decreases the reactivity) in the upper part of the core. When the flow increases, it induces exactly the opposite effect. Older BWRs maintain constant coolant flow, letting the core power set the void fraction necessary for criticality (this self-adjusting ability is an important stability and control feature of BWRs). As far as other fission products are concerned, they also build up in the reactor core during operation. They are mostly long-lived isotopes and reach a maximum concentration just before the fuel element is discharged. An exception is ^{149}Sm. This produces transients similar to ^{135}Xe. The magnitude is, however, much smaller.

The situation for reactor start-up after shutdown or scram is worth a reference. Because ^{135}I has a shorter half-life than ^{135}Xe, the xenon concentration quickly increases to a peak value. This happens because the iodine decays to xenon at a rate that is initially greater than the decay of the xenon, and xenon is no longer being lost to neutron capture. The buildup reaches its peak within 4 to 12 h after shutdown and then slowly decays. In large power reactors xenon transients following shutdown can be minimized by slowly reducing the power to burn out some of the xenon. It may be mentioned that xenon transients are smaller in power reactors than in research or other special reactors because the neutron flux is lower. Because of the fact that xenon concentration grows after shutdown, it is often desirable to start the reactor up as soon as possible, particularly after a scram (rapid shutdown). If this is done quickly, say within half an hour or so, then the xenon will not have yet built up to appreciable levels. Moreover, start-up after a brief shutdown is somewhat easier because the neutron level in the core is still high. In case a longer outage is required, the xenon concentration can be partly reduced by operating the reactor at low power levels prior to shutdown.

REFERENCES

1. **Shultz, M. A.,** *Control of Nuclear Reactors and Power Plants,* 2nd ed., McGraw Hill, New York, 1961.
2. **Tipton, C. R.,** *Reactor Handbook,* Vol. 1, New York, 1964.
3. **Glasstone, S.,** *Source Book on Atomic Energy,* 2nd ed., D Van Nostrand, Princeton, NJ, 1968, 279, 567.
4. **Murgatroyd, R. A. and Kelly, B. T.,** Technology and assessment of neutron absorbing materials, *At. Energy Rev.,* 15(1), 3, 1977.
5. **Anderson, W. K. and Theilacker, J. S.,** Neutron Absorber Materials for Reactor Control, U.S. Government Printing Office, Washington, D.C., 1962.
6. Neutron Cross Sections, Vol. 1, 3rd ed., BNL 325, Brookhaven National Laboratory, Upton, NY, 1973.
7. **Radkowsky, A.,** Theory and application of burnable poisons, in Proc. 2nd U.N. Int. Conf. Peaceful Uses of Atomic Energy, Vol. 13, United Nations, Geneva, 1958, 426.
8. **Cohen, P. and Graves, H. W., Jr.,** Chemical Shim Control for Power Reactors, 22(5), 1964.
9. **Gallagher, J. N., Jr. et al.,** Startup Experiment Program for the Yankee Reactor, YAEC-184, 1981.
10. **Thei, J. A.,** Fluid poison control of boiling water reactors, *Nucleonics,* 16(5), 82, 1958.
11. **Absalom, S. T.,** Boron, Bureau of Mines, U.S. Department of the Interior, 1979, 17.
12. **Hobb, D. Z., Campbell, T. T., and Block, F. E.,** Preparing Boron by Fused-Salt Electrolysis, Rep. Inv. 5893, U. S. Department of the Interior, Bureau of Mines, 1961, 17.
13. **Lipp, A.,** Boron carbide production, properties, application (transl.) *Tech. Rundsch.,* 14, 28, and 33, 1965; 7, 1966.
14. **Bose, D. K., Nair, K. U., and Gupta, C. K.,** Production of high purity boron carbide, *High Temp. Mater. Processes,* 7(2 and 3), 133, 1986.
15. **Pederson, E. S.,** *Nuclear Power,* Vol. 1, Ann Arbor Science, Ann Arbor, MI, 1978, 135.
16. **Thomas, D. E. and Hayes, E. T., Eds.,** The Metallurgy of Hafnium, Naval Reactors, Division of Reactor Development, U.S. Atomic Energy Commission, 1960.
17. **Chi, N., Vergnolle, J., and Remy, A.,** Production of nuclear hafnium by the Van Arkel process, *J. Nucl. Mater.,* 12(3), 310, 1964.
18. **Sharma, I. G. and Gupta, C. K.,** Studies on Electrorefining of Calciothermic Hafnium, *J. Nucl. Mater.,* 74, 19, 1978.
19. **Beudry, B. J. and Gschneider, K. A., Jr.,** Preparation and basic properties of the rare earth metals, in *Handbook on the Physics and Chemistry of Rare Earths,* Vol. 1, Metals, Gschneidner, K. A., Jr. and Eyring, L., Eds., North-Holland, Amsterdam, 1978, chap. 2.
20. **Gschneidner, K. A., Jr. Ed.,** *Industrial Applications of Rare Earth Elements,* ACS Symp. Ser. 164, American Chemical Society, Washington, D.C., 1987.
21. **Murthy, T. K. S. and Gupta, C. K.,** Rare earth resources, their extraction and application, in *Science and Technology of Rare Earth Materials,* Subbarao, E. C. and Wallace, W. E., Eds., Academic Press, New York, 1980, 3.
22. Radiation Protection Procedures, Safety Ser. International Atomic Energy Agency, No. 38, Vienna, 1973.
23. **Friedlander, G. and Kennedy, J. W.,** *Nuclear and Radiation Chemistry,* John Wiley & Sons, New York, 1955, chap. 7.
24. **Glasstone, S.,** *Principles of Nuclear Reactor Engineering,* D Van Nostrand, Princeton, NJ, 1955.
25. **Rockwell, T.,** Reactor Shielding Design Manual, Rep. TID-7004, U.S. Atomic Energy Commission, Washington, D.C., 1956.
26. **Goldstein, H.,** The Attenuation of Gamma Rays and Neutrons in Reactor Shields, U.S. Government Printing Office, Washington, D.C., 1957.
27. **Grotenhuis, M.,** Lecture Notes on Reactor Shielding, Rep. ANL-6000, U.S. Atomic Energy Commission, 1962.
28. **Blizard, E. P., Ed.,** *Reactor Handbook, Vol. 3 (Part B), Shielding,* Interscience, New York, 1962.
29. **Ma, B. M.,** Heat generation and temperature distribution in cylindrical reactor pressure vessels, *Nucl. Eng. Des.,* 11, 416, 1970.

30. **Jaeger, R. G., Ed.,** *Engineering Compendium on Radiation Shielding,* Springer-Verlag, Berlin, 1968.
31. **Schaeffer, N. M., Ed.,** Reactor Shielding for Nuclear Engineers, Rep. TID-25951, U.S. Atomic Energy Commission, Washington D.C., 1973.
32. **Weissman, J.,** *Elements of Nuclear Reactor Design,* Elsevier, New York, 1977.
33. **Jaeger, R. G.,** *Engineering Compendium on Radiation Shielding, Vol. 111, Shield Design and Engineering,* Springer-Verlag, Berlin, 1975.
34. **Davies, N. S.,** *Reactor Handbook,* Interscience, New York, 1964.
35. *J. Am. Concrete Inst.,* 29, 965, 1958, *Concrete for Reactor Shielding,* 2nd ed., American Concrete Institution, 1962.
36. **Tipton, C. R., Ed.,** *Reactor Handbook,* Vol. 1, *Materials,* Interscience, New York, 1960.
37. **Rudiger, O. et al.,** B_4C-Cu-Verbundwer Kstoffe als Neutronenabsorber fur Abschalstabe, *Tech. Mitt. Krupp,* 22, 41/47, 1964.
38. **Suri, A. K. and Gupta, C. K.,** Studies on the fabrication of aluminum bonded boron carbide rings, *J. Nucl. Mater.,* 74, 297, 1978.
39. **Szeless, A.,** Performance of Nuclear Power Plants, *Power Eng.,* 86, 54, 1982.
40. **Masche, G.,** Systems Summary of Westinghouse Pressurized Water Nuclear Power Plant, Westinghouse Electric Corporation, 1973.
41. **Corcoran, et al.,** The critical safety functions and plant operation, *Nucl. Technol.,* 55, 690, 1981.
42. **Skarpelos, J.,** Factors influencing BWR water quality, *Trans. Am. Nucl. Soc.,* 43, 321, 1982.

Chapter 5

RADIOACTIVE WASTE MANAGEMENT

I. INTRODUCTION

The electricity generation industry, like all industries, produces wastes. Very large amounts of waste are generated from the burning of coal. A large power station emits over 100,000 tons of noxious material into the atmosphere and produces nearly 1 million tons of ash each year. Tall stacks ensure that atmospheric emissions are widely dispersed and thus diluted to very low concentrations. Some ash is used commercially, and the remainder is disposed of, generally to landfill sites. In comparison, the nuclear power industry is characterized by the large amount of energy available from relatively small amounts of fuel; 1 ton of uranium in modern nuclear stations produces as much electricity as 25,000 tons of coal. It is thus seen that the resultant quantities of waste are relatively small. However, many of the constituents are radioactive. These emit radiation and must be managed so as not to endanger man and other living organisms. A recent publication has put into perspective the health risks associated with the disposal of radioactive waste.[1]

In many industries the effects on workers, the population, and the environment have been recognized only gradually over a period of time, during which successively improved techniques have been introduced. This has not been the pattern of events in the nuclear industry. From the start, the nuclear industry has always devoted such attention to the development of safe waste management. Those involved have been particularly conscious of the possible hazards associated with radioactive materials. Extremely high standards of protection of individuals and the environment have been instituted. The tight control has ensured that exposure of the public to radiation from all industrial sources, including the nuclear industry, has always been below the recommended limits. Such sources contribute only a small fraction of the total radiation to which men are exposed. Most of this comes from outer space; from naturally occurring radioactive materials in the earth, in buildings, and in people's bodies; and from medical uses of radiation. The literature provides an exhaustive account of the radiations.[2]

In order to manage the radioactive wastes, we have to decide on several key points, for example, whether to disperse them into the atmosphere or the sea or to isolate them from man; whether to process sludge by dewatering; and whether to remove activity from solid waste, thus producing a secondary waste. There are many other decisions about treatment, packaging, storage, and disposal to be made on the basis of technical information, and taking into account national and international laws and regulations. It is, however, not sufficient to manage waste efficiently; it is also essential to respond to the natural concern of the general public and to satisfy them that the wastes are and will be managed safely. Thus, the management of radioactive waste, which for the first 20 years or so of the nuclear age was regarded as a matter of good housekeeping, has now broadened almost into a discipline in its own right, embracing subjects as diverse as radiation biology, corrosion, chemical engineering, and computer modeling.

This chapter covers the management of the radioactive wastes arising from every stage of the nuclear fuel cycle from mining the uranium to reprocessing the spent fuel after its removal from a reactor. The chapter describes current practices and the methods being developed to ensure that waste management can be carried out with no significant threat to current or future generations. The presentation does not include management of waste arising from decommissioning of nuclear reactor plants.

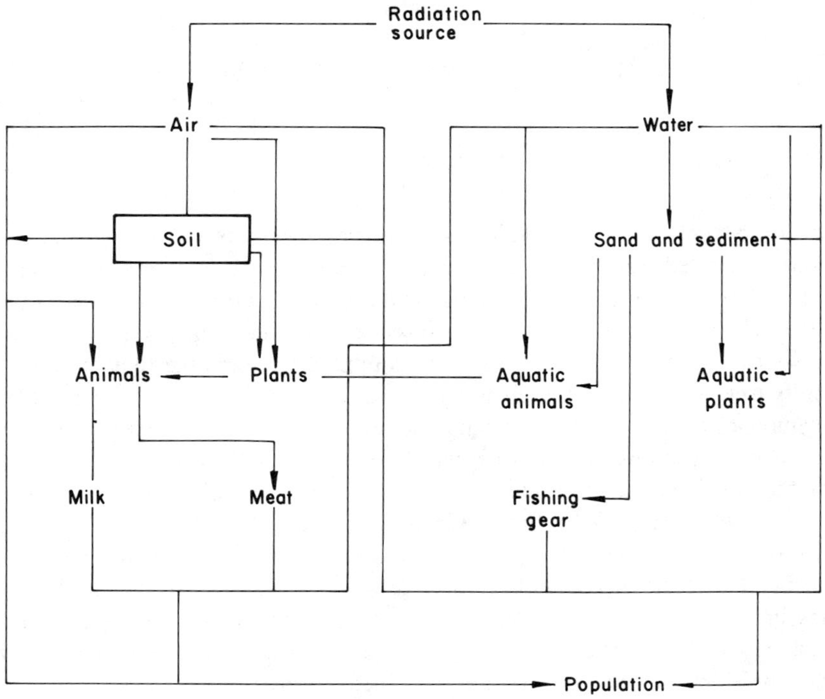

FIGURE 1. Radiation pathways.

II. WASTE MANAGEMENT OBJECTIVES AND PRINCIPLES

The objectives of radioactive waste management must be (1) to comply with the principles of radiological protection for present and future generations, (2) to minimize any impact on future generations to the maximum extent practicable, (3) to preserve the quality of the natural environment, and (4) to avoid preempting present or future exploitation of natural resources. There is wide international agreement in respect to these objectives of waste management.

The principles of radiological protection that are applied throughout the world are based on the recommendations of the International Commission on Radiological Protection. The commission sets upper limits of radiation exposure, or dose, for workers and the public. Limits for radiation workers are set by comparison with the risks of other occupations having high standards of safety. Limits for the public are set by comparison with risks in everyday life. For all except minor disposals, a detailed analysis is carried out of all the possible pathways by which exposure of people might result. The few pathways that will contribute most to radiation exposure are referred to as critical pathways. An idea is given in Figure 1 of radiation pathways. The levels of radionuclides in the key pathways are monitored by the nuclear industry and the authorizing governmental departments, and the results are regularly published.

Among the ways in which the objectives of radioactive waste management can be translated into a practicable scheme, the important ones are (1) to minimize waste arising at source, (2) to dilute and disperse low-level waste to the environment, (3) to retain in store until decay of activity to low levels, and (4) to isolate the long-lived radioactive components.

Regarding minimizing waste arising at the source, it may first be pointed out that the actual volume of the hazardous components of the waste — the radionuclides — is very small. The bulk of the waste consists of common nonradioactive materials such as would be found in any laboratory and industrial operation, though these will be carried with radionuclides. In order to simplify the management of the waste, it is very desirable to maintain a high standard of housekeeping and keep the bulk of the waste low to the extent possible. Operators, for instance, can aim to avoid introducing any unnecessary inactive material into the active area of the reactor or reprocessing plant. Use of reagents that create problems in dealing with the waste should be avoided. It is desirable that the radioactive plant and equipment design lay adequate emphasis on high reliability, since maintenance and replacement inevitably generate waste, much of which is bulky. The operations that are not necessary on radioactive waste should be avoided. Until defining the form in which wastes are to be finally disposed off, it is better to provide safe storage in an interim condition than to have to rework or backtrack at some future date, thereby adding to the bulk of waste and incurring the risk of additional radiation dose to operating personnel.

The safety aspect of the procedure of dilution and dispersion to the environment relies on dilution being sufficiently large and rapid that the concentration of activity gives rise to no appreciable risk to humans by any route. It is used only in those situations where the total quantity of activity is low. It is recognized as a very important route, since it permits the disposal of large volumes of waste that contain little or no activity.

The approach involving retention in storage until radioactive decay can be particularly useful in dealing with wastes in which all the radionuclides possess relatively short half-lives. In this case it is often advantageous to collect and store such wastes. With the passage of time the activity can be reduced to safe levels at which the waste can be safely dispersed into the environment. In this way, exposure to the operators is avoided if the waste were to be processed, or to the public if the materials were immediately released into the environment.

Isolation of waste implies keeping large quantities of radioactivity isolated from the human environment until the activity has decayed to a level at which there is no significant risk. One method of isolation is storage. The term storage means retention of radioactive waste in engineered facilities under supervision. It is, however, practicable as long as the stores are maintained. A decision can finally be made to dispose of such waste in a permanent manner that does not depend for its safety on even the minimal degree of supervision and maintenance that a store would require. The term disposal in radioactive waste management is understood as the release or emplacement of waste materials without the intention of being able to retrieve them. As a first step toward isolation, the waste is immobilized, which means conversion to a suitably stable, insoluble, monolithic, dust-free, and nonflammable solid form. Isolation is then further accomplished by appropriately locating the wastes, so that natural barriers will also be present to impede an accidental return of activity to humans.

III. WASTE SOURCES AND CLASSIFICATION

A. Sources

The first radioactive wastes are probably those emanating from the manufacture of gas mantles and the painting of luminous marks on watch dials. These were, however, not initially recognized as presenting a special problem. The first substantial quantities of radioactive waste arose from the manufacture of nuclear weapons, but at present the sources of waste are very diverse and include (1) uranium ore mining, (2) nuclear fuel manufacture, (3) the operation of reactors, (4) spent fuel reprocessing, (5) the decommissioning of nuclear facilities, and (6) the separation of radioactive isotopes and their use in medicine, manufacturing industry, universities and research establishments, etc. The main wastes arising at a power station during the operating life of a reactor are (1) components which are removed during refueling or maintenance, which are mainly activated solids, e.g., stainless steel containing ^{60}Cd and ^{63}Ni, etc.; (2) materials such as filters and ion-exchange resins which are used to clean the reactor coolant and ventilation gases, which will be contaminated with

Table 1
CLASSIFICATION OF RADIOACTIVE WASTE

Type	Ci/cm³	Ci/gal	Ci/g
LLW	MPC[a] to 0.003 μCi	MPC to 10 μCi	< 10 nCi
ILW	0.003—100 μCi	10 μCi to 0.4 Ci	—
HLW	Above 100 μCi	Above 0.4 Ci	—
Transuranic waste	—	—	> 10 nCi of transuranic elements

[a] Maximum permissible concentration of a radioisotope in water or air.

fission products and activated corrosion products from circuits containing liquid coolants; (3) slightly radioactive gases and liquids which can be discharged into the atmosphere, rivers, or seas; and (4) cleaning materials and clothing, which can be disposed of by shallow land burial.

B. Waste Classification

Radioactive waste may be categorized[3] according to (1) the type of radiation given off (alpha, beta, or gamma); (2) the level of radiation (dose rate); and (3) the physical and chemical forms of the waste. Another way by which radioactive waste is classified is as follows: (1) high-level waste (HLW); (2) intermediate-level waste (ILW); (3) low-level waste (LLW); and (4) gaseous waste. These classifications are based on the type and degree of activity of the waste. They are defined in terms of the concentration of radioactivity they contain, as given in Table 1. In recent years the intermediate category has become largely disused and put together with LLW, while HLWs sometimes also contain transuranic elements.

C. General

The origins of HLW and LLW are quite different and distinct. HLWs are fission products produced when nuclei are fissioned in reactor fuel. This waste may take the form of separated fission products that are the waste stream from a spent fuel reprocessing plant, or, if the so-called throwaway fuel cycle should be adopted, the entire spent fuel assemblies become HLW, including their content of unfissioned ^{235}U and plutonium. HLW is relatively small in volume, but stays highly radioactive for many years.

The transuranic wastes come from fuel reprocessing, although nearly all are from military weapons production. These wastes are primarily alpha particle emitters. Figure 2 generally depicts waste arising in a nuclear fuel cycle. As with all mining activities, huge amounts of waste material are produced at uranium mines. The radioactive content is small compared with wastes from other main points at which waste originates in the nuclear fuel cycle. Some constituents have, however, long lives. All that has been done is to move some material from one part of the earth's crust to another, removing the uranium in the process. However, the fact that the material has been brought to the surface and finely ground increases the possibility of its being dispersed and reaching man. Further, the radioactive gas radon, a decay product of uranium, is able to escape more easily from disturbed material. The ideal procedure would be to return the wastes to the original mines, but this is seldom economical. In practice, the wastes need to be stabilized, generally by earth cover and revegetation, to prevent wind and water erosion and to ensure that the public health hazard is maintained at a very low level.

In fact, the total quantity of uranium mine wastes associated with the generation of a given quantity of electricity is small compared with the coal mining spoil associated with the same quantity of electricity. The introduction of fast reactors, which can ultimately

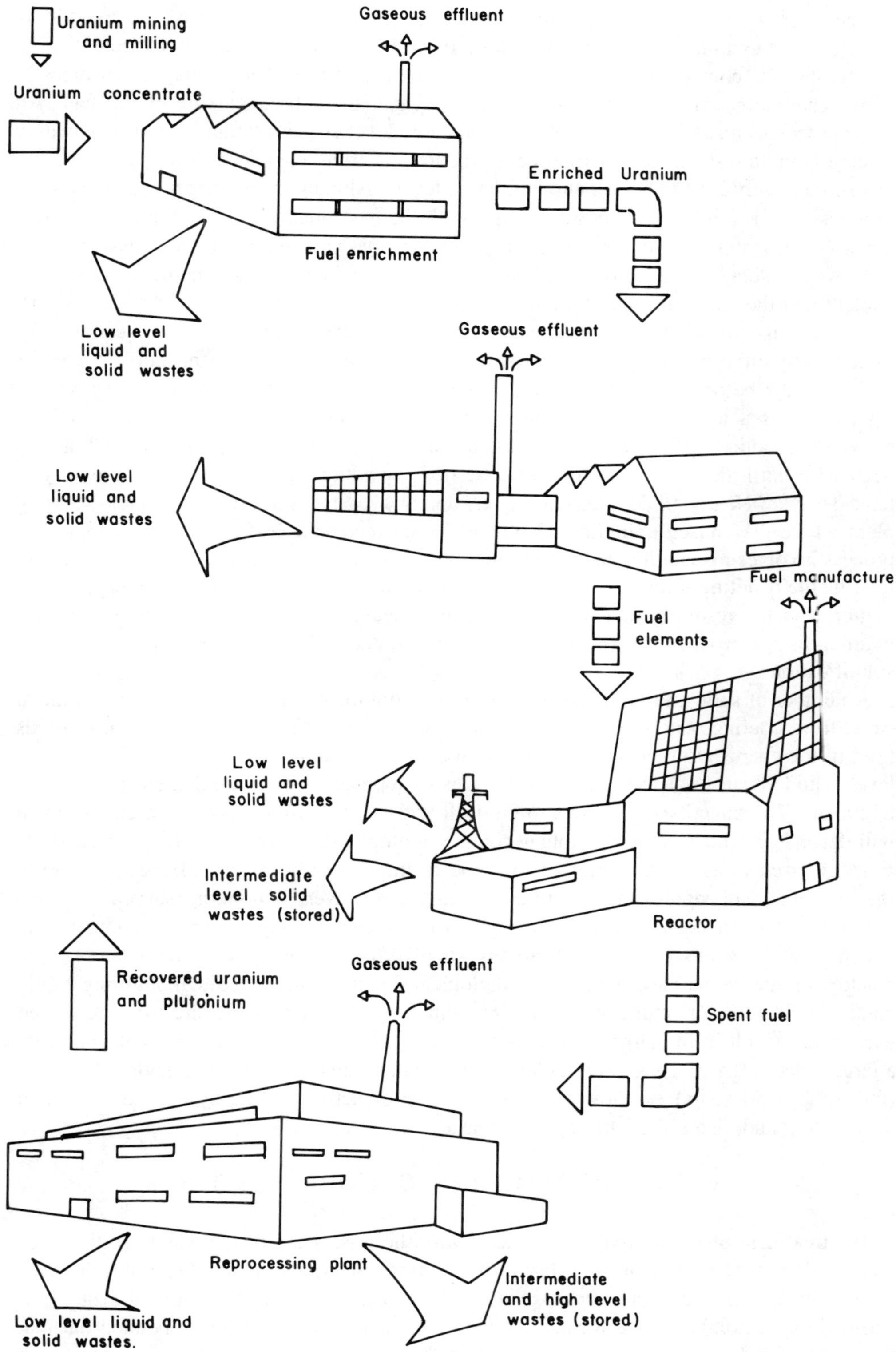

FIGURE 2. Sources of various radioactivity types in the nuclear fuel cycle.

extract over 50 times more energy from each ton of uranium, would require even less mining activity. The uranium, concentrate extracted from the ore needs to be purified, enriched to the fissile ^{235}U component, fabricated into fuel rods, and placed into small metal tubes or cans. These function as sealed containers throughout the subsequent life of the fuel. All these processes result in the production of some low-level effluent streams containing mainly traces of uranium and its radioactive decay products in various chemical forms. These effluent streams add very little to local natural radiation levels. Although reactors generate the fission products and actinides that contribute by far the largest quantities of radioactive wastes, virtually all of this is contained in the fuel and its only extracted at the reprocessing plant.

However, some wastes also result from activation reactions between the neutrons and the coolant and the structures surrounding the fuel. In addition, occasional mechanical or corrosion damage to the fuel's metal cans can result in the release of small quantities of radioactivity either in the reactor or in the cooling pounds where the spent fuel is stored for some months before being transported to the reprocessing plant. Some of this activity is discharged as a low-level effluent stream from the reactor site. A large fraction of the activation products will remain in the mechanical components of the reactor and will not be dealt with until the reactors are decommissioned, by which time much of the activity will have decayed. Nearly all the radioactivity created in a reactor is transferred to a reprocessing plant where the unused uranium and plutonium are recovered to fuel future reactors. The process involves dismantling the metal cans, dissolving the contents in acid, and chemically treating the resulting solution. The gaseous fission products are filtered to remove particulate matter, and the resulting low-activity gases are released. The solution containing the remaining fission products and the actinides is concentrated and stored in cooled tanks pending solidification.

A number of solid and liquid wastes also arise from the dismantling and storing of metal cans, from cooling waters, and from the treating of released gases and the various liquids used in the chemical processes. These ILW are stored pending further treatment. The low-level liquid wastes that arise at various stages of reprocessing are discharged to sea after treatment. The quantities of wastes to be dealt with after facilities have been closed down will depend on which of the possible decommissioning options are pursued. Stage 1, sometimes referred to as 'mothballing', leaves the facility virtually intact and does not lead to the production of significant quantities of waste. However, continuing surveillance and maintenance are necessary. It may be decided to advance to stage 2, or 'entombment', in which everything other than the relatively small primary containment (for example, the reactor pressure vessel) and the concrete biological shield is removed. Site monitoring would continue. Stage 3, or return to a condition suitable for release of the site for unrestricted use, would result in the largest volumes of wastes, typically many thousands of tons from a large reactor. There are good radiological and engineering grounds for delaying this stage (for perhaps 50 years) to allow the bulk of the radioactivity to decay and so ease final dismantling and disposal of the resulting waste.

IV. TREATMENT OF RADIOACTIVE WASTE[3-7]

The treatment of radioactive wastes deals with the processing of the wastes so that they are made easier to handle and less dispersable during transportation and disposal; also, they are less mobile after disposal. In most cases, the method involved is for immobilization of wastes in a monolithic waste form, such as glass, cement, etc. It may be pointed out that treatment of radioactive waste often results in the volume of the final wastes being much larger than that of the waste being treated. The treatment processes are used because they impart to the waste better physical characteristics and because it is often possible to secure concentration of most of the radioactivity in a relatively small fraction of the treated wastes, which is less expensive and easier to handle. It is the intent of this section to identify and describe the treatment options available for the types of wastes for which an appraisal has been given in the foregoing section. The search for improved processes for treatment of radioactive wastes and their disposal is continuing in various nuclear laboratories around the world, as reflected in some of the recent publications.[8,9]

A. High Level Waste

HLWs originate almost entirely from the fuel elements themselves. It is a concentrated waste that results from the reprocessing of spent nuclear fuel and accounts for more than 95% of the total radioactivity which results from the nuclear industry's waste products. It is easily conceivable from the mode by which HLWs are generated that there will be two types: one can be placed in the category of highly active solid waste and the other in the category of highly active liquid waste. In the field of waste management, it may be mentioned that the treatment of HLWs is by far the most extensively investigated topic.

B. Highly Active Solid Waste

The major highly active solid waste, generated during reprocessing operations is the fuel element cladding which has either been removed from the fuel mechanically (for example, Magnox) or remains at the end of the acid leaching process as applied to oxide fuels clad in stainless steel or zircaloy. The volume of fuel cladding waste generated lies in the range of 0.25 to 0.6 m^3/t of fuel processed. In the case of zircaloy cladding, which can be expected to form the major fraction of this waste in the period from 1975 to 2000, about 0.33 m^3 (about 330 kg) of cladding may be expected to be produced per tonne of fuel processed. Each cubic meter of this cladding will be associated with about 10,000 Ci of fission product activity after 1 year of postirradiation cooling (about 2500 Ci/m^3 after 10 years of cooling).[10]

At present, all cladding waste is stored (Magnox and zircaloy are stored under water to avoid a fire hazard), and the following methods of treatment are under investigation for conditioning such waste for reuse, long-term storage, or disposal: packaging after a cooling period of perhaps 10 years; decontamination using aqueous reagents or molten salts; densification by compaction or melting (maximum densification of zircaloy is achieved by melting); and reduction of specific surface areas (casting will provide a reduction of the specific surface area of zircaloy cladding from 100 m^2 to 1.3 m^2/t of fuel processed).

C. Highly Active Liquid Waste

Highly active liquid waste is the liquor remaining after the fuel has dissolved in nitric acid, and the uranium and plutonium values have been extracted. The liquid contains 99% of all the remaining nonvolatile fission products, almost all of the higher actinides from the irradiated fuel, small amounts of unextracted uranium and plutonium (typically they are about 0.1 to 0.5% of the original contents in the fuel), some inactive constituents derived from dissolution of small amounts of fuel cladding, and corrosion products from plant vessels. Although a relatively small volume of liquid is involved, it requires special consideration because of the high level of radioactivity and very high heat output. Naturally, the radioactivity of such wastes diminishes constantly, despite the fact that each radioactive component produces daughter isotopes, which themselves are also radioactive. For example, Table 2 gives the characteristics of fission products and actinides in fuel irradiated in CANDU reactors to 7.5 MW-day/kg uranium.[11]

In reprocessing operations about 5 m^3 of highly active liquid waste is produced per tonne of fuel reprocessed. In the case of high-burnup oxide fuel, this can be concentrated to about 0.25 m^3/t by evaporation for storage in the liquid form.

The basic principles relating to the management of highly reactive liquid waste which are now widely accepted are the following: (1) present methods for conditioning and storage

Table 2
ACTIVITIES OF SELECTED FISSION PRODUCTS AND ACTINIDES IN IRRADIATED FUEL[11]

Radionuclide	Mode of decay	Half-life years	Activity (Ci/kg U) At discharge	After 1 year	After 10 years
			Fission Products		
^{3}H	β	12.3	0.17	0.16	0.10
^{85}Kr	β,γ	10.7	2.22	2.19	1.23
^{89}Sr	β,γ	0.14	443	3.95	9.8×10^{-20}
^{90}Sr	β	29	17.5	16.0	12.9
^{91}Y	β,γ	0.16	578	77	1.1×10^{-16}
^{95}Zr	β,γ	0.18	825	17.3	1.3×10^{-14}
^{95}Nb	β,γ	0.10	802	36.6	2.9×10^{-14}
^{99}Te	β,γ	2.1×10^{5}	3.4×10^{-3}	3.4×10^{-3}	3.4×10^{-3}
^{106}Ru	β	1.0	182	101	0.21
^{129}I	β,γ	1.6×10^{7}	7×10^{-6}	7.9×10^{-6}	7.9×10^{-6}
^{131}I	β,γ	0.02	525	1.2×10^{-11}	0
^{134}Cs	β,γ	2.17	16.9	11.3	0.55
^{135}Cs	β	2.3×10^{6}	4.5×10^{-5}	3.8×10^{-5}	3.8×10^{-5}
^{137}Cs	β,γ	30.2	25.3	24.8	20.2
^{144}Ce	β,γ	0.78	424	181	0.06
^{147}Pm	β,γ	2.6	58.9	50.7	4.7
			Actinides		
^{237}Np	α,γ	2.1×10^{6}	2.1×10^{-5}	2.1×10^{-5}	2.2×10^{-5}
^{238}Pu	α,γ	87.7	7.2×10^{-2}	8.3×10^{-2}	8×10^{-2}
^{239}Pu	α,γ	2.4×10^{4}	0.15	0.15	0.15
^{240}Pu	α,γ	0.24	0.24	0.24	0.24
^{241}Pu	β,γ	14.7	22.9	21.8	14.2
^{241}Am	α,γ	432	11.5×10^{-3}	4.7×10^{-2}	0.3
^{243}Am	α,γ	7380	5.3×10^{-4}	5.3×10^{-4}	5.3×10^{-4}
^{242}Cm	α,γ	0.45	2.58	0.44	8.9×10^{-6}
^{244}Cm	α,γ	18.1	1.6×10^{-2}	1.5×10^{-2}	1.1×10^{-2}

are entirely satisfactory for the protection of the human environment in the time scale of several decades; (2) in the longer term, additional study is required to provide the most effective system for the isolation of this waste from the human environment for the very long periods of time that are necessary; (3) large-scale transportation of liquid HLW is to be avoided; (4) high-level liquid wastes should be solidified at the earliest practicable time for safe extended storage; (5) calcination is not an ideal form of conditioning for long-term storage or final disposal; vitrification is preferred; and (6) in the long term, final disposal is envisaged.

The basic sequence of operations in planning for the highly active liquid waste as adopted in most countries with nuclear power program is (1) interim storage in liquid form; (2) solidification and packaging after a period of cooling; (3) engineered storage of solidified and packaged waste for further cooling; and (4) final disposal of the solidified and packaged waste, after a period of cooling, into geological formations or the ocean.

D. Highly Active Liquid Storage

The waste is stored in stainless steel tanks. The stainless steel tanks are contained in concrete vaults with stainless steel to provide further containment in the highly improbable event that the primary container should fail. The space between the tanks and the vaults is

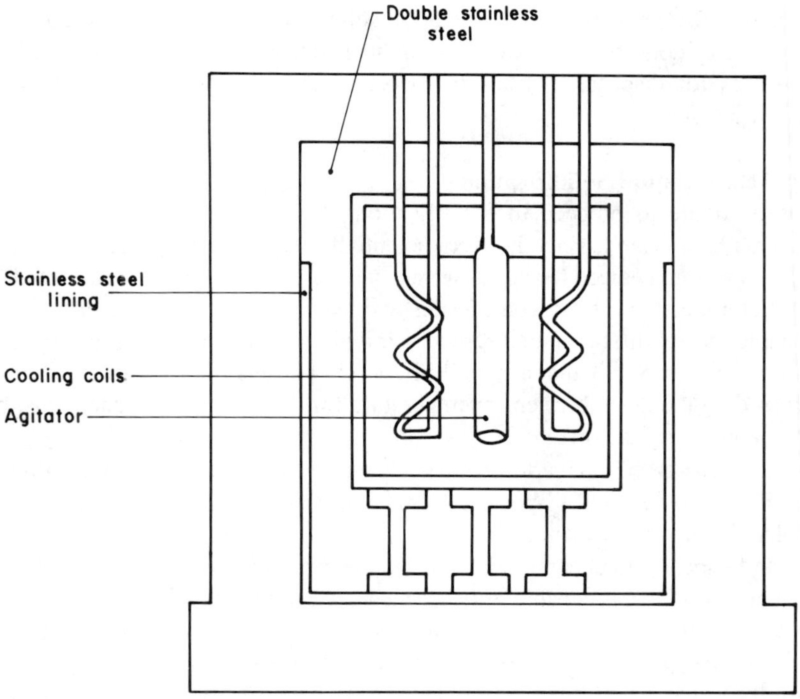

FIGURE 3. Storage tank for high-level liquid wastes.

monitored, and provision is made for transferring the contents to spare tanks should the need arise. Heat is removed by several independent sets of cooling coils. Storage of highly active liquid waste in stainless steel tanks has been practiced for over 25 years, and the record has been extremely good.

Figure 3 shows a schematic of a storage tank for high-level liquid wastes. The reinforced concrete vaults, typically 2 m thick, protect the operators from direct radiation. Provided cooling is maintained, there are essentially no radiological consequences. In the extremely unlikely event of a total loss of coolant, which is estimated to have a probability of occurrence of 1 in 1 million for each year of operation, it would take hours for the contents to boil and days for them to evaporate, allowing ample time to take remedial measures. While absolute safety can, of course, never be guaranteed, the risk of a major release is clearly small enough for the storage method to be deemed acceptable, pending conversion of the liquid waste into a stable solid. The liquid form of waste storage has the advantage of operational flexibility and leaves open the widest range of options for future treatment and disposal, but such a practice can be regarded as no more than an interim measure. Although experience has shown that the practice is safe, the tanks and cooling systems need supervision and eventual replacement.

Although the storage of high-level radioactive liquid wastes in underground tanks has been fairly satisfactory, it has been recognized from the beginning that this method was an expedient and that it has serious drawbacks. To begin with, the tanks occupy considerable space, and although they have been located in remote areas, the demand for storage space is continually increasing. There is always some danger of leakage. The stuctural and chemical integrity of tanks residing for long periods in storage in the presence of long-lived radioactive fission products present considerable concern. Excessive heating due to absorption of alpha and gamma radiations may even cause the stored highly active liquid in the tank to boil. There is a wide international consensus that these liquids should be converted to a solid. Ultimate storage of HLWs as liquids is now considered obsolete, and new techniques are being developed for depositing such wastes as solids. In this approach the advantages are (1) reduced waste mobility in the event of an accident, (2) less supervision, (3) ability to transport wastes for disposal, (4) ability for some further reduction in volume, and (5) economic advantages.

E. Highly Active Liquid Solidification

It seems desirable to proceed to solidification of liquid waste immediately after it is produced. This is indicated, since it offers potentially more safety in storage and a possible economy by avoiding the need for liquid storage tanks. However, the desirability of retaining some liquid storage stems from considering two aspects. One is that it provides a buffer stock to enable the solidification and reprocessing plants to be operated flexibily at a higher overall efficiency. Second is that a period of cooling permits the use of larger glass blocks without exceeding the desirable center temperature limits. A careful balance must be struck between the competing factors.

The waste needs to be converted to a solid whose properties, in conjunction with the characteristics of the selected disposal system, will ensure that the radioactive constituents remain isolated from the environment. The requirements of the solid form should be that it (1) has a good capacity to accommodate all the elements in waste and cope with variations in waste composition; (2) is capable of being made safely and reliably in a remotely operated and maintained plant; (3) has good resistance to radiation damage; (4) has a high thermal conductivity to dissipate the heat produced by radioactive decay; (5) has a good resistance to leaching by water, since this controls the rate at which activity could disperse; and (6) has good mechanical integrity. The following sections enumerate briefly the waste treatment processes.

1. Drying and Calcination Processes

The calcination processes convert the concentrated liquid HLW with a dry oxide powder. The various systems that have been operated include (1) a fluidized bed calcination system, (2) pot calcination, (3) a rotary kiln calciner, (4) spray calcination, and (5) wiped/scraped film evaporator.

In the fluidized bed calcination route, the liquid HLW, along with atomizing air, oxygen, kerosene, and fluidizing air, are injected into a chamber where the liquids are evaporated and the resulting solids in air suspension move through the chamber until they grow sufficiently large to descend to the bottom. There they are removed to a canister where they are baked to eliminate residual nitrates. In the pot calcination process, liquid HLW is injected into a metal cylinder where it is heated to 900°C. The liquids are vaporized, and the oxides remain in the cylinder. The cylinder is continuously fed with liquid HLW until it is filled with the desired quantity of the calcined product. In the rotary kiln calcination process, use is made of a slightly sloped, externally heated, and rotated calciner into the high end of which waste is injected. In traversing to the lower end, the calciner input waste material converts to a finely divided oxide powder which is collected in a canister which is baked at 900°C to remove nitrate residuals. In the spray calcination process, the liquid HLW is atomized into a heated chamber where it is calcined, and drops down into the storage canister and the furnace to bake out residual nitrates. In the wiped/scraped film evaporator process, the calcination takes place in the storage canister. The difference between the pot calcination process and this process is that this one uses an evaporator with a wiper rotating at 1600 rpm for the purpose of concentrating the incoming waste of a thick, flowable slurry or a free-flowing solid prior to the high-temperature calcination.

2. Vitrification Processes

The term vitrification means incorporation of radioactive waste oxides into glass.[12,13] The

Table 3
TYPICAL COMPOSITIONS OF CALCINED HLW AND BOROSILICATE GLASS[14]

Category	Constituent	kg/MTIHM[a]	Category	Constituent	kg/MTIHM[a]
		Calcined HLW			
Fission products	Rb_2O	0.354	Inerts	Na_2O	—
	SrO	1.059		Fe_2O_3	1.511
	Y_2O_3	0.598		Cr_2O_3	0.345
	ZrO_2	4.944		NiO	0.141
	MoO_3	5.176		P_2O_5	0.672
	Tc_2O_7	1.291		MnO_2	—
	RuO_2	2.972		Gd_2O_3	—
	Rh_2O_3	0.480		Zircaloy-2 fines (ZrO_2)	—
	PdO	1.483			
	Ag_2O	0.088			
	CdO	0.097	Actinides	U_3O_8	1.169
	TeO_2	0.725		NpO_2	0.865
	Cs_2O	2.880		PuO_2	0.010
	BaO	1.567		Am_2O_3	0.181
	La_2O_3	1.480		Cm_2O_3	0.040
	CeO_2	3.323			
	Pr_6O_{11}	1.482			
	Nd_2O_3	4.522			
	Pm_2O_3	0.123			
	Sm_2O_3	0.924			
	Eu_2O_3	0.200			
	Gd_2O_3	0.137	Total		40.84

Borosilicate Glass

Constituent	Wt-%
SiO_2	27.3
B_2O_3	11.1
Na_2O	4.0
K_2O	4.0
ZnO	21.3
CaO	1.5
MgO	1.5
SrO	1.5
BaO	1.5
Waste	26.3

[a] Metric tonnes initial heavy metal.

process has been demonstrated to meet all the requirements for high-level radioactivity incorporated in solid form as pointed out earlier. Considerable research has been done to show that the best type of glass for waste immobilization is borosilicate glass. Glass can take on a wide range of elements because it has no very definite structure. Compositions have been found that will accommodate all the nongaseous radionuclides and trap elements in waste and that can contain up to 25 wt% of the waste oxides. Typical compositions of calcined HLW and borosilicate glass are given in Table 3.

The melting points of glasses are in the range of 900 to 1000°C. This is well within the experience of the existing glass-producing industry and acceptable in radioactive plant design.

A number of other attributes of the borosilicate glasses warrant mention. They have proved very resistant to the effects of heat and radiation. A phenomenon known as devitrification takes place for all glasses when held at elevated temperature in the dry state over a long period. The borosilicate glasses as developed for waste retention purposes do not exhibit recrystallization readily, and, when made to do so by holding for a long period above 500°C, the crystallinity produced does not reduce the strength of the glass block significantly. In some cases it may even induce an increase in strength. The glass block retains its coherence with no alteration of its leach-resistant characteristics. It has been found to show irradiation resistance. Published work is replete with findings that radiation doses found necessary to affect borosilicate glass adversely will never be reached in the lifetime of the radioactivity incorporated in it. The leach resistance of borosilicate glasses is very good, and by any normal standards the material can be regarded as insoluble.

The standing of borosilicate glass in its functional role for taking care of HLW is undoubtedly good. However, alternative processes to calcination and vitrification of HLW to a number of other waste forms have been or are being studied in various laboratories. These include ceramics, glass-ceramic composite materials such as glass pellets embedded in a metal matrix, and synthetic crystalline minerals.[12,15,16] The objective underlying this continuing research and development is to look for materials with even greater thermal stability and resistance to leaching. An interesting concept is the incorporation of beads of glass or ceramic oxide containing the oxide in a metallic lead matrix. The metal matrix provides an additional leach-resistant barrier, possesses good thermal conductivity, and thereby facilitates dispersal of heat produced inside because of the radioactive decay process. Another idea being actively pursued is the preparation of high-silica porous glass beads. These beads are soaked in liquid waste and fired for immobilizing the activity. Finally, the beads are given a coat of an inactive layer of silica and subjected to a firing process. The leach resistance of the final product is exceptionally good, as would be expected of high-silica glasses, which are noted for their high melting points. Another concept is the conversion of wastes to ceramic forms composed of highly insoluble crystalline substances. A further development of this idea is SYNROC,[17] in which waste elements are incorporated into a mixture of three structures which are known to be stable entities in many ancient rocks. In general, the incorporation of these materials with HLW is very attractive. Their resistance to leaching or hydrothermal alteration is generally very good. None of the concepts has, however, been taken beyond the laboratory investigation stage. The developmental efforts have often been with synthetic wastes and not with real ones. It is also not known with certainty whether the materials can meet satisfactorily all the requirements set out earlier. It may be mentioned in particular that all the concepts entail considerable process complexity. Some have applied temperatures and pressures considerably higher than desirable in a production plant handling radioactive materials. It is entirely feasible and possible that one of the concepts may supercede glasses eventually, and while the possibilities should continue to be explored, it would not be wise to delay the solidification of wastes by a good technique in the expectation of having or developing a better one. As examples of the general procedures given for the treatment of highly active waste, a brief outline of laboratory and industrial-scale conversion processes is given in the following sections.

a. Laboratory Development Process

The process briefly described here refers to Sellafield wastes.[18,19] A schematic flowchart of the process is shown in Figure 4. The process is basically a continuous one. It starts with calcination in a sloping stainless steel tube rotating in an electrically heated furnace capable of operating up to 850°C. The solution is fed into the upper end of the calciner, and subjected to a progressive process of evaporation and denitration before leaving the lower end as a finely divided powder. The next step is the fusion of calcine with base glass, and this takes

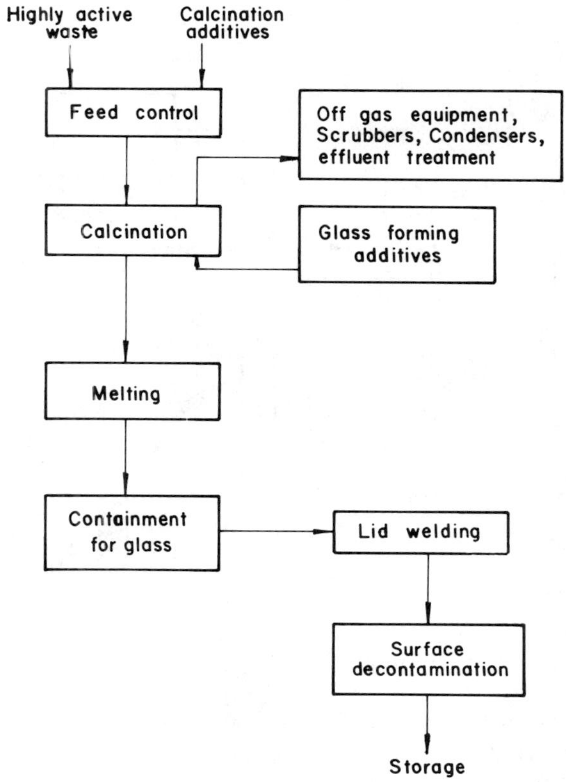

FIGURE 4. Schematic flowchart of the continuous vitrification process.

place in a directly heated, metallic induction furnace operating at a wall temperature of approximately 1100°C. During dissolution and digestion of the calcine, any residual nitrate is decomposed. When the material in the melter reaches a specified level, the glass product is drained off into a stainless steel container. The glass container is allowed to cool, closed with a welded lid, and decontaminated using high-pressure water jets before storage.

b. Industrial-Scale Conversion

The process for vitrifying highly active liquid waste takes different forms that are essentially different engineering approaches to the same type of end product.

The FINGAL (fixation in glass of active liquid) plant in the U.K. features use of a cylindrical vessel in which the glass is manufactured and which is also used as the final storage container. This type of process has come to be known as in-can melting. The liquid waste, together with glass-forming materials, is fed continuously into a heated stainless steel vessel in which the mixture fuses with a homogenous glass. When the vessel is full, it is removed from the furnace, cooled, and sealed, and becomes the storage vessel. A unique feature of FINGAL is the method employed for active gas trapping. The active gases were carried out of the process vessel with the off-gases and passed through a filter installed in a second stainless steel cylinder. This cylinder, with a filter, was used in the next glassmaking operation cycle so that the filter was incorporated in the glass. The FINGAL glass produced blocks of about 50 kg in weight.

In the early 1970s, British Nuclear Fuels, in collaboration with the U.K. Atomic Energy Agency, started the development of HARVEST (highly active residues vitrification engineering study) as a process suitable for routine industrial operation. The principle of

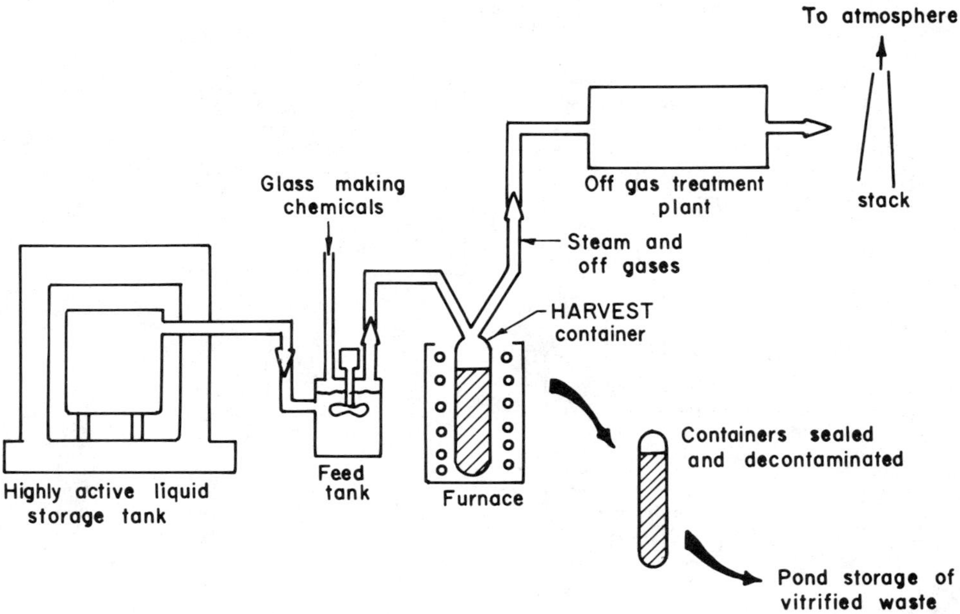

FIGURE 5. HARVEST process for storage of nuclear wastes.

HARVEST[20] is shown in Figure 5. It is based on the FINGAL process, but the scale is increased and the internal dust filter omitted, the dust being collected in a separate condenser system, from which all radioactivity would be concentrated and recycled. Both the processes are characterized by the emphasis on engineering simplicity. They are batch processes. They have no moving parts. The glassmaking vessel has to endure only one cycle at temperature, and this is the most demanding of the stages in terms of potential for failure during operation.

In France, a plant called Atelier de Vitrification Marcoule (AVM)[21] has been operating at Marcoule on a commercial scale since 1978, converting long-store Magnox wastes to glass. The principle of the process is shown in Figure 6. The process makes glass in two stages continuously. In the first stage, liquid waste is dried to a free-flowing powder in a rotary kiln; and in the second, the dried waste is fed, together with glass powder, into the glassmaking furnace, where melting takes place. The molten glass is cast periodically into the storage container, which is cooled and closed with a welded lid. The engineering is more complex. The glass-producing furnace has to last for much longer periods than in the in-can melting route. The advantage is that this type of plant has greater potential for scale-up to a larger production rate per unit.

F. Intermediate- and Low-Level Wastes

Generally speaking, the treatment of ILWs is exactly the same as that for LLWs. It depends essentially on the chemical and physical characteristics of the waste under consideration. Volume reduction and immobilization are generally the two aspects of ILW/LLW treatment.

1. Volume Reduction

In the consideration of volume reduction of ILW/LLW, it is useful to categorize these wastes into noncombustible solid and liquid wastes, and combustible wastes.

a. Noncombustible Solid and Liquid Wastes

Noncombustible solid wastes consist of glove boxes, machinery piping, glassware, high-

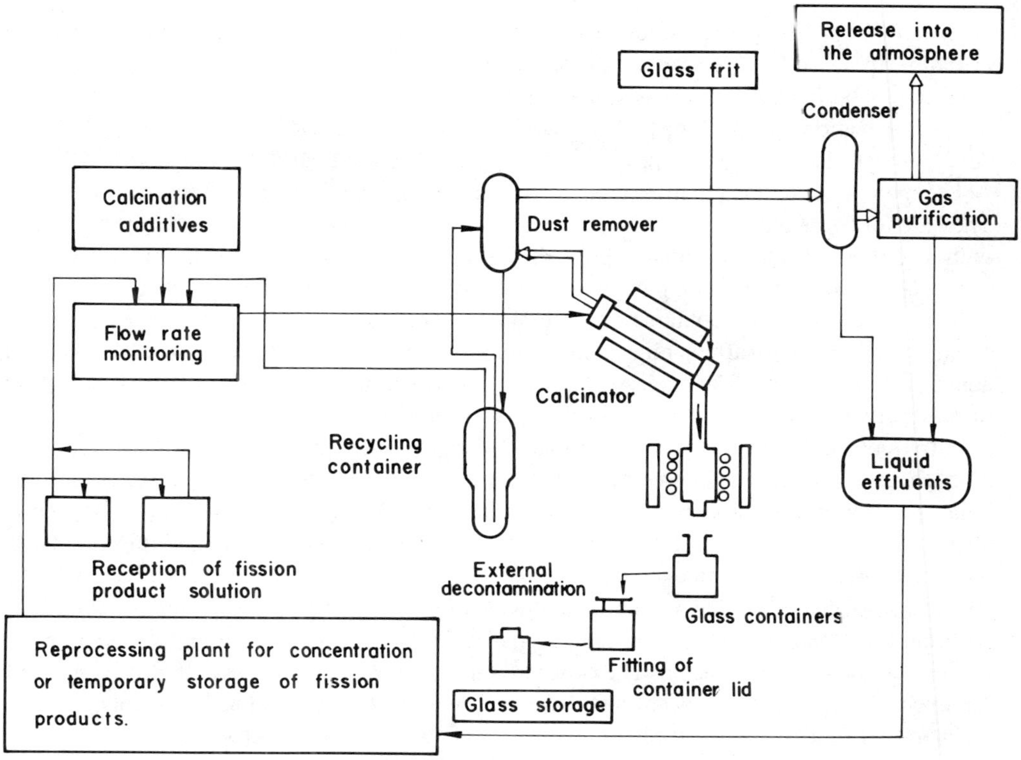

FIGURE 6. French AVM process at Marcoule.

efficiency particulate air (HEPA) filters, and miscellenous trash. The options available for volume reduction for these wastes are (1) no reduction, (2) mechanical disassembly, (3) crushing (or compaction), (4) melting, and (5) dissolution. The no-reduction option is generally used for such materials as massive pieces of metal and floor sweepings where no significant volume reduction can be expected by further treatment. The mechanical disassembly method involves dismantling failed equipment to reduce its volume or to facilitate the handling of a large piece of equipment. An overall waste volume reduction by about 20% is achievable by the mechanical disassembly option. Crushing (or compaction) is a very commonly used volume reduction process for materials such as glassware and HEPA filters. Compaction is also used frequently for combustible solid materials if these are not to be incinerated. Crushing can be expected to secure reduction of the preimmobilization waste volume by about 60%. The melting option involves melting of disassembled metallic wastes (failed equipment and piping) and casting into metal ingots. The dissolution option consists of chemical reaction of the waste with a molten sodium carbonate solution and air to oxidize the wastes. In the process, sodium carbonate acts as a solvent for oxidized wastes.

The types of materials included in the category of noncombustible liquid wastes are generally aqueous wastes containing varying amounts of chemical salts and radioactive isotopes. The proposed volume reduction options for these wastes include (1) evaporation, (2) calcination, (3) concentration using membranes, (4) filtration and centrifugation, and (5) ion exchange. Evaporation is extensively employed in reactor plants for volume reduction of ILW/LLW streams. Evaporation can reduce volume by as much as 90%, depending on the solute concentration in the initial waste. Evaporation is also applicable to concentrate the wastes in such a way that they can be combined with a more concentrated waste stream. For example, concentrated LLW can be combined with ILW streams. ILW can be combined with HLW. Evaporation can also be put to use to prepare a lower-volume feed stream for other ILW/LLW volume reduction processes. The LLW evaporator vapor is sufficiently low in radioactivity and hence can be released into the environment. The ILW evaporator condensate can be reused as process water.

The calcination option of the ILW without combination with the HLW has received limited development. The calcination processes have been described earlier in the case of HLW.

The membrane options that have been considered are reverse osmosis and electrodialysis. They are for LLW only. The reverse osmosis process pressurizes an aqueous waste in contact with a semipermeable membrane above the osmotic pressure for that membrane. The consequence is the transport of pure water through the membrane along with a few of the radioactive constituents contained in the water. The electrodialysis process employs an electrical potential to affect migration of ionic species through a membrane. A purified aqueous stream results. The process is routinely used in desalination applications. Application to radioactive waste has only been in development work.

Filtration and centrifugation, a familiar option, first involves waste solution treatment with chemical additives to precipitate the solutes in the waste. The solid materials thus obtained are removed by using standard filtration or centrifugation technology. Filtration is also commonly used as a pretreatment prior to ion-exchange processing, and this is done to prevent plugging of the ion-exchange bed.

The ion-exchange option involves treatment of ILW/LLW to remove ionic radioactive constituents from the waste by exchanging the radioactive ions in the liquid phase for those in the solid ion-exchange media. The concentrated waste from this process is either a saturated ion-exchange resin or a concentrated aqueous solution formed in the ion-exchange resin regeneration process. Reactor plants routinely use ion-exchange resins.

b. Combustible Wastes

Combustible waste treatment processes involve oxidation of wastes to ashes. These are then treated as noncombustible solid waste. Scrubbing of the off-gases from the oxidation process results in other secondary wastes. Oxidation of combustible radioactive wastes has been practiced for many years. To name a few oxidation processes, mention may be made of the excess air incinerator, controlled air incinerator, pyrolysis incinerator, fluidized bed incinertor, rotary kiln incinerator, molten salt oxidation, acid digestion, etc.

2. Immobilization

The immobilization process used for ILW and LLW are much more conventional than those being used for HLW, and this is because the former generate essentially no heat from radioactive decay. Since ILW and LLW have been produced for years at reactor plants, at previously operated reprocessing plants, and by defense programs, a number of technologies for immobilizing these wastes are presently being used on a commercial scale. The characteristics of the immobilization technologies involving use of absorbents, salt matrix, urea-formaldehyde resins, cements, and bitumen presently in use are dealt with in the following paragraphs.

Materials that are difficult to solidify by other means can be treated with absorbents. Some of the typical absorbent materials are various clays, Portland cement, silica gel, vermiculite, plaster of Paris, and calcium sulfate dessicants. The liquids are fixed by simply mixing them with the absorber until it has been nearly saturated. The absorbed liquids are only immobilized physically. They can be displaced by other liquids and are realtively dispersable due to the particulate nature of the absorbents.

Salt matrix immobilization can be carried out by either of the two methods. The first is to evaporate aqueous ILW or LLW to the extent where bottom fractions from the evaportor crystallize into a solid upon cooling. The second method is to cast the solid salt cake from the molten salt. This type of product would be produced by the salt removed from the molten salt oxidation process (combustible wastes are burned with air in Na_2CO_3 and Na_2SO_4, and inorganic products, metallic oxides, ash, sodium chloride, and fissile material are maintained in the melt). The product is a monolithic salt cake which is not readily dispersible, but which is readily water soluble.

Urea-formaldehyde resins are materials used to immobilize liquid and particulate solid wastes. The resin is a cross-linking polymerization product of the reactants urea and formaldehyde, and consists of a mixture of monomethylol urea and dimethylol urea, depending on the ratio of reactants. The reaction between the constituents is initiated by using a weak acid (sodium bisulfate or phosphoric acid) at a pH between 1.0 and 2.0. The cross-linked structure of the resin traps the aqueous waste within the solid resin matrix.

Hydraulic cements (usually Portland cement) constitute one of the most popular ILW/LLW immobilization techniques in the U.S., along with urea-formaldehyde resin, described above. The procedures for cement use for the purpose of immobilization are basically the same as those for urea-formaldehyde resin except that catalysts are not needed. The improvement of the immobilization characteristics of cements can be brought about by certain additives, among which are (1) sodium silicate, which improves setting properties and increases allowable concentration; (2) fly ash, which improves strontium retention; (3) attapulgite, which aids in suspending the cement; (4) grundite, which improves cesium retention; (5) delta gluconolactone, which retards setting; and (6) tributyl phosphate, which prevents foaming during mixing. A relatively recent innovation with this immobilization process is to impregnate the monolithic cement/waste surface with a styrene monomer. The monomer is polymerized at 50 to 70°C to polystyrene. This gives rise to an essentially impermeable layer on the outer surface, which also improves the strength of the waste form. The durability and resistance to chemical attack are also improved.

Bitumen, because of its chemical inertness, insolubility in water, and good coating properties, is a popular material for immobilizing ILW and LLW. It is used by heating to 150 to 230°C and mixing in wastes to be immobilized. The high temperatures result in evaporation of water in the wastes, which yields a smaller volume of bitumen, but also results in a residual low-level stream that must still be treated.

G. Gaseous Wastes

There are four principal volatile elements which have isotopes having half-lives long enough to require the recovery and isolation of the elements. These elements and the isotopes of interest are iodine (^{129}I), tritium (^{3}H), carbon (^{14}C), and krypton (^{85}Kr). The recovery of these elements, along with such details as their identification and sources, is summarized in Table 4. The paragraphs below outline the treatment of the recovered wastes to render them less mobile.

Iodine waste treatment depends primarily on the process employed for the recovery of iodine initially. In cases where silver-loaded or other metal-loaded absorbents are used, it is likely that iodine will not be recovered from these materials. The most likely treatments would simply be to package the solid iodine-laden absorbents or to immobilize the absorbents with such materials as cement, urea-formaldehyde, or bitumen. In the case of other iodine-recovery processes, the iodine will most likely be recovered as a solid material, as recorded in Table 4. These solids may be packaged without further treatment, packaged with low-temperature immobilization absorbents such as cement, or converted to a different solid material. Regarding the latter option, one of the more promising solid materials is barium iodate, $Ba(IO_3)_2$. An immobilized iodine waste containing about 9% iodine as barium iodate seems to have favorable leach properties. The leach rate is reduced by an additional factor of 3 by the addition of butyl stearate to the cement to an extent of about 1%.

The proposed tritium waste treatment options, as with iodine wastes, include a wide variety of processes designed to reduce the mobility of the tritium and thus restrict the amount

Table 4
SOURCES AND RECOVERY PROCESSES FOR VARIOUS GASEOUS WASTES

Iodine	**Identification**: Semivolatile fission product element. The isotope of interest for fuel management purposes is ^{129} I, which is a low-energy beta emitter with a half-life of 15.9 million years. **Sources:** In the reprocessing plant iodine in off-gases is in two distinctly different forms, elemental iodine (I_2) and organic iodides (e.g., methyl iodide). **Recovery processes:** (1) Scrubbing with aqueous solutions. (a) Iodex processes: scrubbing with 20 — 22 *M* HNO_3. The product is solid $HG(IO_3)_2$, which is convertible to $Ba(IO_3)_2$ for disposal (c) Caustic: scrubbing with KOH solution. The product is solid K_2CO_3-KI-KIO_3-KOH. (2) Trapping on silver-containing sorbents. (a) Silver-exchanged zeolites: sorption on zeolites in which silver has replaced the sodium normally present in the crystal. The product is the alumino-silicate zeolite plus the sorbed iodine and silver. (b) Silver nitrate-impregnated silicic acid: sorption on silic acid loaded with silver nitrate, which is the reactive material. The product is the silicic acid plus the sorbed iodine and silver compounds. The processes described are effective with elemental iodine. Only Iodex and trapping on silver-containing sorbents are totally effective on organic iodides.
Tritium	**Identification:** Volatile (when elemental) or semivolatile (as tritiated water) hydrogen isotope. It is a low-energy beta emitter with a half-life of 12.3 years. **Sources:** Produced in several ways, including as a fission products; as a low-energy neutron activation product of any lithium impurities in the fuel; as a high-energy activation product of many structural materials; as a neutron activation product of naturally occurring deuterium in coolant water in nuclear reactors. **Recovery processes:** Voloxidation process under development to remove tritium from spent fuel. The process involves oxidation of high-density UO_2 to lower-density U_3O_8 with an attendant phase change, resulting in disintegration of the fuel material and the release of tritium not combined with the cladding. This process, after catalytic oxidation of the elemental tritium, will result in a relatively concentrated stream of HTO. The concentrated tritiated water vapor can then be moved by refrigeration or by molecular sieves.
Carbon	**Identification:** Since generally present as CO_2 it may be treated as volatile species. Isotope of interest is ^{14}C, which is a low-energy beta emitter with a half-life of 5730 years. **Sources:** Produced in the fuel principally by activation of nitrogen impurities. Reprocessing plant off-gases carry ^{14}C-contaminated carbon dioxide. **Recovery processes:** Proposed processes for recovering ^{14}C-contaminated CO_2 from the reprocessing plant off-gases include (1) caustic scrubbing with NaOH or $Ca(OH)_2$, (2) sorption on molecular sieves, and (3) absorption in liquid fluorocarbon. All produce calcium carbonate as the final product because a calcium hydro-oxide scrub cycle is used to convert the CO_2 recovered by other means to the carbonate.
Krypton	**Identification:** Volatile noble gas. Isotope of interest is ^{85}Kr, which is a beta-gamma emitter with a half-life of 10.7 years. **Sources:** Produced as a fission product in the fuel. **Recovery processes:** Kr being nonreactive chemically, its recovery processes rely on physical processes. Processes that can recover are (1) cryogenic distillation, (2) selective sorption, and (3) diffusion through permselective membranes. Among these, diffusion is not being considered seriously at present. The distillation process yields a relatively concentrated Kr combined with some stable fission product Xe (Kr product is about 6%). The sorption processes include both adsorption on activated carbon or molecular sieves, and absorption in fluorocarbon (it is the same as that used to recover ^{14}C).

of tritium-hydrogen exchange that would happen if the waste were to come into contact with water after disposal. It should be feasible to immobilize tritium effectively for its entire radioactive life, since it will have decayed to innocuous levels in about 100 years. The simplest of treatments uses the tritium waste, which is usually tritiated water, as the aqueous agent in making hydraulic cement. The cement can either be a standard cement, a high alumina cement, or a polymer-impregnated cement. A more complex tritium waste treatment is to make the tritium into a relatively insoluble hydrogenous compound which may either be packaged directly or be immobilized in cement during packaging. Some of the compounds that have been investigated include polyacetylene, polyacrylonitrile, polystyrene, and metallic hydrides.

The carbon waste treatment (assumed to be in the form ^{14}C-contaminated $CaCo_3$), has not been studied to any extent. However, there are only two options that have been considered for treating the calcium carbonate. They are direct packaging without further treatment, and mixing the calcium carbonate and then packaging. It may be mentioned that both of these are only concepts. They have not been investigated for leach resistance, integrity, or the effects of carbonate on cement.

Noble gas waste treatment presents a unique problem. The noble gases are volatile under all conditions expected during disposal, and they form only a few marginally stable compounds. Treatment of these gases, it is understandable, is based therefore on physical rather than chemical methods. The most straightforward of the processes is to store them pressurized in cylinders similar to those used to contain welding and laboratory gases. Since most of the ^{85}Kr will decay to the stable daughter rubidium in 100 years, this mode of treatment is expected to be adequate as long as the storing of the cylinders is done under controlled conditions. The material of the cylinder should be made to withstand high pressure and somewhat elevated temperatures, and the corrosive effects of the rubidium daughter. The second approach to containing the noble gases is known as zeolite encapsulation. Krypton is provided with sufficient activation energy by application of sufficient pressure and temperature to diffuse into the interstitial cavities of zeolite. The zeolite is subsequently brought to ambient temperatures and pressures, and under these conditions krypton atoms are effectively trapped. Elimination of having to monitor the waste in a controlled environment and elimination of the possibility of a sudden release of noble gases contained in a pressurized cylinder are the two advantages that can be cited in favor of this process. The disadvantages include requirements of high temperatures and pressures, the need for separation of krypton from xenon, and the possibility of krypton release by moisture.

H. Miscellaneous Wastes

In the category of miscellaneous wastes, two specific types are being considered. The first type is uranium mill tailings. The very large volume of mill tailings effectively precludes the types of waste treatment processes used for other waste types. It is, however, a fortunate situation that the very dilute nature of these wastes makes this type of treatment unnecessary. Some treatment is required for preventing (1) public access to this material, (2) their becoming airborne from erosion, (3) the leaching of ^{226}Ra, and (5) emanation of radon gas. This can generally be accomplished by covering the tailing pile with an earth overburden. This takes care of public access and delays release of 3.8-day ^{222}Ra sufficiently that virtually all of it decays to its nongaseous daughter prior to its release into the atmosphere. A layer of asphalt before piling on the earth overburden can provide additional radon retardation.

The second type of waste in the miscellaneous category is spent fuel. The treatment of as-produced spent fuel, which possesses many characteristics of HLW, will most probably not be the same as that used for HLW described earlier. The reasons for this are many. The spent fuel is a solid material, whereas most of the as-produced HLW is a liquid. The spent fuel elements contain volatile fission product gases (krypton, iodine, etc.) under relatively high pressure. The zircaloy cladding is not likely to withstand the elevated temperatures that go along with the vitrification processes. The presence of all the ^{235}U and plutonium makes criticality a potential problem. It may be mentioned that various treatment processes are under intensive study, and at the present time it is not possible to pinpoint the option that will finally be chosen for spent fuel treatment.

V. RADIOACTIVE WASTE DISPOSAL

The objective in disposal is to place the waste in a situation such that, with no further action by man, the radionuclides cannot return to the environment in concentrations regarded as hazardous. The schemes for disposal of ILWs and HLWs are outlined in the following paragraphs.

A. Intermediate-Level Waste Disposal

ILWs may be segregated into the following two streams, which have different disposal requirements: stream 1, wastes containing activated materials, fission products, or both, and only minor quantities of actinides (no long-lived activity). For stream 1, the repository design requirements are very similar to those for HLWs insofar as radionuclide paths back to man are concerned. The important difference is that heat generation in the first 100 years is very much less, and this aspect modifies the design constraints. The requirements of stability of rock, low water movement, and ability to accept climatic and topographic changes over long periods of time are still of prime importance. For stream 2, it might be possible to use underground caverns excavated at a shallower depth than for stream 1.

1. Shallow Burial

The shallow land burial of radioactive wastes involves placement of the waste at relatively shallow depths (as compared to geologic disposal) with no provisions of retrievability at a later date. Burial in shallow trenches is a well-established technique for the disoposal of low-level solid wastes. ILWs have a higher level of activity and require more protection to ensure that they remain isolated for the necessary period. This is typically 150 years for the short-lived wastes (stream 2). Their disposal has been carried out routinely for a number of years in the U.S. and in France, where they are packaged, placed in concrete-lined trenches, and covered with more concrete. In one design being developed in U.K., the wastes would be immobilized in concrete or resin and sealed into steel drums. These drums would be packed in concrete casks which would be placed in reinforced concrete trenches some 18 m deep. When the trenches have been filled to a depth of 8 m, they would be covered with a reinforced contrete slab of 1-m thickness. This would then be covered by a 3-m layer of compacted clay, a 1-m thick reinforced concrete instrusion shield, and a final 5 m of clay. Such sites would require monitoring for up to about 150 years. A schematic of the disposal of ILWs in deep trenches is shown in Figure 7.

2. Deep Undergroud Burial

The disposal of ILWs containing significant concentrations of long-lived species such as actinides and some fission products (stream 1) can only be allowed if adequate containment for many thousands of years can be ensured. By placing these wastes in a suitable geological formation at a depth of about 300 m the possibility of accidental human intervention or natural disturbance can be minimized. In a deep repository the rocks provide a massive barrier between the waste and man's environment, while the depth reduces the possibility of accidental intrusion by future generations. The only conceivable pathway by which the radionuclides in the waste could return to man is movement into groundwater which eventually reaches the surface to enter the food chain. There are several barriers to such movement. The inherent insolubility of the waste form itself, and the container and backfill material with which waste would be surrounded limit the rates at which the radionuclides could be dissolved. By choosing a host geological formation in which water moves extremely slowly and in which radioactive species are readily absorbed and thus retained, it is possible to ensure that no significant levels of radioactivity enter the food chain. Many common geological formations, such as clays, mudstones, and evaporites (for example, bedded salt deposits), possess such properties.

Evidence from natural phenomena supports the experimental findings that many radioactive materials are stongly absorbed by solids and rocks. Particularly striking are the results from the series of natural reactors which occurred some 1800 million years ago in a uranium ore body at Oklo in Gabon, West Africa, in a fairly common rock geology. Nuclear fission processes continuing over several thousand years created several tons of fission products and significant amounts of plutonium. Detailed studies of the deposits have shown that nearly

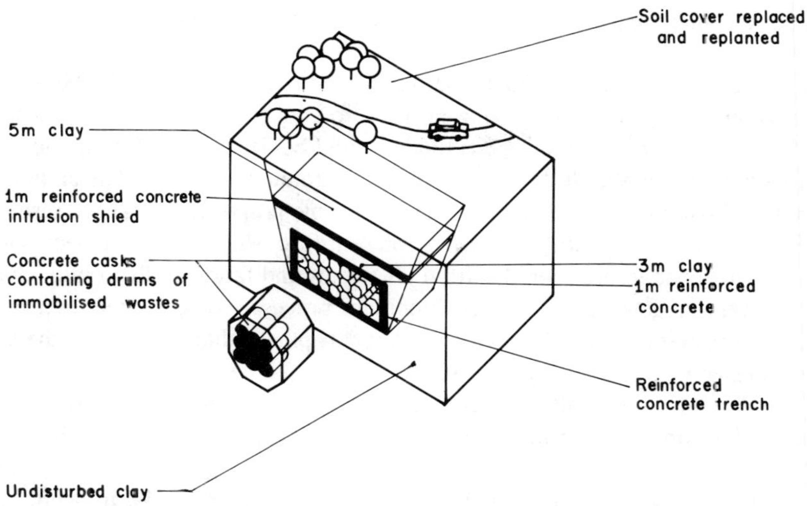

FIGURE 7. Disposal of ILWs in deep trenches.

all the heavy elements, including the plutonium and most of the fission products, remained where they were formed.

There is no way in which the integrity of the waste form or its immediate surroundings can be affected by the ILWs themselves. This is not the case with HLWs, with their initial high radiation level and heat emission. As with shallow burial repositories, site-specific studies need to be carried out to establish the suitability of any particular geological formation for a deep repository. However, studies in many countries indicate that there is every likelihood that repositories in appropriate types of geological formations can be shown to be very safe indeed.

B. Highly Active Waste Disposal

Disposal of highly active waste has received the most attention to date and is the most challenging problem.[22] The following are the disposal options: (1) space disposal, (2) ice sheet disposal, (3) partitioning and nuclear transmutation, (4) geological disposal, and (5) ocean bed disposal

The first two options actually constitute waste placement outside man's environment. The idea of space disposal drew a good deal of attention from the National Aeronautics and Space Administration (NASA) for the U.S. Department of Energy. According to the scheme, selected waste packages would be lifted by a space shuttle in the near-earth orbit, from which the package would be propelled into an appropriate solar orbit that would be stable for millions of years. Apart from the high cost involved, this option has the enormous disadvantage that if there should be a fault with the launching vehicle and it crashed to the earth, quantities of radioactive materials could be spread out in the open, providing radioactive pollution. This method of waste disposal has received opposition virtually from all nuclear power authorities.

Regarding the disposal of packages of waste in continental ice sheets, in particular, in Antartica, the first advantage is the remoteness from human activities. Besides this, there is the simplicity of the technology for emplacement. The properties of ice (self-scaling through plastic flow and impermeable) would appear to contain the radioactivity for thousands of years. There are, however, considerable uncertainties regarding the long-term stability of ice sheets and their interaction with the waste. At present, there is very little information on which to judge the safety of the concept. International agreement exists not to introduce radioactivity into the area. Because of these reasons, this concept of disposal is not being pursued.

Various suggestions have been made that long-lived radionuclides might be destroyed by nuclear transmutation processes, leaving no need to consider any possible long-term environmental effects. In principle, all the actinides could be destroyed by irradiation in fission reactors in a manner parallel to that of the use of plutonium as a fuel. The actinides would be replaced by fission products in substantially the same proportions as those resulting from the burning of uranium and plutonium in our present reactors. It would be necessary, in the first instance, to separate the actinides from all wastes and fabricate them into suitable fuel elements. A very large research and development process would be required before any industrial-scale operation could even be considered as practicable. Moreover, the separation of the actinides neptunium, americium, and curium from other waste to the required degree calls for the use of additional, very complex separation processes, different from those already well developed for uranium and plutonium. This would produce its own set of wastes.

Of the various disposal options listed, the most promising are burial in deep, geological strata on land or under the deep oceans.

1. Geological Disposal[23,24]

Disposal in geological formations under the land has been accepted internationally in principle as a practicable, safe, long-term method of nuclear waste disposal. It is expected that suitable encapsulated waste, placed deep into a carefully chosen geological formation, can be effectively isolated from the biosphere. Guidelines which might be used within the U.K. in the selection of areas containing geological formations potentially suitable for the location of sites for disposal of high-level, heat-emitting radioactive wastes have been published by the Institute of Geological Sciences.[25] The aim is to select a rock formation and area that meets the following criteria: (1) sufficient thickness and area to house the repository, and sufficient depth to isolate it from the surface; (2) simple and determinable hydrogeological conditions of low hydraulic conductivity; (3) a long geological history of stability, away from active faults and abnormally high seismic activity; (4) absence of potential resources that might encourage man's intrusion; (5) absence of previous mining activities; (6) suitable chemical characteristics, including high ion-adsorption or -exchange capacity, and adequate thermal stability; and (8) adequate thermal conductivity. The types of geological formation that have been considered[26,27] for the containment of HLWs are (1) salt; (2) argillaceous deposits such as clays, shales, and slates; and (3) crystalline hard rocks such as granites and gneiss.

Salt formations are quite old, and their very existence indicates the absence of water flow. Salt has a high thermal conductivity to permit heat to be dissipated easily. Under the pressure of overlying rocks, salt creeps so that any fissures produced by earth movement are self-sealing. This property has prevented ingress of surface water, and it also ensures that salt will rapidly close a round the emplaced hot waste container, thereby isolating it effectively from moving groundwater.

Argillaceous formations are sedimentary rocks varying from unconsolidated, plastic clays to the harder rocks such as shales and slates produced when clays have been subjected to great pressure over long periods. The principal attraction of these is their very large ion-exchange capacity, which makes them admirably suitable. They adsorb radioactive species leached from the waste by flowing groundwater. The plastic clays have self-sealing properties. The shales and slates are prone to crack and fissure if subjected to earth movements. The water movement rate through these formations is relatively low. It is virtually nil in plastic clays. The quantity of water inherent in their constitution can be considerable. Clays have the greatest quantity of water, which is progressively less in shales and slates. It is the behavior of this water under thermal influence that constitutes the most important question in relation to the emplacement of hot wastes. These formations, however, appear to have very good potential for plutonium-contaminated and other ILWs with low heat emission.

Crystalline rock formations (granites and gneisses) are quite good absorbers/adsorbers of materials that might be leached out of waste. Granite is not as good a thermal conductor as salt. Heat dissipation will therefore be slower in granite. It is less susceptible to alteration by heating than the clay-based rocks. Being a strong rock, it is suitable for engineering operations. Being brittle, however, it may be subject to cracking if stress is induced by earth movements or the thermal expansion of a large volume of rock.

The design of a repository for HLW depends on the type of rock in which it is to be built and the constraints imposed by the temperature limits to protect the waste and the integrity of the repository. It is necessary to take into account (1) the maximum rock temperature adjacent to freshly emplaced waste containers, which will control the extent to which the walls of the hole or chamber in which waste is placed will break down; (2) the bulk rock temperature (regional effects), which will determine the effect on the stability of the rock formation as a whole; and (3) the temperature of the waste containers and of the glass within them. This will determine the rate of corrosion of the container and the rate of leaching of glass.

2. *Ocean Bed Disposal*[28,29]

Presently, international regulations forbid storage of highly radioactive materials underneath the oceans, although on the face of it such techniques would appear to be more satisfactory than burial on land. The main fact is that such wastes, once on the ocean floor, would be irretrievable. In fact, however, there is little point in ocean bed disposal, as the burial of vitrified nuclear waste in dry land storage is so satisfactory. Nevertheless, two concepts have been put forward for deep ocean disposal of HLW: emplacement of waste canisters in ocean bed sediments and placement of them on the ocean bed.

Disposal in ocean bed sediments is in many ways similar to disposal in geological formations on land, but with the certainty that the formations are permanently water saturated. There are large areas of deep ocean bed where very thick deposits of sediment have remained undisturbed for millions of years. The absence of any water flow through the sediment, together with its good absorptive properties, offers very long-term isolation of radioactive wastes with the additional barrier of enormous dilution of any material which might reach the surface of the sea. The principle is that of multiple barriers between the waste and man's environment. The first barrier is the canister; the second is the solid form of the waste itself (glass or other suitable solids); the third is the sediment; and finally there is the ocean itself, which provides the fourth barrier in dispersing and diluting the activity.

Disposal on the ocean bed, as the phrase implies, is free of the sediment barrier. In this method, the canister and the waste form are similar to those described in the foregoing paragraph. In this case there would be very good heat removal, thus eliminating high temperatures and corrosion rates; hence there would be less need for limiting waste concentrations or for long land storage. The levels of radioactivity would depend on the dispersion process by bottom currents and dilution by mixing. Some radioactivity may be scavenged from the water by adsorption onto suspended particles, which would either be deposited locally or transported elsewhere. This depends, however, on the prevailing conditions.

3. *Wastes in SYNROC*[17,30,31]

Immobilization of high-level nuclear reactor wastes in SYNROC is an alternative to that done in borosilicate glasses. SYNROC is a titanate ceramic composed of three constituent minerals: zirconolite ($CaZrTi_2O_7$), holandite ($BaAl_2Ti_6O_{16}$), and perovskite ($CaTiO_3$). These

Table 5
SYNROC PRODUCTION PROCESS STAGES (In-can Hot-pressing Process)

1. Mixing of SYNROC precursor (TiO_2 + ZrO_2 + CaO + BaO + Al_2O_3) with HLW nitrate solution to form a thick, homogeneous slurry.
2. Spray-drying and rotary kiln operations to calcine the slurry. The calcined product is mixed with 2% metallic Ti.
3. Loading the calcine in a primary canister. The canister is made from stainless steel in a configuration of a bellows so that it can be shortened by a factor of about 3 when subjected to axial loading. A thin perforated metal liner sits inside the canister and space between it, and the liner is filled with a low thermal conductivity powder (e.g., ZrO_2). The opening of the canister is sealed with a steel plug after filling the liner with SYNROC powder.
4. Hot pressing of the primary canister.
5. Stacking of the hot-pressed, collapsed primary SYNROC canister in a final disposal canister, which is fabricated from the highly corrosion-resistant alloy Ni_3Fe. The space between primary canister and disposal canister is filled with molten lead, and the disposal canister is finally sealed.

minerals have the superb capacity to accept all the elements present in HLW into their crystal lattices. The evidence of long-term geologic stability, combined with experimental observations showing that these minerals are extremely resistant to the attack of hydrothermal solutions, suggests that SYNROC would provide a superior method of immobilizing HLW.

When a material containing HLW is finally buried in a geological repository, the principal way by which subsequent contamination of the biosphere might occur is via leaching of the waste form by circulating groundwaters, followed by migration of HLW species into the near-surface environment. The superior leach resistance of SYNROC at elevated temperatures will make it possible to bury it in geologic repositories within a few years after reprocessing, thereby avoiding the necessity for prolonged storage and supervision. It should be possible to immobilize about three times the weight of radwaste in a given volume of SYNROC than is now envisaged for glass. This will generate substantial cost savings. A conceptual flowchart for SYNROC production is summarized in Table 5. The in-can hot-pressing technology by which the process is known and which is under development will cause less radiological hazards and contamination than the melting, refining, and degassing processes implicit in glass production. In addition to the technical and cost advantages as cited, SYNROC possesses a major advantage over glass in terms of public acceptability.

VI. SUMMARY[32-35]

Nuclear fuel generates a range of radioactive wastes. Figure 8 shows the origins of various types of nuclear wastes. The waste produced may be of gas, liquid, or solid form. The wastes arising from different sources are classified as low level, intermediate level, or high level, depending upon radioactivity levels.

The basic objective in the disposal of radioactive waste is the protection of man and the environment from unacceptable harm. There are basically three principles by which radioactive waste is managed: (1) dilution and dispersion of short-lived or very dilute radioactive wastes, (2) concentration and containment of long-lived radioactivity, and (3) delay and decay storage of very short-lived radioactive waste to permit its decay into nonradioactive species. These principles are applied separately or in combination, depending on the nature and concentration of the radioactivity in the waste, its toxicity, its mobility, radioactivity in the waste, radioactive half-life, and the type of radioactive emission (alpha particle, beta particle, gamma ray, or neutron). These characteristics govern the choice of management procedures to be adopted.

Waste with low and intermediate levels of radioactivity is produced in a variety of forms by commercial nuclear power reactors; research reactors; nuclear fuel cycle facilities; and medical, industrial, and research uses of radioisotopes. Because of its radionuclide content

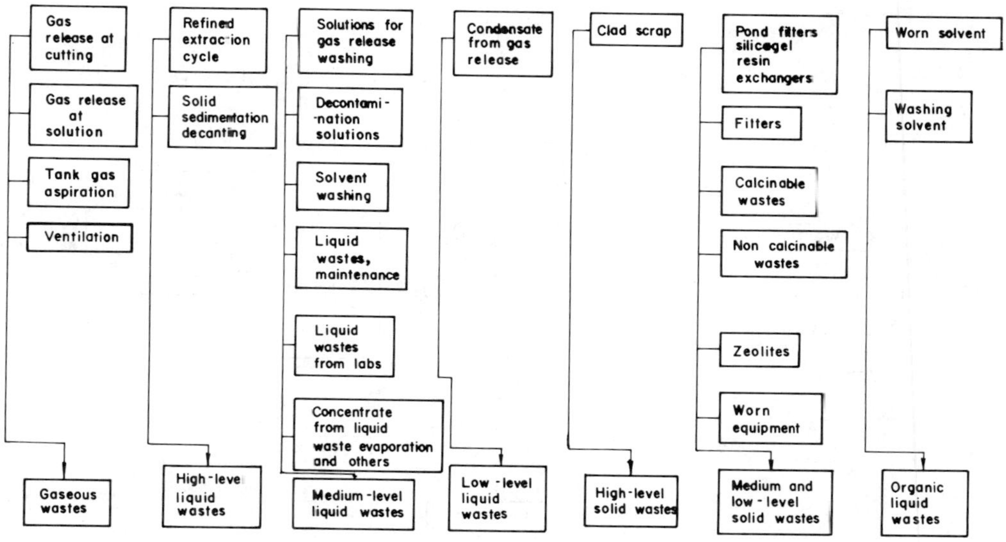

FIGURE 8. Origins of various types of nuclear wastes.

or physical form, it does not have the intense radioactivity and high heat generation rate that demands the unique management measures necessary for HLW. However, some of this waste which is contaminated with long-lived alpha-emitting nuclides has to be given special attention to limit the rate of its return to the biosphere.

Waste containing intermediate levels of alpha-emitting, long-lived materials, e.g., plutonium, is considered to require isolation from the biosphere for longer periods than can be accepted for the shorter-lived beta and gamma emitters in the bulk of the waste. Accordingly, in those countries where this type of waste is generated, the plan is to package the waste and dispose of it in underground repositories, some of which may also be designed to accept HLW.

The uranium mine and mill operations generate the largest tonnages of solid LLW. Some particularly bulky items can result from the repair and decommissioning of nuclear facilities. The bulk of LLW is buried in suitably sited shallow earth trenches. The volume of this waste is often reduced by compaction or incineration before transport to the burial site.

ILW contains much the same beta- and gamma- emitting radionuclides as does LLW, but at a higher level of specific activity such that it requires shielding to permit safe handling. This type of waste has been disposed of in shallow land sites in some countries with provision of additional barriers, such as encapsulation into concrete or bitumen and addition of contrete linings in the repository or provision of specially engineered structures. The main alternative disposal route for this waste and for LLW waste not placed in shallow trenches is the use of dry abandoned mines or mined cavities at depths of a few hundred meters. The dumping of packaged LLW and ILW, usually incorporated in concrete in steel drums, into the deep ocean has also been practiced for several decades.

HLW consists of the fission products and unwanted actinide elements contained in spent nuclear fuel. It is produced, usually as a concentrated liquid, in the reprocessing of the fuel. However, spent fuel itself can be regarded as HLW if a decision is made to dispose of it without reprocessing. In some countries, for instance, Belgium, France, India, Japan, the U.K., and the U.S., HLW solutions from commercial nuclear fuel reprocessing have been stored in stainless steel tanks for periods up to 25 years. The tanks require cooling to remove fission product heat, secondary containment to restrain any leakage from entering the environment, additional spare tanks for transfer of solution in the event of tank failure, and

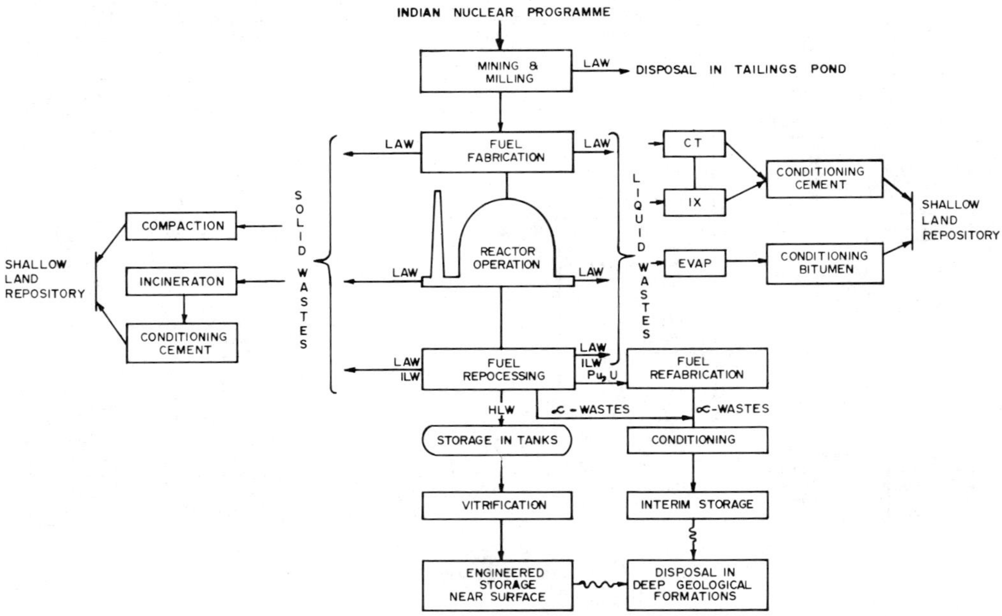

FIGURE 9. Schematic of radioactive waste management programs.

constant supervision. Storage of HLW solution is an interim measure before solidification of the waste into a stable, practically insoluble glass or ceramic form.

Solidification of HLW solution involves evaporation of nitric acid and water, and calcination of residual nitrates of fission products and actinides into oxides, into a stable waste form. Several engineering alternatives have been developed for the evaporation and calcination operations, and for fusing, casting, or pressing the waste forms. These operations must be carried out remotely owing to the high level of radioactivity involved.

Vitrification is the immobilization of waste in glass. The only established engineering-scale process is the use of borosilicate glass. Glass is an inert and durable material. So long as it is dry, there is no natural mechanism by which the contents can disperse and contaminate the environment. France has developed a process successfully to a commercial scale at its plant at Marcoule. Vitrification has acquired commercial status and is being followed in many countries in the world.

Alternatives to borosilicate glass with the prospect of improved retention of radioactive waste have been studied over many years, particularly in the U.S. The most promising alternative is now recognized to be a synthetic rock known as SYNROC, which is being developed by the Australian Atomic Energy Commission and the Australian National University, where it was first developed by A.E. Ringwood. SYNROC combines three refractory and leach-resistant minerals which are capable of accepting most of the high-level radioactive waste constituents into their crystal lattices. These minerals are known to have been stable in a variety of geological and geochemical environments for many many years. The main advantages claimed for SYNROC over borosilicate glass are its thermodynamic stability; its greater resistance to leaching by water, especially at high temperatures; and its potential for allowing greater flexibility in disposal schemes.

The ultimate disposal of the bulk of the radioactive wastes of the nuclear power industry deep underground will effectively close the nuclear fuel cycle, which starts with mining of uranium, often deep underground. A schematic of radwaste management programs in India is shown in Figure 9 as a general illustration.

No country disposes of high-level radioactive wastes underground at present. Many countries have active research programs involving *in situ* experiments and geological investigations for the establishment of demonstration and commercial high-level repositories. Geological media under study include bedded and domed salt deposits, crystalline rocks, basalt, clay, and tuff.

Disposal in the deep ocean is a safe method used for dealing with some of the relatively small quantities of low-level radioactive waste produced by the nuclear power industry and the radiochemical industry. Oceanographic research is needed to exploit the vast potentialities which the sea bed can provide in the management and disposal of wastes.

As conclusions to the management of radwaste, one can record the following:

1. The wastes from nuclear industry are relatively small in volume.
2. Waste management complies with established radiological protection principles.
3. LLWs are discharged into the environment with no significant radiological consequences.
4. ILWs will be packaged and disposed of in engineered repositories near the surface or deep underground.
5. Technology exists to vitrify and store HLWs for an unlimited time.
6. The ability to store vitrified wastes for long periods implies that there is no immediate need for a disposal route for these wastes. Feasibility exists for deep underground disposal. Further research is proceeding for other options so that disposal may be implemented when required.
7. More geographic and oceanographic data will be required before optimum choices can be made.
8. The SYNROC route to waste immobilization appears potentially attractive. SYNROC survives a wide range of geochemical and geological environments for periods exceeding that necessary for radwaste to decay to safe levels.

REFERENCES

1. **Marshall, L.,** Nuclear waste and Nimby, *Atom,* 356, 2, June 1986.
2. **Wade, B.,** Radiation and nuclear power, *Atom,* 301, 289, November 1981.
3. **Turner, A. D. and Dell, R. M.,** Electrochemistry and radioactive wastes, *Atom,* 327, 14, January 1984.
4. **Collins, J. C., Ed.,** *Radioactive Wastes: Their Treatment and Disposal,* John Wiley & Sons, New York, 1960.
5. **Glueckauf, E., Ed.,** *Atomic Energy Waste: Its Nature, Use and Disposal,* Interscience, New York, 1961.
6. **Amphlett, C. B.,** *Treatment and Disposal of Radioactive Wastes,* Pergamon Press, New York, 1961.
7. **Mawson, C. A.,** Processing of Radioactive Wastes, IAEA Rev. Ser. No. 18, International Atomic Energy Agency, Vienna, 1961.
8. **Cooper, M.,** Radwaste immobilization, *Atom,* 362, 2, December 1986.
9. **Fellingham, L. R.,** Radioactive waste management R & D at Harwell, *Atom,* 351, 18, January 1986.
10. **Franklin, N. L.,** Irradiated fuel cycle, British Nuclear Fuels Limited, paper presented at the European Nuclear Conf. Nuclear Energy Maturity, Paris, April 21 to 25, 1975.
11. Management of Radioactive Fuel Wastes, Atomic Energy of Canada Research Company, Pinawa, Canada, 1978.
12. **Grover, J. R.,** High-Level Waste Solidification — Why We Chose Glass, UKAEA Rep. No. AERE 9432, U.K. Atomic Energy Agency, 1979.
13. IAEA, Techniques for the Solidification of High-level Wastes, Tech. Rep. Ser. No. 176, International Atomic Energy Agency, Vienna, 1977.
14. Annual Report on the Characterization of High-level Waste Glasses, Report PNL-2625, Battele Pacific Northwest Laboratory, Seattle, WA, 1978.

15. National Technical Information Service, Proc. Int. Conf. Ceramics in Nuclear Waste Managements, CONF-790420, U.S. Department of Commerce, Springfield, VA, 1979.
16. IAEA, Characteristics of Solidified High-Level Waste Products, Tech. Rep. Ser. No. 187, International Atomic Energy Agency, Vienna, 1979.
17. **Ringwood, A. E., Kerson, S. E., Ware, N. G., Hibberson, W., and Major, A.,** Immobilization of high level nuclear reactor wastes in SYNROC, *Nature (London),* 278, 219, 1979.
18. **Smith, W.,** Vitrification of Sellafield wastes, *Atom,* 341, March 1985.
19. **Larkin, M. J.,** Development of highly active waste conditioning at Sellafield, *Atom,* 361, 25, November 1986.
20. **Clelland, D. W.,** A review of European high-level waste solidification technology, in Proc. Int. Symp. Management of Nuclear Wastes from the LWR Fuel Cycle, Rep. No. CONF 760701, National Technical Information Service, Springfield, VA, 1976, 137.
21. **Costi, J. A., Jouan, A. F., Papult, C., and Porteau, C. H.,** Vitrification of high-level waste solutions at Marcoule, France, in *Proc. European Nuclear Conf.,* Vulkan Verlag, Essen, West Germany, 1979.
22. **Marshall, W.,** The disposal of high-level nuclear wastes, *Atom,* 300, 262, 1981.
23. **Chapman, N., Gray, D., and Mather, J.,** Nuclear waste disposal: the geological aspects, *Atom,* 261, 186, July 1978.
24. An ICSU Committee, The geology of nuclear waste disposal, *Atom,* 338, December 1984.
25. **Gray, D. A.,** Disposal of Highly Active Solid Radioactive into Geological Formations — Relevant Geological Criteria for the United Kingdom, Institute of Geological Sciences Rep. No. 76/12, Her Majesty's Stationery Office, London, 1976.
26. IAEA, Site Selection Factors for Repositories of Solid High-Level and Alpha-Bearing Wastes in Geological Formations, Tech. Rep. Ser. No. 177, International Atomic Energy Agency, Vienna, 1977.
27. **Chapman, N. A.,** Geochemical Considerations in the Choice of a Host Rock for the Disposal of High-Level Radioactive Waste, Institute of Geological Sciences Rep. No. 79/14, Her Majesty's Stationery Office, London, 1979.
28. **Lewis, J. B.,** The case for deep sea disposal, *Atom,* 317, 49, March 1983.
29. **Lewis, J.,** Sea disposal, *Atom,* 299, 222, September 1981.
30. **Ringwood, A. E.,** *Safe Disposal of High-Level Nuclear Reactor Waste: A New Strategy,* Australian National University Press, Canberra, 1978, 64.
31. **Ringwood, A. E., Kenson, S. E., Ware, N. G., Hibberson, W., and Major, A.,** The SYNROC process: a geochemical approach to nuclear waste immobilization, *Geochem. J.,* 13, 141, 1979.
32. **Saunders, P. A. H.,** *The Management of Radioactive Wastes,* 2nd ed., Information Services Branch, U. K. Atomic Energy Agency, 1983.
33. **Saunders, P. A. H.,** *The Effects and Control of Radiation,* U.K. Atomic Energy Agency, 1982.
34. **Roberts, L. E. J.,** Radioactive waste disposal — policy and perspectives, *Nucl. Energy,* 18(2), 85, 1979.
35. *Nucl. Energy,* 21(4), 235, 1982.

Chapter 6

FUSION REACTOR MATERIALS

I. INTRODUCTION

The pivotal role of energy consumption in modern society is too well known to need any elaboration. There is also a great deal of awareness the world over today of the potential shortages arising from the pending exhaustion of fossil fuel resources, and of the social stresses that result from sharply rising fuel prices and the localized nature of the reserves. In this situation nuclear energy has a definite place.

One source of nuclear energy is fission. Fusion is the other source. It has been widely acclaimed that fusion is a promising energy option for the future. Even though it still requires more energy to create nuclear fusion reactions than is produced, immense energy would become available if the requisite scientific and engineering advances could be achieved. Nuclear fusion is the process whereby the nuclei of light atoms fuse together to release large amounts of energy. It takes place in high-temperature matter, as in the sun. Net energy production from nuclear fusion has been achieved, but only transiently, in thermonuclear detonations. A much more controlled energy release is, of course, the important goal for commercial applications. It has been estimated that a fusion reactor could produce 2000 MW of electricity continuously using only 1 t of fuel per year, compared with 7 million t needed to produce the same output from a coal-fired station. Controlled nuclear fusion is potentially a final solution to the energy problem, a feature which it has in common with both nuclear energy from fission breeder reactors and solar energy (both direct and indirect forms such as wind, waves, and thermal gradients).

Nuclear fusion, like fission, offers the prospect of continuous central electricity generation. Environmentally it appears to have potential advantages over fission because the fuel used is more abundant and no fission products are produced, and fusion itself does not involve any fissile or transuranic elements. The final product of a fusion reaction is helium, which is both nontoxic and nonradioactive. There are, therefore, no long-term disposal problems. On the other hand, during the operation of a fusion reactor plant there is a high neutron flux, which produces radioactive isotopes when impinging on structural materials. If a fusion plant is ever shut down, the radioactive substances would produce the same problems as radioactive wastes produced by nuclear fission.

Fusion research is being intensively pursued internationally.[1,2] As it stands now, the prospects appear bright for this important new nuclear energy source to contribute to meeting the world's energy needs. This chapter aims to provide a general outline of nuclear fusion energy with emphasis on technology and materials for fusion reactors.

II. FUSION ENERGY SOURCES[3-17]

Fusion is essentially the reverse of the fission process. In the former pairs of relatively loosely bound light nuclei fuse, while in the latter the heavy elements split. Both processes lead to the formation of more lightly bound nuclei (as compared with the reactants) closer to the medium weight range. The resultant decrease in the total mass appears as energy. Fission is the process fundamental to the operation of nuclear power reactors. The self-sustaining fission processes are essentially manmade phenomena.

Self-sustaining nuclear reactions are much more difficult to achieve than their chemical counterparts. In the fusion of light elements, pairs of atomic nuclei, which by their nature are all positively charged and are thus constrained by electrostatic repulsive forces (the

Table 1
FOUR POTENTIALLY USEFUL FUSION REACTIONS

Reaction	Shorthand notation	Reaction energy (MeV)	Threshold plasma temperature (keV)	Max energy gain per fusion
${}_1D^2 + {}_1T^3 \rightarrow {}_2He^4 + {}_0n^1$	DT	17.6	10	1800
${}_1D^2 + {}_1D^2 \rightarrow {}_2He^3 + {}_0n^1$	DD	3.2	50	70
${}_1D^2 + {}_1D^2 \rightarrow {}_1T^3 + {}_1P^1$	DD	4.0	50	80
${}_1D^2 + {}_2He^3 \rightarrow {}_2He^4 + {}_1P^1$	D-^{3}He	18.3	100	180
${}_5B^{11} + {}_1P^1 \rightarrow {}_2He^4$	^{11}B-P	8.7	300	30

coulomb barrier) from approaching each other closely, have to be brought into intimate contact so that the very close-range nuclear attractive forces can take over and bind the two nuclei into one. To accomplish this, much energy has to be supplied to the nuclei to accelerate them to high velocities.

In the central core of stars the temperature is high enough for the random velocities of the nuclei to be adequate for self-sustaining fusion reactions to proceed. Because the kinetic energy necessary to achieve penetration of the coulomb barrier is thermally derived, the process is known as thermonuclear fusion. Use of the phrase controlled thermonuclear fusion emphasizes that net energy production must be based on chain reactions driven by thermal energy rather than by a neutron population. Neutrons do play a secondary role in fusion.

A. Fusion Reactions

Numerous chemical elements of low atomic weight have one or two isotopes that undergo nuclear fusion reactions either between like nuclei or with other isotopes of the same or different elements. Four potentially useful fusion reactions are identified in Table 1. Reaction equations and energies are included for each entry. The reactant nuclides are hydrogen (${}_1H^1$ or ${}_1P^1$), deuterium (${}_1H^2$ or ${}_1D^2$), tritium (${}_1H^3$ or ${}_1T^3$), helium-3 (${}_2He^3$), and boron-11 (${}_5B^{11}$), with neutrons and alpha particles (${}_2He^4$) also included among the products.

To enable the fusion reaction between, for example, deuterium and tritium to occur, it is necessary for the nuclei to have enormous velocities in order to overcome the mutual electrostatic repulsion forces that exist between the positively charged nuclei. In practice it means that a nuclear fusion reaction cannot occur if the temperature of the combining gases is below 10^8 K. A fusion reaction normally only takes place at 1 to 2 $\times$ 10^8 K, when a sufficient number of nuclei have velocities large enough to overcome the electrostatic repulsion forces. The atoms at such temperatures are said structurally to be in plasma state.

The four states of matter are solid, liquid, gaseous, and plasma. The conversion from one state to the next, in the above order, is dependent upon the kinetic energy of the fundamental particles of the state (Figure 1). A solid becomes liquid when the kinetic energy of its molecules exceeds the binding energy of the solid crystals. When the kinetic energy of the molecules in a liquid increases sufficiently to overcome the Van der Waals forces, the liquid changes to a gas. When the kinetic energy exceeds the ionizing potential of the atoms, the particles become ionized, and we have a mixture of negatively charged electrons and positively charged ions. This is the plasma state of matter. A plasma may be weakly ionized, in which case it is a mixture of electrons, ions, and nonionized particles; or completely ionized, in which case it is a collection of electrons and ions only.

The plasma state, in the context of a fusion reaction, is where the atoms are completely ionized with the nuclei and electrons coexisting as separate entities in a sea of charged particles. A second criterion for the occurrence of thermonuclear fusion is the plasma number density. Fusion cannot normally take place if this is less than 2 to 3 $\times$ 10^{14} nuclei per cubic

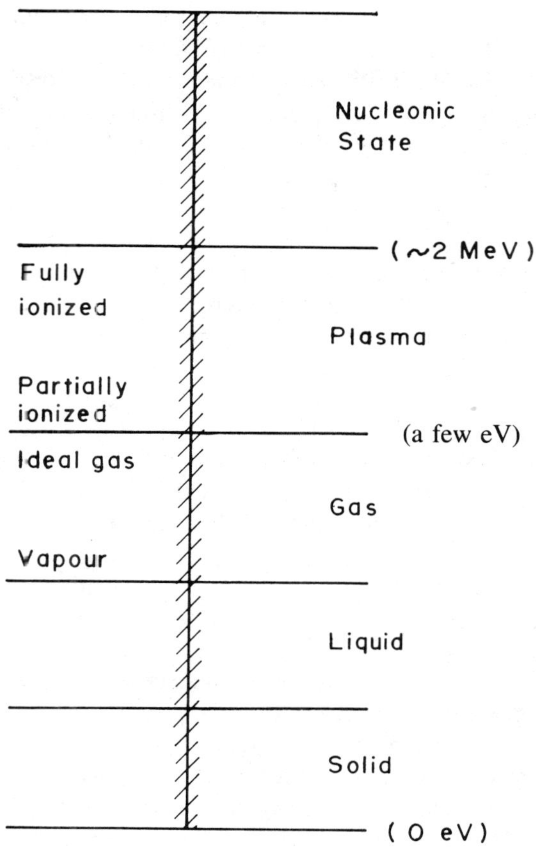

FIGURE 1. The states of matter and particle energy.

centimeter. Finally, nuclear fusion can only occur if both the minimum fusion temperature and the minimum plasma density are maintained for about 0.5 to 1 s (energy confinement time). These three criteria are called Lawson's criteria, after the British physicist J. D. Lawson, who published a paper on the subject of nuclear fusion in 1957.

The first goal of controlled fusion is to exceed break-even, i.e., to produce more energy than that required to cause the reactions. The frequently used measures of break-even requirements for fusion are Lawson's criteria, and, as stated, they are expressed in terms of plasma temperature T and the product of the plasma particle density n and the confinement time T_E. The most usual statement of Lawson's criteria for the deuterium-tritium fusion example given is that break-even requires an 8.6-keV (100×10^6K) temperature at $nT_E \approx 10^{14}$ cm^{-3} s.

It is shown in Table 1 that further data are included: approximate threshold energies for each of the reactions, and maximum energy gain per fusion (which is simply the ratio of the exothermal reaction energy to the average plasma energy). The most accessible reaction around which the first generation of fusion reactors is being developed is deuterium-tritium (D-T, or simply DT). It has the lowest threshold energy and a large energy gain. This reaction can ignite at 100 M°C with $nT_E = 10^{14}$ cm^{-3} s. The presence of deuterium suggests that some DD reactions also occur, but with much lower probability. They, in turn, produce $_2He^3$, which reacts with other deuterium. Thus, the first four tabulated reactions contribute to DT fusion.

The neutrons produced in DT fusion are both an advantage and a disadvantage. On one hand they make a positive contribution in providing a method for producing the tritium fuel, while on the other the 14.7-MeV energy of the neutrons complicates energy conversion, and leads to potential neutron activation and damage problems for the structural materials. Other reactions, such as the DD reactions (which are based on more readily available fuel and give off a smaller fraction of their energy to neutrons) or the ^{11}B-P reaction (which produces no neutrons and does not yield nuclei that undergo secondary reactions), offer advantages, which, however, are offset by their higher threshold energies. In other words, they require higher values of temperature and of nT_E. The energy threshold, it may be seen from the table, is a factor 30 higher than that for the DT reaction. Despite the drawbacks associated with this reaction, it is the reference in current thermonuclear fusion technology. The success with this reaction might in the future open the way for DD and ultimately ^{11}B-P systems. With this in mind, the balance of this chapter addresses DT fusion without much reference to other concepts.

B. Fusion Fuels

From the foregoing analysis, it can be generally said that the two reactions of most interest in the near future are by isotopes of the same element, hydrogen. They are the reactions between pairs of deuterium nuclei and, in particular, the reaction between pairs of deuterium and tritium nuclei.

The nucleus of the hydrogen atom is a single proton, a positively charged basic particle of matter. The nucleus of the deuterium, or heavy hydrogen atom, whose weight is very nearly twice that of a hydrogen atom, is a proton bound to a neutron; it occurs in heavy water (D_2O) as 1 molecule in 6500 of all the earth's surface water (or equivalent to 1 kg deuterium in 30,000 kg water). The nucleus of tritium is a proton in combination with two neutrons. It is radioactive, with a relatively short half-life of 12.3 years, and so only occurs in nature as transient material at very low concentration. Tritium can be manufactured through nuclear transmutation from naturally occurring chemical elements sufficiently abundant to provide fusion fuel. The alpha particle that is formed has a helium nucleus with two protons and two neutrons; one free neutron is given off. It is the collision of the free neutrons with other nuclei that generates heat as the reaction proceeds.

As has been pointed out, priority has been assigned to the fusion of deuterium with tritium, this being much more readily promoted and sustained than the DD reaction, which requires attainment of much higher temperatures.

There is a fusion cycle that has several similarities to that considered previously for fission systems. Isotope enrichment, breeding, and recycle steps are important components. These terms are, however, not always employed in fusion technology. Isotopic separation is accomplished by the well-established process employed to produce heavy water. The tritium component of the fuel is bred by transmuting the nuclei of suitable materials, the preferred candidate being lithium. The following reaction pair is employed: ${}_3Li^7 + {}_0n^1 \rightarrow {}_1T^3 + {}_2He^4 + {}_0n^1$ (fast neutron), and ${}_3Li^6 + {}_0n^1 \rightarrow {}_1T^3 + {}_2He^4 + 4.8$ MeV (thermal neutron). The neutron from DT fusion would react with ^{7}Li to give one ^{3}T. The neutron from the latter reaction would be thermalized to allow production of a second ^{3}T by interaction with ^{6}Li.

A process which can be said to be equivalent to breeding occurs when the average number of ^{3}T nuclei produced per fusion exceeds unity. As it is known that a fusion reaction is thermally driven rather than driven by neutrons, the fission breeding requirement of more than two neutrons per reaction is not applicable in this case. The DT fuel cycle with ^{3}T breeding is depicted in Figure 2. Tritium breeding would most likely be facilitated by using a lithium blanket around the outside of the fusion reaction region (as a fertile blanket is employed with a fissile core in a breeder reactor). Lithium in a molten salt form could serve the additional functions of heat removal and, because of its low mass, neutron moderation to enhance the ^{6}Li reaction. The tritium and helium gases produced by the reactions are readily separated from the salt and from each other. Deuterium and lithium are abundantly

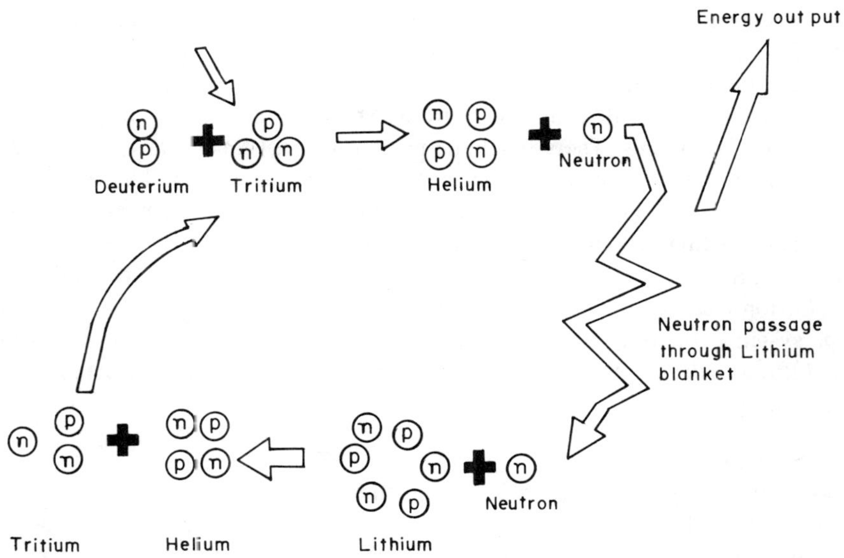

FIGURE 2. DT fusion with tritium bred from lithium.

available. An eventual transition to DD fusion removes lithium requirements completely. In this way the supply of fuel for fusion can be said to be limitless.

The DT reaction is not without technological problems. The neutron resulting from the reaction is emitted at a very high energy. It carries with it as kinetic energy four fifths of the total energy released. These energetic neutrons are exceptionally potent in causing radiation damage in solids and induce radioactivity in virtually all the materials used in the construction of the reactor. Tritium is radioactive and gaseous, and, being essentially a hydrogen atom, it leaks from containers more readily than most other substances. Because it also has to be present through a great part of the reactor system, it is difficult to contain, transfer, and process. The permissible tritium concentration is very low, owing to its radioactivity and ready ability to enter biological systems through its incorporation into water molecules.

III. CONFINEMENT

The high temperature requirement demands that the plasmas be prevented from contacting material surfaces. There are also additional requirements in order to achieve the specified density and confinement time as set out in Lawson's criterion. All these required conditions can, in principle, be fulfilled in two ways: magnetic confinement and inertial confinement.

A. Magnetic Confinement[18-20]

The plasma, as has been said, basically consists of charged particles. Being an electrical conductor, plasma will therefore interact with electric and magnetic fields. The direct confinement by external electric fields of a plasma whose density is high enough to be of thermonuclear interest is not possible because plasma in bulk is almost exactly neutral, to better than one part in a million. Thus, the external electric forces on the oppositely charged nuclei and electrons just cancel out. Such an objection does not apply to the use of magnetic fields for confinement. The magnetic confinement is based on the fact that in a magnetic field the charged particles no longer move outward in straight lines, but are constrained to perform helical orbits in the direction of the magnetic field (this applies to particles of both positive and negative charges, the fundamental constituents of plasma). It is thus possible to devise suitable configurations of magnetic fields which will confine plamsa for times long enough for it to react. The pressure of the plasma particles is compared to the pressure from the magnetic field through the parameter beta, defined as the ratio of plasma particle pressure to magnetic field pressure. Beta may vary from zero for overpowering magnetic fields to unity for balanced pressures. There would be no confinement for a beta value above unity. There exist a larger number of fusion concepts employing magnetic fields for confinement and/or heating of plasmas. Differences are based on magnetic field strength and configuration, heating method, fueling procedures, and operating mode. Three substantially different concepts are (1) TOKAMAK a Russian acronym for toroidal magnetic chamber roughly translated as TO, toroidal; KA, chamber; and MAK, magnetic; (2) magnetic mirrors; and (3) pinch-type systems. Many of the other concepts have some similarities to one or more of those listed. Prior to dealing with magnetic confinement, it is appropriate to make a reference to plasma heating processes.

1. Plasma Heating

The choice of heating method is an important subject of fusion research. The heating of up to several hundred cubic meters of plasma, typical of magnetic confinement designs, to the 100-million-degree temperature necessary for thermonuclear fusion requires an input of several hundred megawatts of energy, which may consume a substantial fraction of the gross electric output of the plant. The four favored methods, used individually or in combination are ohmic heating, neutral beam injection, radio-frequency waves, and adiabatic compression.

a. Ohmic Heating

In ohmic heating, an electric current is induced to flow in the plasma itself by electrical transformer action. This causes resistive heating quite similar to that in metallic conductors. It also generates a magnetic field, which may also be advantageous for the confinement system. However, the effectiveness of ohmic heating is limited by reduction of the resistivity of the plasma with increasing temperature.

b. Neutral Beam Injection Heating

The injection of energetic particles into the plasma is a proven method of raising its temperature. In neutral beam injection heating, a stream of atoms of one or both of the reactant particles is injected into the plasma at very high velocity (a few thousand kilometers per second). The electrically neutral atoms collide with plasma ions, sharing their tenfold greater energy with them and becoming part of the plasma; indeed, this is a possible method for refueling the plasma. The enormous velocities required can only be produced by electrical acceleration of the particles to be injected, so they first have to be ionized, positively or negatively.

The simplest and most developed technique uses positive ions. After acceleration, the beam of energetic charged particles is reneutralized to the greatest practicable extent by passing it through a region of low-pressure gas to provide a surplus of electrons which attach to the positive ions. It is essential to reneutralize the accelerated ions so that they can pass through the powerful magnetic confinement field and enter the plasma. Unfortunately, this complex process is only about 10% efficient when used for large plasma volumes; and, since very large amounts of power have to be injected, the losses would be very serious in a fusion power plant. The much more difficult technique of negative ion acceleration may greatly improve this situation.

c. Radio-Frequency Waves

There is much optimism that radio-frequency wave heating may be a better method, but

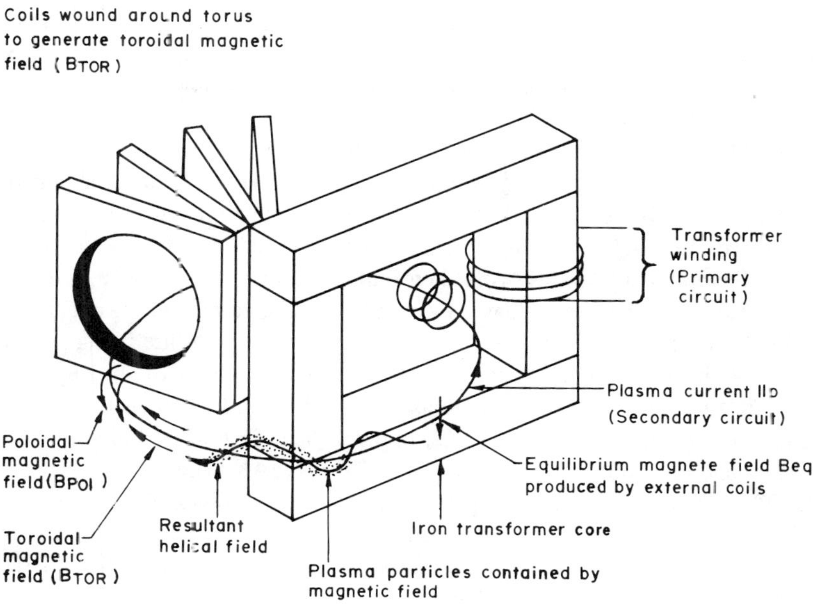

FIGURE 3. Schematic of the TOKAMAK configuration.

it is somewhat less developed. Very high-frequency electromagnetic energy is launched from wave guides or antennae directly into the plasma, the frequency being chosen to resonate with the rotations of one or the other of the charged particles around the field lines. In this way the energy may be transferred quite directly to the further acceleration of the plasma particles.

d. Adiabatic Compression

Adiabatic compression can be performed by rapidly strengthening the confinement field and may be of value in experimental devices as an auxiliary technique. It does not at present appear to be a major heating technique for fusion reactor concepts.

2. TOKAMAK[21-24]

There are essentially two main magnetic field configuations for magnetic confinement for fusion reactions. One is the closed line, with toroidal geometry, and the other is the open line, with cylindrical geometry. Both configurations have fundamental advantages and drawbacks. Confining the present discussion to the closed-end system, it may be said that there is an absence of end losses of plasma in this system. As drawbacks of the system, mention may be made of the nonavoidance of radial field gradients which have adverse effects on particle orbits and also of the lower accessibility of the center of the confined plasma.

The bases for plasma confinement magnetically are shown schematically in Figure 3. The core of the system is the doughnut-shaped vacuum vessel, the torus, in which the introduced gas is heated to form a plasma by passing a large electric current through it. The metal-walled torus is wrapped with a number of equally spaced D-shaped field coils to produce a toroidal magnetic field in the direction shown by the figure. The primary coils induce a current in the plasma to generate a poloidal field. The combination of toroidal and poloidal fields is a helical or spiral magnetic field which provides the cage that prevents the hot plasma from hitting the walls of the vacuum vessel. This complex system of magnetic fields is called a TOKAMAK. In typical systems, the toroidal field has a strength about ten times greater than that of the poloidal field. Since the TOKAMAK is a low-beta device, the fusion energy per unit volume is low, and the TOKAMAK sizes must be relatively large. Beta can be increased somewhat by fattening the doughnut (i.e., reducing the ratio of the major radius to the minor radius) and by using a noncircular plasma cross section. Overall, it is necessary to balance decreased system size (increased beta) against confinement stability (favored by low beta).

The TOKAMAK devices in operation around the world represent more than half of the effort toward controlled thermonuclear fusion. TOKAMAK were first developed in the Soviet Union. The first achievement in this field was in the I. V. Kurchatov Institute of Atomic Energy in Moscow in 1969 under the guidance of L.A. Artsimovich,[25] with the model T-3 TOKAMAK. Several TOKAMAK were then built in the U.S., and these easily surpassed the Russian results with respect to basic Lawson's criteria (temperature, density, and confinement time). At present, several research reactors have been or are being constructed in various parts of the world to test heat take off and the operation of the superconducting magnet system. One of these is the joint European torus (JET), which is jointly operated by fusion research laboratories in the U.K., France, West Germany, Italy, Belgium, Denmark, Sweden, and Holland.[26] The most advanced TOKAMAK in current use is the Princeton large torus (PLT), which was put into operation at Princeton University in 1976. Two larger and even better devices currently being constructed in the U.S. are the TOKAMAK fusion test reactor (TFTR) and the international TOKAMAK reactor (INTOR). Some characteristics of these and a few other TOKAMAK facilities are given in Table 2.

3. *Magnetic Mirrors*

Magnetic mirrors are high-beta devices and generally employ an open-line cylindrical geometrical system for magnetic confinement. The magnetic field lines are allowed to leave the confinement region. The particles are, however, prevented from escaping from the ends of the machine along these lines by making the magnetic field stronger at the ends. As the charged particles approach the ends, the strong magnetic fields there force them to spiral faster in smaller radii so that their velocity components along the lines of forces are reduced. If all the kinetic energy of the particles along the lines is converted to energy of spiral motion, the particles are reflected; hence the term magnetic mirror. The simple-mirror reactor would operate in a driven steady-state mode, compared to the cyclic operation of a TOKAMAK. The plasma burn would be maintained by continuous injection of fuel into the plasma, while spent fuel (unreacted fuel and reaction products) is removed. Natural plasma end losses provide an inherent spent fuel removal mechanism. Neutral beam injection provides both a start-up heat source and a refueling mechanism.

The main advantages of a simple mirror reactor are steady-state operation, high power density, and automatic impurity control through end losses. The disadvantages are high leakage and low energy gain. Excessive plasma losses make the simple mirror an undesirable container. These can be reduced by employing a minimum-beta mirror configuration where the magnetic field increases from the center outward in all directions.

In all of the mirror concepts, the magnetic fields and fuel injection would need to be maintained on a continuous basis. As is also true for the TOKAMAK, superconducting magnets are a requirement if total power requirements are not to be excessive.

4. *Pinch-Type Systems*

Magnetic pinch confinement takes advantage of the fact that the plasma, composed of electrically charged ions and electrons, is an excellent conductor of electricity. The electrical conductivity of a plasma is proportional to $T^{1.5}$, where T is the absolute temperature. (At ignition temperatures, the plasma conductivity is some hundred times that of copper at room temperature.) It is possible, therefore, to pass large electrical currents within a plasma.

An electrical current in a conductor (such as a plasma) generates its own magnetic lines of force, which encircle the current. As in simple electromagnetic devices where the righthand

Table 2
SOME CHARACTERISTICS OF TOKAMAK FACILITIES

Facility	Plasma current	Torus outer radius (m)	Plasma radius (m)	Toroidal magnetic field (T)	Ionic temp (keV)	Plasma density ($10^{20}m^{-3}$)	Lifetime (μsec)	Remarks
JET (western Europe)	3000 (5000)	2.96	1.25/2.1	2.7 (3.5)	5—10	0.3	300—2000	In operation
PLT (U.S.)	500	1.32	0.41	3.2	7.1	1	100	In operation
TFTR (U.S.)	2500	2.48	0.85	5.2	15	0.5	300	In operation
INTOR (international)	6400	5.2	1.2/1.9	5.5	10	1.4	1400	Project[36]
T-10 (U.S.S.R.)	650	1.5	0.39	5.0	1.1	0.6	60	In operation
TFR-600 (France)	600	0.98	0.22	6.0	2	2	35	In operation

rule applies, if the current is downward, the lines of force are clockwise. The magnetic field, in turn, interacts with the fast-moving plasma particles, exerting radial inward forces upon them, and therefore acts to constrict or confine the plasma to the center of the tube. This is the so-called pinch effect.

The pinch-type magnetic confinement systems differ substantially from both the TOKAMAK and the mirrors by operating in a pulsed mode. They rely on shock-heating the plasma by application of a very fast-rising magnetic field. There are essentially two types of concepts: the Z pinch concept and the theta pinch concept.

a. Z-Pinch System

The Z pinch is among the simplest of concepts for magnetic confinement fusion because the necessary magnetic field would be produced by a current carried in the plasma itself rather than by external coils. The plasma is confined in a torus, and the current is induced by making the plasma essentially a one-turn secondary of a large transformer. A poloidal magnetic field would be produced by the current. A very rapid rise in the primary current would cause the magnetic field to squeeze the plasma, shock-heat it, and provide confinement. The resulting plasma burn would produce an energy pulse. Thus, power applications would have to be based on frequent repetition of the sequence.

b. Theta-Pinch System

The theta-pinch system involves a cylindrical containment region filled with low-temperature plasma. A capacitor bank discharging to the single-turn coil around the region would create a circumferential current, sharply increasing the magnetic field parallel to the axis. A cylindrical sheath about the surface of the plasma would be formed by the field. The plasma would then be driven inward. Heating of plasma would first be accomplished by shock from the sheath motion and later by compressiin as the magnetic field continues to increase (at a decreasing rate). Consequent to the field attaining its maximum strength, the plasma would exist in a dense state until escaping through the ends. Formation of linear sections into a toroidal shape facilitates reduction in end losses.

Between the two pinch systems, the Z pinch concept of magnetic confinement has received maximum attention. Extremely high plasma densities ($n \approx 10^{21}$ cm^{-3}) are possible in the concept. In such a situation, a confinement time as short as $T_E \approx 10^{-7}$ s would satisfy Lawson's criterion for break-even every production. The problems for either of the pinch concepts include (1) inherent instability; (2) highly energetic pulsing, which components need to withstand; and (3) chamber walls which could hold up to the cyclic stresses caused by the short time scale repetitions.

B. Inertial Confinement[27-36]

Besides the various types of magnetic confinement, among which the TOKAMAK is outstanding and the most advanced, research is also being pursued to develop the inertial confinement, among which laser fusion is the most studied type.

A simplified representation of the physical phenomenon of the DT reaction by laser fusion is shown in Figure 4. The DT mixture is manufactured in the form of tiny frozen pellets, which are suspended in air. A high-intensity laser beam is then employed to provide an irradiation of 1 kJ of energy to the particle per nanosecond, i.e., the laser beam must have a power of no less than 1 million MW. The external layer of the DT particle evaporates to form a gas, and, as it escapes, it produces strong implosion forces on the DT particle. This is compressed to a density about 1000 times that of the original solid particle. Under the action of a sufficiently powerful laser beam, a minimum temperature of 10^8 K can be achieved. Implosion forces enable this temperature level to be maintained for the necessary confinement time and also to maintain the needed plasma density. Once fusion takes place,

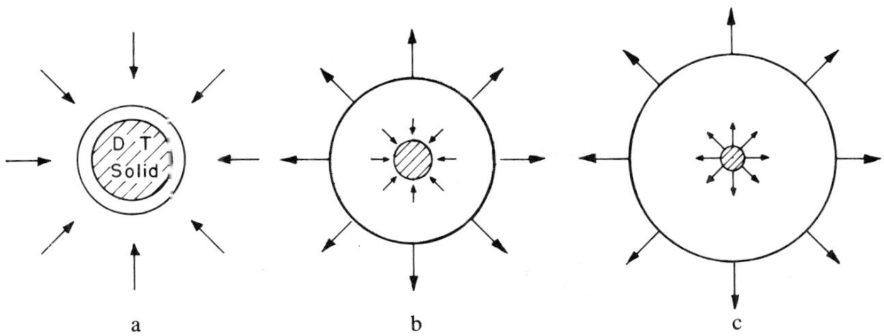

FIGURE 4. Laser fusion reaction sequence. (a) Solid DT fuel is heated at 1 kJ/nsec; (b) the external layer evaporates, implosion occurs to reduce the volume to 0.1% of its original, and the pellet is heated to 10^8 K; (c) thermonuclear microexplosion takes place.

the pellet resembles a miniature hydrogen bomb. It gives out a relatively large, but controllable amount of thermal energy.

On the international scene, considerable progress in inertial confinement has been achieved with laser installations at a number of laboratories in the U.S. (Livermore, CA, Los Alamos, NM, and Rochester, NY); Japan (Osaka University); and the U.S.S.R. (Lebedev Institute). The main problems are the provision of a laser beam with a capacity of 1 million MW. It is true that such an intensity of electrical output is only needed for 1 s, but the problems faced are still very daunting because of the difficulties of satisfying all three of Lawson's criteria. The central issue of producing high compression and high temperature simultaneously will be tackled by the operation of new laser facilities such as the Nova Installation at Livermore, due to become operational later in the decade.

C. Ion Beam Focus

Ion beam focus for fusion is essentially an inertial confinement. The approach is to freeze a mixture of deuterium and tritium with tiny, hard pellets a few millimeters in diameter at a temperature of about 4 K. They are then imploded by concentrating laser beams, as pointed out earlier, or beams of ions on each pellet in succession from various directions, so that the pellet is bombarded from all sides. In either way, enormous amounts of energy can be delivered, causing the outer layer of the pellet to vaporize so violently that it behaves like a rocket motor, imparting powerful thrust inward as well as outward. In other words, heat energy is carried into high-density material to what is known as the ablation front, where hot, dense matter blows off to lower-density regions. This imparts momentum to the rest of the material, accelerating it inward so that it compresses the fuel to create very high densities up to 100 g/cm^3 or up to 1000 times the density of ordinary matter; it also generates tremendous heat. Since the density is much higher in this case as compared with the magnetic confinement system, the time of containment needed to achieve a self-sustaining reaction is very much less, of the order of 1 ns. A topological representation of the situation described is shown in Figure 5.

A simplified block diagram showing the essential components of an inertial confinement nuclear power station based on ion-beam fusion is shown in Figure 6. There is good reason to hope that a reactor using ion beams to achieve fusion could be much more efficient, and thus more economical to run, than one using laser beams. Only 5% of the energy that powers the most efficient laser that engineers can hope to design emerges in the laser's beam. This is an enormous obstacle to overcome before the fusion can be of commercial promise. With beams of ions, by contrast, 30% of the energy pumped into the particle accelerator emerges in the beam. So the technique promises a sixfold advantage in efficiency over the laser in driving a fusion reaction. Another advantage is that, because the ion beam is so much more efficient, the fusion reaction is a great deal less violent.

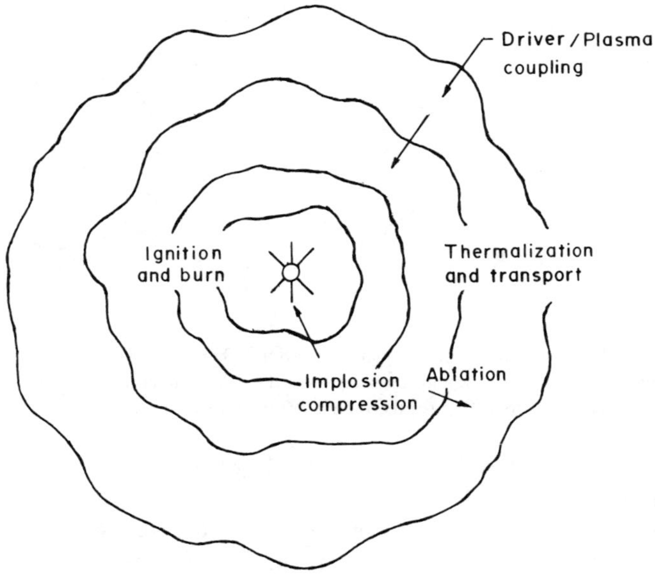

FIGURE 5. Topological representation of the inertial confinement method of fusion.

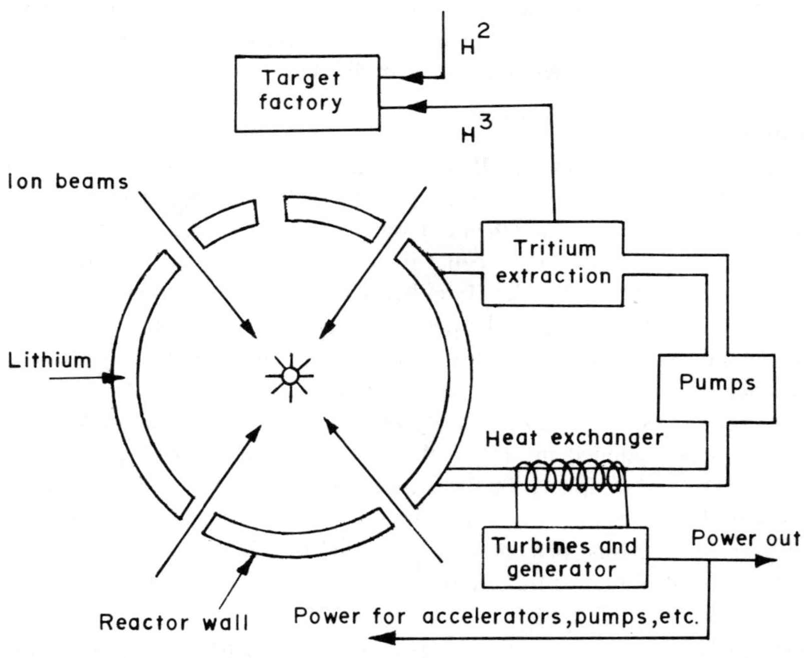

FIGURE 6. Essential component of an inertial confinement nuclear power station based on ion-beam fusion.

With so many points in favor of using ion-beam fusion, development of this method is still lagging behind. One reason is that a great deal of money, effort, and prestige has already been devoted to magnetic confinement fusion system development (TOKAMAK, JET), because it was the first to show promise. Another reason is that powerful lasers are already in use in laboratory tools. However, there is a need for critical experiments to be performed.

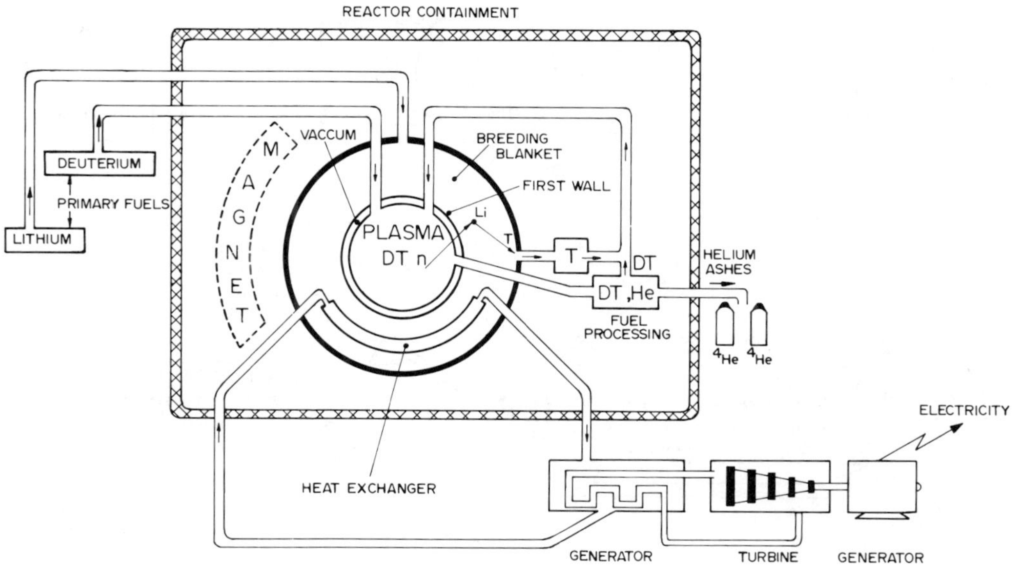

FIGURE 7. Conceptual layout of a DT fusion power reactor.

Vital questions that must be answered concern focusing the beams of ions, which need to be concentrated from all directions on the tiny fuel pellets. For the self-sustaining reaction to ignite, the beam must be held steady on target for about 30 ns. A crucial experiment to investigate such problems will be performed by the U.K. Rutherford Appleton laboratory, using the new accelerator known as the spallation neutron source (SNS). With the huge efficiency advantage offered by ion beams over lasers, if the crucial experiments as planned go well and encourage the setting up of ion-beam experiments on the present scale of magnetic and laser research, then the ion-beam-driven fusion technology might one day take over to provide fusion power.

IV. THERMONUCLEAR FUSION REACTOR

A conceptual fusion reactor plant is shown in Figure 7. The fusion fuel plasma is confined magnetically within a container (vacuum chamber). The fusion mechanism itself may also be based on inertial confinement. The remainder of the system would have somewhat similar features for either mechanism. The present reference relates to a conceptual DT fusion reactor. The fusion plasma in the center emits its energy primarily in the form of helium ions and neutrons which would strike the first wall. A blanket surrounds the vacuum chamber and would contain molten lithium salt for heat removal from the first wall and for breeding tritium. Deuterium and tritium fuel would need to be introduced into the chamber, while spent fuel material would be discharged. The coolants for use in the primary coolant loop and the secondary coolant loop of a nuclear fusion power plant will be different, which is shown in Figure 7. The magnetic coils are placed on the outside. Finally, the reactor is surrounded by suitable biological shielding. The fuel and energy pathways and a cross section of the thermonuclear reactor with various materials in place are given in Figures 8 and 9, respectively.

V. FUEL MATERIALS FOR FUSION

Primary interest in controlled thermonuclear fusion research has centered around mag-

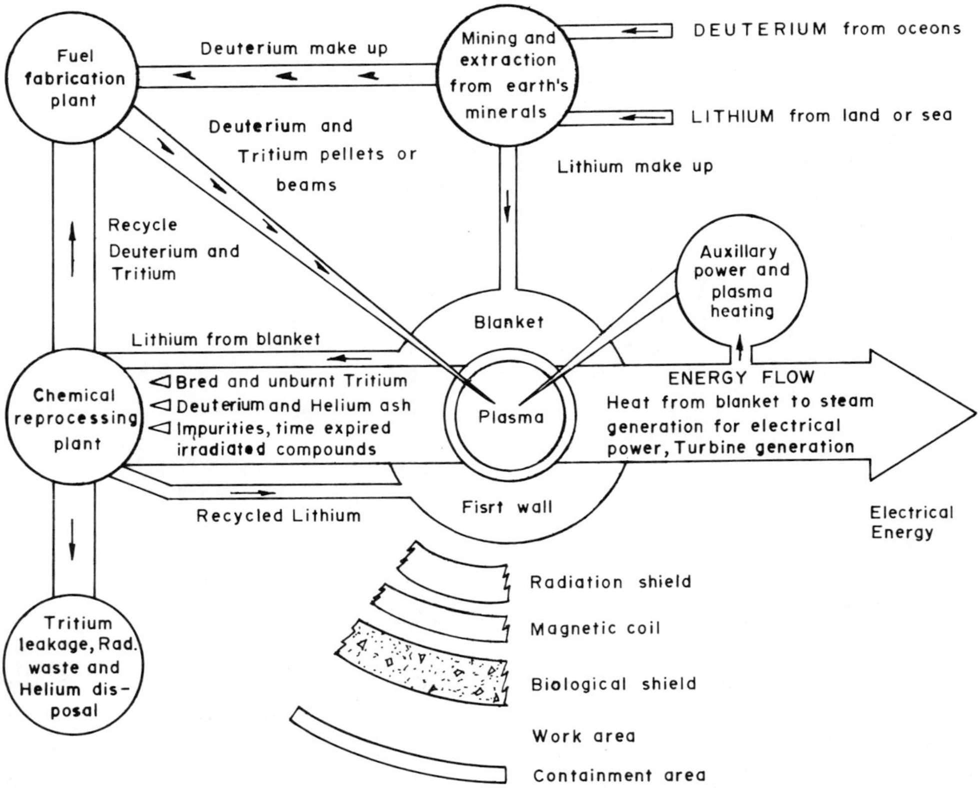

FIGURE 8. Fuel and energy pathways in a fusion reactor.

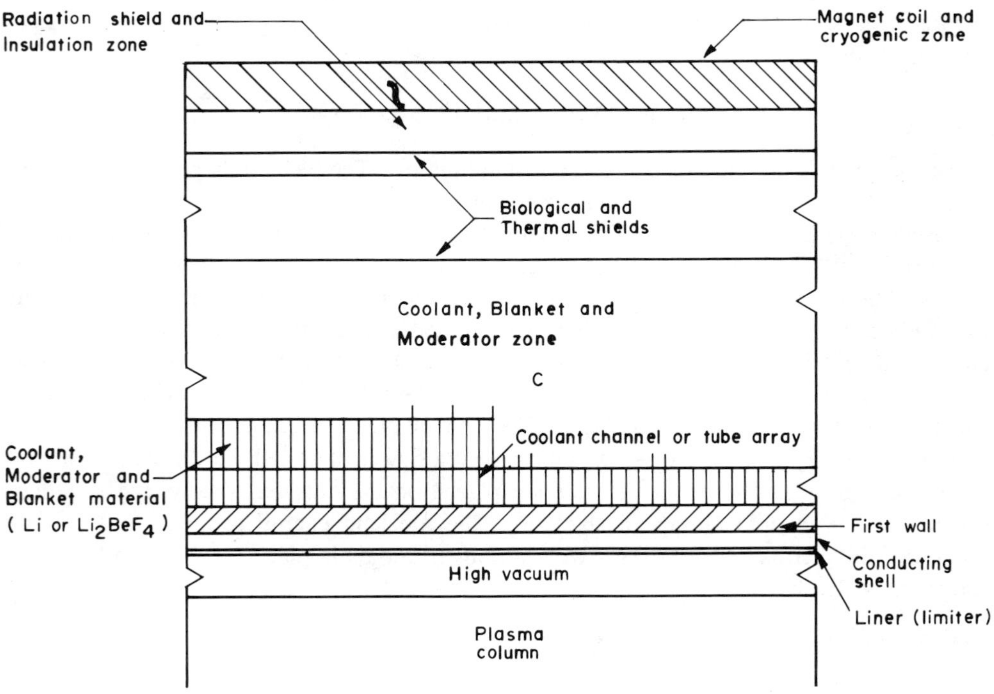

FIGURE 9. Schematic section of a thermonuclear reactor.

netically confined DT plasma as the basis of a power-producing reactor. Many favorable points have already been stated as to why a DT-fueled fusion reactor is preferable to all other types. Electrically neutral neutrons produced by the fusion reactions escape the magnetic confinement region inside the material container of plasma and pass into the blanket (which surrounds the hot DT plasma core, the seat of plasma reactions). It is here that these energetic neutrons, which carry off 80% of the energy liberated in the DT fusion reaction, have their kinetic energy converted to thermal energy. This is then carried off by a coolant to heat steam which is required to turn turbogenerators in a standard steam power generation cycle. The remaining 20% of reaction energy liberated is in the form of charged alpha particles which remain trapped in the magnetic confinement region and maintain the plasma temperature. Thus, the fusion reactor core is basically a source of heat, and indeed there are other possible applications (e.g., the production of fissile fuels, the transmutation of fission reactor waste, and the productin of synthetic fuels) apart from electricity production. It is also in the blanket that tritium breeding will take place. Lithium must be present there so that neutrons will react with it, and the tritium which has been produced must be extracted to refuel the combustion chamber. Both primary fuels of the DT reactor — deuterium and lithium — are plentiful, geographically widespread, and cheap. The possibility of exploiting fusion fuel such as DD and other advanced fuel such as $P + B^{11}$ with their inexhaustible supply of cheap fuel and the neutrons they produce are fewer or nil and lower in energy is the basic motivation for research in these areas. The system studies show, however, that neither is likely to yield an economically acceptable power in the two presently pursued fusion reactor concepts involving magnetic and inertial confinements.

VI. MODERATOR, REFLECTOR, AND BLANKET MATERIALS

Among the nuclear fission reactor materials, moderator (for thermal reactor) and blanket (for fast reactor) materials are not compatiable for working together. The nuclear requirements for blanket materials in a fast reactor are quite different from those for moderators and reflectors in a thermal reactor. The nuclear properties required for the blanket in fast breeder reactors are (1) nuclear fertile fuel, (2) large neutron absorption cross section, and (3) ease of fuel reprocessing. The blankets in fast reactors should also have reasonable good heat transfer characteristics to dissipate the thermal energy during neutron capture and subsequent radiation emission in fast reactors. In a nuclear fusion reactor, e.g., a DT fusion reactor, however, the moderator, reflector, and blanket materials can work together. For instance, liquid ^{6}Li can moderate and slow down the energetic fusion neutrons, reflect it, and finally absorb it to carry out the breeding reaction for tritium production. In other words, the liquid ^{6}Li can perform functions of moderator, reflector, and blanket materials in a DT reactor simultaneously. The liquid lithium is also a very important coolant. In the case of a hybrid fusion-fission reactor, the energetic neutrons passing through the moderating, slowing-down, and thermalization processes may be absorbed in the fertile material: depleted uranium or thorium.

From the foregoing, it is apparent that lithium is one of the important materials in the repertoire of fusion reactor materials. The lithium blanket surrounding the vacuum chamber has multiple functions: (1) it acts as a moderator and neutron absorber that receives the high-energy neutrons of the DT reaction and converts their kinetic energy into heat, (2) it acts as the heat source for the power plant, (3) it acts as the tritium breeder, (4) it absorbs the gamma radiation from the neutron inelastic scattering reactions, and (5) it is a first shield for magnetic coils from irradiation damage and excessive heat loads. The material of the blanket could be pure lithium metal, a lithium-bearing fused salt such as flibe, or a mixture of the two.

A. Lithium

Lithium is the lightest metal, its density being approximately half that of water. The most important minerals containing lithium are lepidolite, spoduman, petalite, and ambligonite. Lithium is obtained by electrolysis from its molten chloride. It reacts with water. Since it has the highest specific heat of all solid metals, it may be efficiently used in heat exchange. Lithium has a high electric conductivity in the molten state, which makes possible its recycling by means of electromagnetic pumps. Both lithium and its salts in the molten state are highly corrosive. Lithium enrichment in one of its isotopes can be performed through various methods. By isotopic exchange between ^7Li and ^6Li, the enrichment in ^6Li may reach 16%.

Lithium can be used in reactors as liquid compounds (melts) or as solid compounds. The most common molten salt that can be used for cooling and tritium breeding is the eutectic compound Li_2BeF_4. This compound offers advantages over liquid lithium because of its less corrosive nature. As for neutron absorption, the use of this compound is, again, preferable leading to a lower demand of cooling agent — as low as 20% in comparison with plain lithium. A more convenient extraction of tritium from the compound can be cited as a further point in its favor. The compound modifies its chemistry as a result of nuclear reactions with neutrons ($^6LiF + n \rightarrow {}^4He + TF$, $^7LiF + n \rightarrow {}^4He + TF + n$, and $BeF_2 + n \rightarrow 2n + 2\,{}^4He + F_2$). Another salt mixture to be considered for cooling and breeding is LiCl-LiF. As far as Li concentration, this compound is more convenient, but raises some questions related to its higher melting temperature and to the production of ^{35}S isotope [^{35}Cl (n,p) ^{35}S]. Isotope ^{35}S is beta active, with an energy of 0.17 MeV and a half-life of 87.1 d, its activation cross section for thermal neutrons being 0.13 b. Among the solid compounds of lithium considered for breeding, the most important ones are Li_2O and $LiAlO_2$. The breeding ratio of tritium is superior in Li_2O as compared to $LiAlO_2$. From a thermal stability point of view, $LiAlO_2$ is more stable than Li_2O. Instability is expressed in the thermal decomposition of these materials, accompanied by a predominant production of gases (O_2), besides Li, at temperatures from 700 through 1400°C. The need to use the tritium thus produced and tritium's low solubility in these materials determine their disposition in an inhomogeneous structure (for instance, included in graphite or distributed in the porous compound Li_2O-Al_2O_3). This calls for a deeper knowledge of the reactions that may occur. In the Li-Al-O system, only the $LiAlO_2$ looks like a good choice from the consideration of the amount of lithium required. The component aluminum in the compound, on irradiation, forms nuclei of ^{28}Al of short half-life. Consequently, this compound raises no problems of lasting radioactive pollution. Unfortunately, the compound has a low thermal conductivity, which may lead to temperature gradients on the blanket.

Corrosion induced by liquid metals when coming in contact with structural materials stands out as an important technological problem. The use of liquid metals as cooling agents in fission reactors has contributed significantly to knowledge in this area. It is well known that usual structural materials (Ni, Cr, Fe, and Nb) as well as ordinary nonmetallic impurities (C, N, H, and O) are generally more soluble in lithium than in liquid sodium, and this raises difficult problems of compatibility of materials with lithium. From the data available, the maximum admissible temperatures for the use of structural materials in liquid lithium are as follows: Ni alloys, 450°C; ferritic steel, 500°C; carbon steel, 550°C; austenitic steel, 670°C; and Nb-1% Zr, 900°C. Several studies consider that the maximum temperatures should be much lower (for instance, according to these, austenitic steel should not be used at temperatures over 500°C). In general, molten salt compounds of lithium seem to be less corrosive and therefore more compatible with structural materials. Another phenomenon may also arise for compounds such as Li_2BeF_4. This compound is a powerful deoxidizing agent, and this implies that it will act as a cleansing agent of any oxide layer of the structural materials, thereby rendering them less protective against corrosion. The compound, if transferred through a powerful magnetic field, may be chemically destabilized as a consequence of the induced electrical effects — a fact that is accompanied by an increase in its corrosion potential. Finally, although lithium is inert to graphite, it cannot be used together with other refractory structural materials, since it may carbonize them through carbon transfer.

B. Tritium Breeding and Neutron Multiplication

This topic is important enough to be worth coverage as a separate section in the present text. As a disadvantage of the DT fuel cycle, the absence of large quantities of naturally occurring tritium may be brought to attention. The need is, therefore, to breed this component of the fuel. It is bred in the lithium blanket. All natural lithium is composed of 7.42% ^{6}Li and 92.58% ^{7}Li. The breeding reactions, as pointed out earlier, are $^6L + n \rightarrow {}^4He + T + 4.8$ MeV and $^7Li + n + 2.5\ MeV \rightarrow {}^4He + T + n$. While ^{7}Li has higher cross sections for the 14-MeV, newly born neutrons, it has lower cross sections for moderated neutrons. Neutrons are moderated by lithium to energies at which cross sections of ^{6}Li predominate.

The lithium reactions given produce neutrons of their own. Each produces one triton (tritium nucleus) for each neutron absorbed. The DT fusion reaction, unlike fission reactions, produces but one neutron. Some of these neutrons are lost through parasitic capture and leakage, as in a fission reactor. It is also necessary to breed more than one triton per triton consumed because of the loss of tritium by radioactive beta decay, and because of losses in plant hold-up tanks and in recovery and recycling operations. Tritium breeding ratios (tritons bred per triton consumed) of more than 1.15 are said to be necessary. This points to the necessity of finding a means of neutron multiplication through some (n,2n) reaction.

Neutron multiplication can be accomplished through nonfissile neutron multiplication by bombarding several metals with fast neutrons produced in the DT fusion reaction. One promising metal is beryllium. The reaction is $^9Be + n \rightarrow 2{}^4He + 2n$. The cross section for this reaction is about 0.6 for neutrons in the 5- to 15-MeV range, but drops to zero below 2 MeV. A beryllium layer may then be placed next to the vacuum chamber surrounding the plasma to take advantage of the energetic neutrons. Other metals with high (n,2n) reactions are W, Pb, Ta, Mo, Fe, and Cu as well as ^{7}Li itself. The presence of the latter in the blanket may make it unnecessary to add a layer of beryllium or other material.

VII. COOLANT AND SHIELDING MATERIALS

Good heat transfer and thermal properties are the basic requirements of a fusion reactor coolant. Therefore, the coolant should have high thermal conductivity, large specific heat, low melting point, high boiling point, low density, low viscosity, and little corrosiveness. Moreover, the coolant must be compatible with the structural, channel, and piping materials of the nuclear fusion reactor. The primary requirements and basic materials selected for the fusion reactor coolant materials are given in Table 3. The variety of materials shown include lithium, a lithium-bearing fused salt compound such as flibe (Li_2BeF_4), helium, and water.

Liquid lithium has the advantage of combining cooling and breeding functions. One problem arises with the isotopic content of lithium. In fission nuclear reactors, where neutron economy is of paramount importance, one can use as cooling agents only those lithium compounds that are ^{6}Li free, since this particular isotope is a powerful absorber of thermal neutrons; thus, ^{7}Li isotope is preferred. With fusion reactors, the odds are reversed; ^{6}Li is essential for breeding tritium. In pumping liquid conductive coolant lithium across magnetic fields, there occur magnetohydrodynamic pumping losses. From analysis on several proposed systems, one finds these losses to be as high as 18% of the gross electrical output of a 1000-MWe plant in the case of lithium and 13% for flibe. With lithium, there is also a compatibility problem with many structural materials. Stainless steels undergo thermal gradient mass transfer in flowing lithium above 500°C. Refractory metals such as niobium and vanadium alloys exhibit compatibility with lithium to such higher temperatures, but these materials must be protected from the atmosphere and from impurities in the blanket. Flibe has a

Table 3
PRIMARY REQUIREMENTS AND BASIC MATERIALS FOR FUSION REACTOR COOLANT

Primary requirement	Basic materials
Good heat transfer properties	Li, Li_2BeF_4 (flibe)
Low melting point, high boiling point	He
Low density, low pumping power	H_2O
Low neutron (absorption)-induced radioactivity	
High thermal and irradiation stability	
Least corrosiveness and safe handling	

comparatively high melting point (460°C). As an electrolyte, it affords the potential for galvanic attack under the influence of a magnetic field. The compatibility of flibe strongly depends on the ratio of lithium to beryllium fluoride contained in the salt. If the oxidation potential of the salt can be buffered with chromium, for instance, the more reactive element of the first wall, flibe can be used with stainless steels up to about 600°C. Flibe is compatible with nickel-based alloys up to about 800°C and with molybdenum alloys to even higher temperatures.

Flibe is, unfortunately, a marginal breeding material; and if it is used for both breeding and cooling, most designs require additional beryllium as a neutron multiplier. Regarding the other coolants, helium of commercial purity is not compatible with niobium or vanadium alloys, but is compatible with molybdenum, stainless steels, and nickel-based alloys up to at least 800°C. The major drawback of helium is that it is a relatively low-density coolant, and either large coolant ducts or high pressures are required. Stresses from the high-pressure gas plus high thermal stresses could lead to an early failure of coolant ducts in the blanket. Water could be used as a coolant, but it requires high pressure, and it would be difficult to remove any tritium that reached the coolant stream. The reaction between water and lithium in the event of a pipe failure is also of concern. Water probably would not be used with refractory metals, but perhaps with aluminum, stainless steels, or nickel-based alloys.

The purpose of the shield is to prevent excessive energy deposition in the superconducting magnets. These magnets will operate at a temperature near 4 K, and about 1 kW of electrical energy is required to remove 1 W of deposited energy. Thus, the shield must be effective if the fusion device is to be a net producer of energy. The shielding materials for use in a nuclear fusion reactor fall into three groups according to their purposes: (1) heavy and moderately heavy elements to attenuate the gamma rays and X-rays, and to slow down fast neutrons and ions; (2) light elements or hydrogenous substances to scatter and slow down neutrons and ions; and (3) light materials containing boron to capture neutrons with no secondary gamma-ray emission. The primary requirements and basic materials for thermal and biological shieldings for fusion reactors are given in Table 4.

VIII. STRUCTURAL MATERIALS[36-40]

Included in this group are all materials used in the construction of the combustion chamber, thermal insulation, and reinforcement structure as well as the materials meant to protect certain parts in the blanket and cooling system against corrosion or erosion.

Fusion material problems in general, and the problem of first wall in particular (defined as that solid structure which directly faces the plasma), draw a high level of concern. Several reasons account for this. In contrast to fission reactors, the radiation environment in fusion reactors is not very well established. This is especially true for the laser fusion route to fusion energy development. There are at present no completely developed selected test facilities that can simulate the entire radiation environment of a first-wall structure. In addition

Table 4
PRIMARY REQUIREMENTS AND BASIC MATERIALS FOR FUSION REACTOR SHIELDING

Primary requirements	Basic materials
Attenuating primary gamma and X-rays	Pb, Bi, W
Producing little secondary gamma rays	LiPb, Li_2BeF_4, B, boron compounds
Slowing down fast neutrons and ions	Boral, Fe, iron mineral
Absorbing slow ions and thermal neutrons	Concrete

to these uncertain features, one comes across a wider range of materials requirements than is the case in fission reactors.

A. Plasma First-Wall Materials

The most difficult service conditions pertain to the materials of the first wall that contains the thermonuclear plasma in the vacuum chamber. The probable environment of a DT fuel-cycle-based, controlled thermonuclear reactor first wall will primarily be (1) fast neutron flux and energetic ion flux; (2) nuclear radiation energy loading; (3) an operating temperature on the first wall surface between 600 and 800°C; (4) a high vacuum in the range of 10^{-4} to 10^{-7} torr between the outer surface of the plasma and the inner surface of the first wall; (5) irradiation surface sputtering, blistering, and erosion on the inner surface of the first wall to increase the impurities in the plasma and to deteriorate the heat transfer properties of the first wall; (6) helium production, helium embrittlement, bubble formation, and irradiation swelling produced by nuclear transmutation of thermonuclear neutrons from the plasma; (7) thermal expansion, thermal stock, and thermal cycle fatigue and cracking; (8) irradiation creep, increase in yield and ultimate strength, and reduction in ductility due to irradiation hardening and impure isotope production; and (9) enhancement of oxidation and radiation corrosion in the intense irradiation environment to cause stress fatigue crack and SCC at the inner and outer surfaces of the first wall.

Based on such a probable environment to which the first wall is exposed, the required important nuclear, physical, thermal, and mechanical properties; safety considerations; and economic factors are briefly indicated as (1) the necessity of coping with intensive fast neutron and energetic ion fluxes imposed on the first wall; (2) small neutron capture cross section; (3) low nuclear transmutation radioisotope production (especially helium); (4) low induced radioactivity rate and radioisotope decay heating rate; (5) short radioisotope decay heating time; (6) low gamma transition and decay heat; (7) low induced surface sputtering, blistering, and erosion rates; (8) low permeability, solubility, and diffusivity to deuterium, tritium, and helium ions; (9) low irradiation swelling and creep rates; (10) high melting point; (11) moderate thermal expansion coefficient, and high thermal conductivity and diffusivity; (12) high mechanical strength and ductility; (13) high corrosion resistance; (14) good fabricability and machinability; (15) high thermal irradiation stability, and operating experience, reliability and safety; and (16) availability and low cost. The choice or selection of the most probable or appropriate material from the candidate structural materials for a first wall will actually be a compromise based on the material requirement factors as outlined. The materials that come under the purview of selection are (1) refractory metals and their alloys, (2) iron alloys, (3) nickel alloys, (4) aluminum-based materials, and (5) carbon-based materials.

1. Refractory Metals and their Alloys

The list of chief refractory metals includes niobium, vanadium, molybdenum, zirconium, and titanium. The extractive schemes of these metals from their various resources have been excellently reviewed in a recent publication.[41]

Niobium is a soft, ductile metal. Apart from zirconium, niobium has the lowest neutron capture cross section among the candidate wall materials. One drawback of using niobium as a structural material comes with its activation, accompanied by production of radioisotopes with long half-lives, which requires long-term storage, much like the radioactive wastes from fission reactors. Niobium seems to be very compatible with liquid lithium. Niobium exhibits poor oxidatioin resistance, and therefore oxygen levels in lithium must remain very low. Oxygen concentrations in excess of some hundred parts per million in weight may lead to a complete penetration of niobium by the lithium within a couple of hours. For thermal insulation, metallic niobium has been suggested, if coated with a thin film of alumina. Niobium alloys exhibit high-temperature strength properties. Examples of niobium alloys of interest as a fusion reactor first-wall structural material include Nb-Ti, Nb-Zr, Zr-Nb-Sn, Nb-W-Ti-Zr, and Nb-Mo-Ti-Zr.

Vanadium is a white, soft, and cold-ductile metal like niobium. Unlike niobium, vanadium can withstand lithium penetration, even under high oxygen concentrations. Vanadium-based alloys provide interesting properties for fusion reactor applications.

Molybdenum is a silvery-white hard metal. It features a high elasticity modulus and oxidizes only at high temperatures. An important alloy, the TZM alloy (99.4% Mo, 0.5% Ti, 0.08% Zr, and 0.01% C) has received attention. It is a high-fracture-strength material, even at high temperatures. It oxidizes at high temperatures, which calls for protection by surface coating or by reinforcing the outer surface by other materials such as stainless steel. At high temperatures this material is compatible with both helium and liquid lithium. The alloy is good in withstanding neutron irradiation.

Titanium in alloyed forms has received attention. From the standpoint of machinability and fabricability, operating experience, reliability, availability, and low cost, titanium alloy is a good material to choose. The chief merits of titanium alloys are a low density and a high resistance to corrosion. In fact, titanium alloys enjoy a reputation for outstanding corrosion resistance properties, even when subjected to the effect of seawater or other corrosive agents.

2. Steels

Alloyed steels and high-alloyed steels are basic materials for structure. Their main characteristics have been dealt with in an earlier chapter on fission reactor structural material. Present reference is being directed toward characteristics of special concern when it comes to their use in thermonuclear fusion reactors. The maximum temperature at which high-alloyed steels may still be used is around 1200°C. The relative shortcoming of their being affected by radioactivity implies storage problems for relatively long periods of time, spanning nearly 50 years or so. Austenitic steels with refractory metals (especially chromium) show superior tensile properties at high temperatures. These steels are also corrosion resistant. At extremely high temperatures, even the high-alloyed steels lose their strength. Steels designed for service at high temperatures are described below:

1. For use in the range of 400 to 550°C, steel must have chromium in the range of 1 to 3 wt%, so as to prevent cementite coagulation and graphitization. The chromium content must increase, and addition of carbide-forming elements (Mo, W, and V) must also be made in steels to extend their application range to 450 to 550°C.
2. For use in the range of 500 to 600°C, steels must have more chromium (5 to 15 wt%) and be alloyed with Mo, W, V, or Nb.
3. For use in the range of 600 to 650°C, the austenitic steels are in order.

A general conclusion may be drawn that in fusion reactors steels cannot be employed as structural materials or for reinforcement purposes in extremely high-temperature operating

Table 5
COMPOSITION OF NICKEL ALLOYS (wt%)

Alloy	Ni	Mo	Cr	Fe	Cu	Mn	Si	Ti	C	Co
Hastelloy A	58	20	—	22	—	—	—	—	—	—
Hastelloy N	71	17	7	5	—	—	—	—	—	—
Hastelloy X	45	23—25	22	20	1.5—3	0.5	—	—	<0.15	—
Nimonic	Balance	—	18—21	<5	<0.5	<1	<1	0.2—0.6	<0.15	—
Inconel	72	—	15.5	8	—	1	0.5	—	0.15	—
CONEL	71	—	—	8	—	—	—	3	—	18
MONEL	67	—	—	2	29	2	—	—	—	—

conditions. They are, however, suitable and bound to be used when the temperatures do not exceed 600°C.

3. Nickel-Based Alloys

Unlike the situation with fission reactors, where such materials are ordinarily avoided, nickel alloys may be considered for many fusion reactor projects. The composition of various nickel alloys is given in Table 5.

Nickel alloys, depending on their chemical composition, may be used up to temperatures of 650 to 800°C. A feature of nickel alloys is that they do not present phase transitions. In order to increase alloy strength, formation of dispersed particles such as intermetallic compounds Ni_3 (Ti and Al), of nitrides (TiN), or even of carbides (TiC) in the alloy is sought. For this purpose Cr, Al, and Ti are added. Such alloys may be used up to 700°C.

4. Aluminum-Based Materials

Aluminum-based materials cannot be used at temperatures as high as those tolerated by the other materials outlined above, yet they possess the outstanding advantage of low density. They are also considered in terms of the special service conditions of fusion reactors, featuring intense and rapidly fluctuating magnetic fields. In such situations, aluminum is very convenient indeed. As a structural material the metal-ceramic compound, sintered aluminum product (SAP), obtained by sintering together Al_2O_3 (5 to 25%) and Al (powder) is ordinarily employed. The low activation level and considerable availability throughout the world speak in favor of aluminum materials.

5. Carbon-Based Materials

The eligibility of carbon for employment in fusion reactors is primarily due to its high melting point. Of the structural materials considered, one can name graphite, silicon carbide, and fibrous carbon. Carbon fibers are latecomers in the family of carbon-based materials. They may serve as a carbon curtain, lining the inner wall of the combustion chamber, with the role of limiting the effect of the plasma particles upon the wall. Such a curtain may well capture ions and neutral atoms escaping the plasma, and may also collect all atoms sputtered out from the wall by neutrons. The carbon curtain would reduce the effect of charged particles on the first wall. The induced radioactivity as ^{14}C should be negligible. The curtain is likely to serve under the highest temperature in the reactor, for example, up to about 1800°C. Use of graphite in contact with liquid lithium is out of the question. Use of silicon carbide is also not without problems, due to the associated brittleness and difficulties encountered in manufacturing large-size carbide structural components. There is a question regarding the use of graphite with respect to its behavior under high irradiation. The consequences of irradiation require further research.

IX. MATERIALS FOR MAGNETIC DEVICES[42,43]

This class covers the materials meant for the normal or superconducting coils. Nonsuperconducting, water-cooled copper magnets can provide a conventional means of inducing high magnetic fields for large TOKAMAK using a well-developed, existing technology. The chief function of these magnets is to produce reliable, steady, and unsteady magnetic fields for plasma confinement. The conventional, nonsuperconducting magnet system has been used for near-term fusion experimental devices in laboratories. However, the consumption of electric current, I, and the loss of electric power due to external and internal resistances of the nonsuperconducting magnet system are extremely high, which makes them unsuitable in commercial nuclear fusion power reactors.

A. Superconducting Magnet System

The most used superconducting materials nowadays are NbTi and Nb_3Sn. They are at present commercially available as tape or multifilament wire. In the superconducting magnet design, NbTi can be used up to a maximum magnetic field about 8 to 10 T, and Nb_3Sn up to a maximum magnetic field of approximately 18 T. Experimental data indicate that a superconducting magnet system for ignited fusion power reactors should have stored energies on the order of 10^5 MJ and peak magnetic field intensity from 8 to 16 T. Thus, the present commercial superconducting materials, NbTi and Nb_3Sn, are apparently adequate to generate such high magnetic fields for use.

X. IRRADIATION EFFECTS[44-48]

An appreciation of the damaging radiation environment from a DT plasma can be obtained from reference to the fundamental reaction involving the two species. The reaction is ${}_1D^2 + {}_1T^3 \rightarrow {}_2He^4$ (3.52 MeV) + ${}_0n^1$ (14.08 MeV). In the three fusion devices (TOKAMAK, mirrors, and lasers), there are three major forms of radiation: (1) photons (X-rays ranging from a few electronvolts in magnetically confined plasmas up to 100 to 10000 eV or more from pellets); (2) energetically neutral or charged particles (unburned fuel such as D and T, helium ions, and nonfuel pellet debris ranging from H, Be, C, O, Si, and Fe to as high as U); and (3) neutrons. There are generally two main classes of radiation problems that come under consideration: bulk, and surface or near-surface effects. The former result primarily from high-energy neutron bombardment, and the latter from photon and primary charged particle irradiation. As pointed out, the DT plasma prevents this radiation environment. Perhaps the most serious currently perceived materials problem in the fusion reactor is radiation damage to the structural first wall. Second only to radiation damage to first wall are those materials problems associated with superconducting magnets.

A. First-Wall Material

The first-wall material interaction with 14-MeV neutrons is of considerable importance, compared to that with the 3.5-MeV He ions from the plasma. In an ideal case, all the 3.5-MeV He ions would be retained in the plasma, and their energy all would be used to heat the plasma. However, He ions actually leak out of the plasma and irradiate the surface of the first wall. He ions with a considerable energy are injected into the first wall and stop at the depth around their range from the surface. Then He ions become He gas atoms. If the number of He gas atoms is sufficient, they form, at a certain depth from the surface, He gas bubbles, which result in surface blisters if the gas pressure is large enough. When the He ion irradiation becomes more intense, there occurs the exfoliation of blister skins. If such a severe exfoliation does occur, such materials cannot be used as the first wall. Regarding radiation damage by 14-MeV neutrons, two major problems, namely, swelling and ductility effects, have been identified for fusion reactor first walls.

Irradiation swelling is the volumetric instability induced by void formation, bubble formation, and agglomeration of inert gases, mainly helium, in the wall material. In addition to the normal (n, γ) reactions which cause some elements to be transmutated to others, the high-energy neutrons of a fusion reactor can produce gaseous atoms via (n,p) and (n, α) reactions. Of the two transmutant atoms, solid or gaseous, it has generally been observed that gaseous atoms can have the greater effect on the mechanical behavior of a metallic system. The high gas production rates (helium and hydrogen), coupled with displacement damage, make the fusion reactor environment potentially more severe than that of the more familiar fission reactors, where similar displacement damage occurs but transmutation rates are much lower.

An even more serious problem than swelling is found when one examines the possible efffects of 14-MeV neutron irradiation on the potential fusion materials. A certain amount of ductility is necessary in any large structure in order to provide relief from the consequences of all manufacturing defects, unanticipated temperature and stress excursions, and continually changing materials properties as the reactor is operated. In the extreme case, where there is no ductility available, any stress that exceeds the yield strength would lead a component to fracture with catastrophic results in a large, highly radioactive structure containing hundreds of millions of curies of tritium and activated materials. The ductility loss in metals has been attributed to at least three mechanisms: production of high densities of dislocation loops, formation of second-phase precipitates, and helium gas bubble formation. Of these, the latter is perhaps the most serious. It has been shown that helium levels of as little as 10 ppm can reduce uniform elongation in stainless steel to well below the desired level of 10% at temperatures above 500°C.[49] Since helium production is a nuclear phenomenon characteristic of all the potential materials, the only way to alleviate this effect is to reduce the energies of the neutron spectra. By placing solid or liquid materials between the plasma and first wall, the helium production rates can be lowered by 2 or 3 orders of magnitude.

B. Superconducting Material

Irradiation affects all decisive parameters of the superconducting materials employed in thermonuclear reactors: critical temperature T_c, critical field H_c, and the current's critical density J_c. This is to be expected as a result of the structural disorder induced by radiation in the material. There appears to exist a critical level up to which radiation does not affect T_c; also, it appears that H_c depends mainly on changes in T_c and that the value of J_c is not altered, if J_c is originally higher than 1.3×10^6A/cm^2 (at 5 T).

XI. SUMMARY[2,50]

It is well accepted that the enormous amounts of stellar energy are getting generated through the process of nucleosynthesis by which heavier elements are formed from hydrogen, the dominant constituent of stars. The possibility of achieving such thermonuclear reactions in a controlled way under terrestrial conditions has long held out the attractive prospect of a safe, clean, and unlimited source of power. The earth's surface abounds in suitable light elements with the potential of providing a virtually inexhaustible supply of energy.

Controlled nuclear fusion for energy release has several basic advantages. Notable among these are

1. The basic nuclear fusion fuel, deuterium, is a chemical component of heavy water, which is a part of ordinary water; thus, the fuel resource of nuclear fusion energy is inexhaustible.
2. The fusion reactions produce no appreciable amount of radioactive waste (alpha particles or helium atoms only).

3. The energy release from fusion is much greater than fission with equal weight of basic fuels (1 g of ^{2}D can produce about 4 MWD, and 1 g of ^{235}U can produce fission energy of only about 1 MWD).
4. The inherent safety is a feature of the controlled nuclear fusion reactor. The density of fusion fuel, plasma, is so low that it is not explosive. The reactor shuts down automatically when the plasma stability is off balance.

The conditions necessary to realize controlled thermonuclear reactions are vigorous and extreme, far beyond the realms of normal experience. In essence these reactions involve heating a small mass of hydrogen isotopes to stellar temperatures and sustaining the reaction in virtually complete isolation until a net energy gain accrues. Necessary criteria, called Lawson's criteria, must be met for achieving this. The criteria, as set forth, are that the product of the number density of the reactants n and the time they are kept isolated T_E should exceed a certain threshold value, given by $nT_E > 10^{14}$ cm^{-3} s for the most favorable fusion reaction; and the temperature of the reactants should be at least 100 million K.

Present international research is making advances toward achieving these criteria along different routes. The three routes to controlled nuclear fusion presently being considered are a so-called magnetic bottle (magnetic confinement fusion system), the use of laser beams (inertial confinement fusion system), and the use of ion beams (inertial confinement system).

Of the magnetic confinement system, there are three subsystems: (1) toroidal or closed-end devices, such as torus (or TOKAMAKs); (2) linear or open-end devices, such as magnetic mirrors; and (3) pinch-type systems. In the magnetic confinement nuclear fusion reactor system, the thermonuclear fusion is confined in the magnetic fields — toroidal, poloidal, and vertical — to confine a thermonuclear plasma in a large torus (TOKAMAK), Figure 10. These fields are oriented at right angles to each other. Progress toward the plasma conditions needed for power generation has been steady and rapid, especially in the magnetic confinement systems, of which basically two types are being studied today: mirrors and TOKAMAK. Both rely on magnetic fields to constrain the trajectory of the high-temperature D^+, T^+, and He^{2+} so that they do not touch the first wall and cause it to melt. The mirror concept uses a cylindrical magnetic bottle configuration in which the field is higher at the open ends to prevent leakage of the plasma particles. In the magnetic confinement systems, the plasma is essentially of low density, so the containment time has to be relatively long, of the order of 10 to 20 s, to take the reaction to the point where it sustains itself.

Laser-driven fusion is also being developed rapidly. A general opinion is that it is likely to overhaul the magnetic confinement of plasma as a route to generating power by nuclear fusion and become the chosen technique for the commercial nuclear fusion reactor of the future. A third contender has entered the race: there are now sound reasons for believing that the approach using ion beams to drive fusion may eventually prove to be the simplest and most efficient way to control nuclear fusion and bring it into commercial operation, if certain technical problems can be solved.

There is sufficient worldwide confidence in the successful demonstration of fusion power. In recent years much attention has been given to conceptual fusion reactor designs (Figure 11). Diverse disciplines of plasma physics, nuclear engineering, materials science and technology are being taken into consideration for defining the problems likely to be encountered in the realization of a practical and economic fusion reactor power plant.

The primary components and materials of nuclear fusion reactors consist of nuclear fuels, structural materials, moderator, reflector and blanket materials, coolants, shields, and non-superconducting and superconducting magnet systems.

There are four potentially useful nuclear fusion fuel cycles: (1) DT, (2) DD, (3) D-^{3}He, and (4) ^{11}B-P. Associated with the fuel cycles are the two breeding reactions $^{1}n + {}^{6}Li \rightarrow {}^{4}He + {}^{3}T$ (for the DT fuel cycle) and $^{1}P + {}^{6}Li \rightarrow {}^{4}He + {}^{3}He$ (for the D-^{3}He fuel cycle)

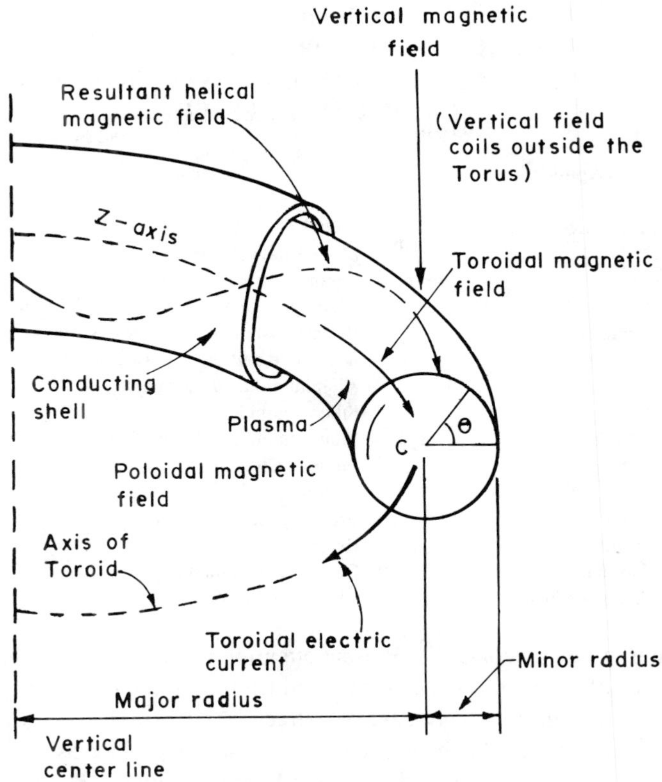

FIGURE 10. Thermonuclear plasma confinement in magnetic fields (TF, toroidal field; PF, poloidal field; VF, vertical field).

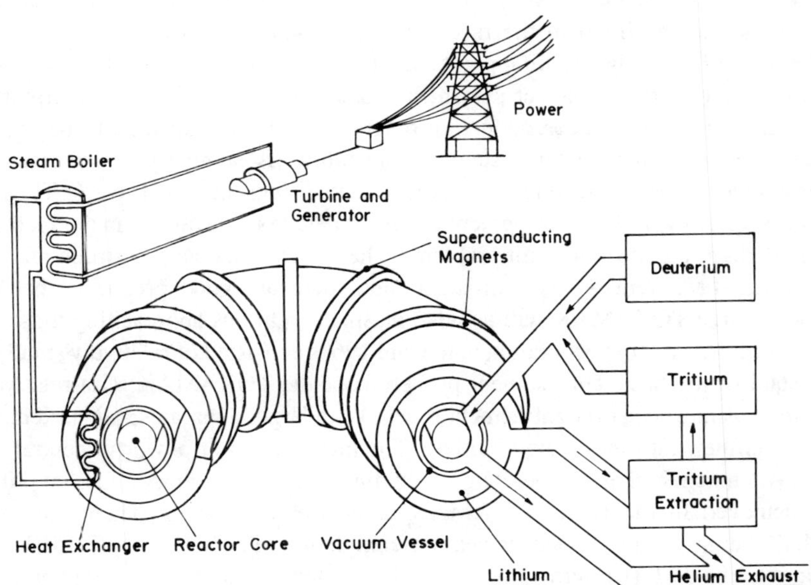

FIGURE 11. Conceptual fusion reactor.

Table 6
SUMMARY OF MAJOR FUSION REACTOR COMPONENTS AND MATERIALS OPTIONS FOR EACH AS PERCEIVED

Application	Materials
Breeding	Liquid lithium
	Molten salts (Li_2BeF_4, LiF)
	Ceramic compounds (Li_2O)
	Aluminum compounds (LiAl, $Li_2Al_2O_4$)
Structural	Refractory-based alloys (Nb, V, Mo)
	Iron-based and nickel-based alloys
	Aluminum-based materials
	Silicon carbide
Cooling	Liquid lithium and potassium
	Molten salts
	Helium
Moderation	Breeding materials, graphite
Neutron multiplication	Beryllium
Normal conductors	Copper and aluminum
Superconductors	NbTi, Nb_3Sn, V_3Ga

for the production of new fusion fuels. Present attention is on the DT fusion fuel cycle. The two heavy isotopes of hydrogen — deuterium and tritium — are present in vast amounts in the world's oceans and are relatively easy to extract.

Among the structural materials, the vacuum or the first wall of the plasma vacuum chamber faces the severest of environments: radiation and high heat fluxes. The candidates for first wall or structural materials are austenitic stainless steel, nickel alloy, niobium alloy, vanadium alloy, and titanium alloy, or combinations of stainless steel-niobium alloy and stainless steel-vanadium alloy.

The coolant-blanket materials are Li and Li-Pb eutectic alloy in liquid form, and $LiFBeF_2$ (or Li_2BeF_4, known as flibe) in the fused or molten salt form. Li_2O, LiAl, $LiAlO_2$, and Li_7Pb_7 (because of their high melting points) are usable as solid form blanket materials only.

The nonsuperconducting magnet system has been used for near-term laboratory fusion experimental devices. The superconducting magnet sytems are expected to be required for commercial fusion power reactor systems in the future. A summary of materials that are being considered for use as fusion reactor components is presented in Table 6. The materials problems associated with the development of fusion reactors constitute a major technological challenge. Perhaps the most difficult problem is the radiation damage to structural first wall.

Progress has been remarkably consistent, and there are good prospects that the new generation of large TOKAMAK will achieve plasma conditions approaching those required in a fusion reactor. The first generation in the late 1960s consisted of relatively small devices, built to establish the basic confinement properties of the TOKAMAK with relatively low-temperature plasma and short confinement times. With larger apparatus, higher temperatures and larger confinement times were obtained. The introduction of powerful neutral injection systems gave much higher temperature, including the record temperature of 80 million degrees, achieved in the U.S. TOKAMAK at Princeton in 1978. There are four large TOKAMAK being commissioned or nearing completion: TFTR (U.S.), T-15 (U.S.S.R.), JT60 (Japan), and JET (western Europe). TFTR started operating in December 1982, followed by JET in mid-1983. Built by the JET Joint Undertaking at Culham in Oxfordshire, U.K., JET is approximately 12 m high and 15 m in diameter — the largest and most powerful TOKAMAK of its generation. TFTR and JET are designed to operate with deuterium and tritium plasma in the latter part of their program, generating several megawatts of heat for a few seconds. JT60 and T-15 are designed to operate with hydrogen plasmas only. A further stage of TOKAMAK must be built to establish the engineering aspects of an electricity-generating fusion device before prototype reactors are built. The collaborative effort between Europe, Japan, the U.S., and the U.S.S.R. in the TOKAMAK design known as INTOR is a step in this direction. Fusion power stations are likely to gain ground in the next century, when the need for further sources of electricity will be even more pressing.

REFERENCES

1. **Pease, R. S.,** International fusion research, *Atom,* 315, 2, 1983.
2. **Ekland, S.,** Status report on fusion, *Atom,* 260, 150, 1978.
3. **Rose, D. J.,** On the Feasibility of Power by Nuclear Fusion, ORNL-TM-2204, May 1968.
4. **El-Wakil, M. M.,** Fusion power, in *Nuclear Energy Conversion,* Intext Educational, London, 1971, 518.
5. **Clarke, J. F.,** The next step in fusion: what it is and how it is being taken, *Science,* 210, 967, 1980.
6. **DeBellis, R. J. and Sabri, Z. A.,** *Fusion Power: Status and Options,* EPRI-ER-510-SR, Electric Power Research Institute, 1977.
7. **Dingee, D. A.,** Fusion power, *Chem. Eng. News,* April 2, p. 32, 1979.
8. **Glasstone, S.,** Controlled Nuclear Fusion, Office of Information Services, U.S. Atomic Energy Commission 1974.
9. **Glasstone, S. and Loverberg, R. H.,** *Controlled Thermonuclear Fusion,* D Van Nostrand, New York, 1960.
10. **Gough, W. C. and Eastland, B. J.,** The prospects for fusion power, *Sci. Am.,* 224(2), 1971.
11. **Kammesh, T.,** *Fusion Reactor Physics: Principles and Technology,* Ann Arbor Science, Ann Arbor, MI, 1975.
12. **Kenton, J.,** Capturing a star: controlled fusion power, *EPRI J.,* December, p. 6, 1977.
13. **Krall, N. A. and Trivelpiece, A. W.,** *Principles of Plasma Physics,* McGraw-Hill, New York, 1973.
14. **Olds, F. C.,** Fusion power developments, *Power Eng.,* 82(11), 46, 1978.
15. **Rose, D. J. and Clark, Jr. M.,** *Plasmas and Controlled Fusion,* MIT Press, Cambridge, MA, 1961.
16. **Rose, D. J. and Feirtag, M.,** The prospects for fusion, *Technol. Rev.,* 79(2), 20, 1976.
17. **Zaleski, P.,** *Nuclear Energy Maturity, Proc. April 1975, Conf. Vol. 12, Enrichment and Fusion,* Academic Press, New York, 1976.
18. **Carlson, G. A., Shultz, K. R., and Smith, A. C., Jr.,** Definition and Conceptual Design of a Small Fusion Reactor, EPRI-ER-1045, Electric Power Research Institute, 1979.
19. **Chen, F. F.,** *Introduction to Plasma Physics,* Plenum Press, New York, 1974.
20. **Chen, F. F.,** Alternate concepts in magnetic concepts, *Phys. Today,* 32(5), 36, 1979.
21. **Coppi, B. and Rem, J.,** The Tokamak approach in fusion research, *Sci. Am.,* 229(2), 61, 1972.
22. **Murakami, M. and Eubank, H. P.,** Recent progress in Tokamak experiments, *Phys. Today,* 32(5), 25, 1979.
23. **Rose, R. P.,** Design Study of a Fusion-Driven Tokamak Hybrid Reactor for Fissile Fuel Production, EPRI-ER-1083, Electric Power Research Institute, 1979.
24. **Steiner, D. and Clarke, J. F.,** The Tokamak: model T fusion reactor, *Science,* 199, 1395, 1978.
25. **Artsimovich, L. A.,** Nuclear Fusion, 12, 215, 1972.
26. **Gibson, A.,** The Jet project, *Atom,* 254, 1977.
27. **Booth, L. A. and Frank, T. G.,** Commercial Applications of Inertial Confinement Fusion, LA-6838-MS, Los Alamos Scientific Laboratory, Los Alamos, NM, 1977.
28. **Emmett, J. L., Nuckolls, J., and Wood, L.,** Fusion power by laser implosion, *Sci. Am.,* 230(6), 24, 1974.
29. **Freiwald, D. A. and Frank, T. G.,** Introduction to Laser Fusion, Los Alamos Scientific Laboratory, Los Alamos, NM, 1975.
30. LLL Fusion, Laser Fusion Program, Lawrence Livermore Laboratory, Livermore, CA, UCRL-52000-77-B, 1977.
31. Laser Fusion Program at Los Alamos, LA-79-29, Los Alamos Scientific Laboratory, Los Alamos, NM, 1979.
32. Sandia, Particle Beam Fusion, Sandia National Laboratories, Albuquerque, NM, 1980.
33. **Stickley, C. M.,** Laser fusion, *Phys. Today,* 31(5), 50, 1978.
34. **Varando, S. G., Mitchner, J. L., and Yonas, G.,** Civilian Applications of Particle-Beam-Initiated Inertial Confinement Fusion Technology, SAND 77-0516, Sandia National Laboratories, Albuquerque, NM, 1977.

35. **Yonas, G.,** Fusion power with particle beams, *Sci. Am.,* 239(5), 50, 1978.
36. IAEA, International Tokamak Reactor, Zero Phase, Report of the International Tokamak Workshop, International Atomic Energy Agency, Vienna, 1979.
37. **Gruen, D. M.,** Materials for thermonuclear fusion reactors, in *Materials Science in Energy Technology,* Libowitz, G. G. and Whittingham, M. S., Eds., Academic Press, New York, 1979, 325.
38. **Gold, R. E. et al.,** Materials technology for fusion, current status and future requirements, *Nucl. Technol. Fusion,* 1, 169, 1981.
39. **McHargue, C. J. and Scott, J. L.,** Materials requirements for fusion reactors, *Met. Trans. A,* 9A, 151, 1978.
40. **Kulcinski, G. L.,** Fusion reactors: their challenge to materials scientists, *Contemp. Phys.,* 20(4), 417, 1979.
41. **Gupta, C. K.,** Extractive metallurgy of niobium, tantalum, and vanadium, *Int. Met. Rev.,* 29(6), 405, 444, 1984.
42. **McHenry, H. I. and Reed, R. P.,** Structural alloys for superconducting magnets in fusion energy system, *Nucl. Eng. Des.,* 58, 219, 1980.
43. **File, J., Knudson, D. S., and Rappie, G. H.,** Non-superconducting magnet structures for near-term large fusion experimental devices, *Nucl. Eng. Des.,* 58(2), 167, 1980.
44. **Hasignti, R. R.,** Radiation damage in metallic materials, *Trans. Jpn. Inst. Met.,* 22(12), 12, 831, 1981.
45. **Ma, B. M.,** Radiation effects on controlled thermonuclear reactor first wall materials with fast neutron fluxes, Int. Conf. Radiation Effects with Tritium Technology for Fusion Reactor, Gatlinburg, TN, October 1 to 3, 1975.
46. **Ma, B. M.,** Irradiation swelling, creep, thermal shock and thermal cycling fatigue analysis of cylindrical controlled thermonuclear reactor first wall, *Nucl. Eng. Des.,* 28, 1, 1974.
47. **Ma, B. M.,** Swelling, Creep, and Thermal Fatigue Analysis for Cylindrical Fusion Reactor Vacuum Wall, invited lecture at Proc. 2nd Int. Conf. Structural Mechanics in Reactor Technology, September 10 to 14, West Berlin, Germany, 1973.
48. **Ma, B. M.,** Irradiation effects on CTR first wall, *Nucl. Eng. Des.,* 39, 203, 1976.
49. **Fish, R. L., Straalsund, J. L., Hunter, C. W., and Holmes, J. J.,** Effects of Radiation on Substructure and Mechanical Properties of Metals and Alloys, ASTM-STP-529, American Society for Testing and Materials, 1973.
50. **Green, B. J.,** The state of controlled thermonuclear fusion research, *Nucl. Eng.,* 25(5), 186, 1984.

INDEX

A

Ablation front, 199
Activation and ordinary water, 67—68
Adiabatic compression, 194, 195
Advanced gas-cooled reactor (AGR), 53, 89, 111
AGR, see Advanced gas-cooled reactor
Albedo, 64
Alkyl amines, 82, see also specific types
Allotropic modification, 20
All-volatile treatment (AVT), 109, 110
Alpha particles, 143
Aluminothermic reduction, 35
Aluminum, 2—3, 33, 207, 209
Amax process, 5
Amines, 82, see also specific types
Ammonia, 78—82, 110
Applications, see also specific types
  of boron carbide, 136—137
  of burnable poisons, 130—131
  of niobium, 36—38
  of stainless steel, 45—51
  of zirconium, 21—23
Austenitic steels, 43, 46, 47, 204, 208, see also specific types
AVT, see All-volatile treatment

B

Balzano process, 5
Beryl, 91
Beryllium, 90—94, 205
  production of, 91
  properties of, 93, 94
  radiation and, 93
Beryllium fluoride, 206
Beta particles, 143—144
Biological shielding, 201, 206
Bismuth, 118, 119
Bithermal exchange, 76, 81
Blanket, 64, see also specific types
  lithium, 192
  materials for, 203—205, see also specific types
BLWR, see Boiling light water reactor
Boiling light water reactor (BLWR), 53, 107
Boiling water reactor (BWR), 32, 50, 53
  burnable poisons and, 130
  coolant systems for, 99—100, 106—107
  experimental, 107
  fuel elements in, 22, 29
  ordinary water and, 67
  power distribution transients and, 157
  pressure vessels in, 51, 52
Boric acid, 108, 132
Boron
  applications of, 136—137
  chemical metallurgy of, 133—136
  as control material, 133—137
  properties of, 136, 137
  as shielding material, 151
Boronsilicate glass, 171
Bremsstrahlung, 144
Burnable neutron poisons, 129—131
BWR, see Boiling water reactor

C

Cadmium alloys, 137—138, see also specific types
Calcination, 170
CANDU, 22, 32, 33, 38, 53, 54
Carbon, 86, 207, 209
  fibrous, 209
  pyrolytic (pyrocarbon), 85
  transport of, 123
Carbon dioxide, 112—116
Carbon steel, 204
Cascade, 70
Cementite, 208
Centrifugal pumps, 120
Chain reaction, 97
Chemical exchange, 72, 76—82
Chemical shim, 131—133
Chemistry and volume control system (CVCS), 108
Chloride SCC, 44
Cladding materials, 1, see also specific types
Clinch River Breeder Reactor (CRBR), 50
Cold-pressing, 92
Columbite-tantalite, 35
Combustible waste, 176, see also specific types
Combustion chamber, 206
Compatibility, 88—89, 121
Compression, 194, 195, see also specific types
Compton effect, 144, 145
Confinement, 193—201, see also specific types
  inertial, 198—199
  ion beam focus type, 199—201
  magnetic, 193—198
  plasma, 210
  time of, 191
Containment, 99
Control
  materials for
    boron-based, 133—137
    comparisons of, 142
    properties of, 128, 134
    rare earth-based, 140—142
  methods of, 125—133, see also specific types
    burnable poisons and, 129—131
    chemical shim in, 131—133
    movable control rods as, 125—129
Controlled nuclear fusion, 189
Controlled thermonuclear fusion, 190
Control rods, see also specific types
  classification of, 126

design of, 127—128
geometries of, 126—127
movable, 125—129
operating criteria for, 127—128
Coolant materials, 97—123, 205—206, see also specific types
characteristics of, 101—105
coolant systems and, see Coolant systems
gaseous coolants, see Gaseous coolants
heavy water as, 106
liquid metal, 118—123
types of, 103, 104, 106
water coolants, 106—110
Coolant systems, 98—101, see also specific types
in BWR, 99—100
in LMFBR, 101
in PHWR, 100—101
in PWR, 98—99
types of, 98
Corrosion, 17, 122, see also specific types
lithium and, 204
radiation and, 112—115
radiolytic, 114
reactor material, 115—116
sodium and, 121—123
stress, 116
thermal, 112
Coulomb barrier, 190
CRBR, see Clinch River Breeder Reactor
Creep strength, 29
Crystallization, 12, see also specific types
CVCS, see Chemistry and volume control system

D

Deaerators, 108
Deep underground burial of waste, 180—181
Demineralization, 108, 110
Depletion, 128
Deuterium, 207
Deuterium-tritium fusion, 191
Direct-cycle coolant system, 98
Distillation, see also specific types
extractive, 15
in heavy water production, 72—73
hydrogen, 72—73
vacuum, 70
water, 73—74
Double coolant system, 98
Double fluorides, 12
Downcomer, 98
Dow process, 5
Duplex stainless steel, 43, 47

E

EGCR, see Experimental gas-cooled reactor
Elastic scattering, 146
Electrolysis, 72, 74—76
Electrolytic reduction, 5, 92
Electromagnetic pumps, 120
Energy, see also specific types
fusion, 189—193
kinetic, 193
threshold, 192
Wigner, 90
Energy decrement, 62
Entombment, 166
Epithermal worth, 128
Erosion, 206
Experimental boiling water reactor (EBWR), 107
Experimental gas-cooled reactor (EGCR), 111
Extraction, see also specific types
of aluminum, 2—3
of magnesium, 4—5
of solvents, 12, 13
of zirconium, 10—17
Extractive distillation, 15
Extractive metallurgy, 34—35

F

Fabrication
niobium and, 38—39
zirconium and, 23—29
Fast breeder reactor (FBR), 53
Fast flux test facility (FFTF), 53—54
FBR, see Fast breeder reactor
Fermi age, 63
Ferritic steel, 43, 46, 116, 204, see also specific types
Fertile materials, 64, see also specific types
FFTF, see Fast flux test facility
Fibrous carbon, 209
Finned fuel elements, 110
First walls, 206—211
Fission, 146, 168, 189
Flibe, 205, 206
Fluorides, 12, see also specific types
Fractional crystallization of double fluorides, 12
Fuel elements, 1, see also specific types
in BWR, 22, 29
classification of, 2
design of, 56
finned, 110
Magnox, 5—9
in PWR, 21, 29
Full-flow condensate demineralization, 110
Fusion, 189
coolants in, 205, 206
cycle of, 192
deuterium-tritium, 191
fuel materials for, 192—193, 201—203, see also specific types
laser, 198
thermonuclear, 190, 201
Fusion energy sources, 189—193
Fusion reactions, 190—192
Fusion reactor materials, 189—215, see also specific types
confinement of, 193—201
inertial, 198—199
ion beam focus type, 199—201

magnetic, 193—198

G

Gamma rays, 144—145, 206
Gas-cooled fast reactor (GCFR), 53, 118
Gas-cooled reactor (GCR), 53, see also specific types
coolant systems for, 104 111
examples of, 113
experimental, 111
Gaseous coolants, 110—118, see also specific types
Gaseous waste, 177—179, see also specific types
GCFR, see Gas-cooled fast reactor
GCR, see Gas-cooled reactor
Geological disposal, 182—183
Girdler sulfide (GS) process, 76
Glass, 171, see also specific types
Graphite, 83—90, 209
compatibility and, 88—89
preparation of, 84—85
properties of, 85—88
radiation and, 89—90
Graphitization, 208
GS, see Girdler sulfide process

H

Hafnium, 138—140
Heat capacity, 104
Heat exchangers, 98
Heat transfer, 102
Heat transfer reactor experiment (HTRE), 111
Heavy elements, 149, see also specific types
Heavy water, 69—83
as coolant, 106
production of, 70—83
chemical exchange in, 72, 76—82
distillation in, 72—74
electrolysis in, 72, 74—76
radiation and, 70
Heavy water reactor (HWR), 53
Helium, 116—118, 207, 210
Hexone, 12
Hideout, 132
High level waste (HLW), 167, 179, see also specific types
Highly active waste, see also specific types
disposal of, 181—184
liquid, 167—168
solidification of, 170—174
storage of, 168—170
High-temperature gas-cooled reactor (HTGR), 50, 53
coolant systems for, 97, 111, 112 116—117
graphite and, 88
High-temperature reactor (HTR), 83
HLW, see High level waste
Holandite, 183
Hot-pressing, 93
HTGR, see High-temperature gas-cooled reactor
HTR, see High-temperature reactor
HTRE, see Heat transfer reactor experiment
HWR, see Heavy water reactor
Hydrazine, 108
Hydride orientation, 27
Hydrogen, 26, 78
ammonia and, 78—82
tensile properties and, 42
Hydrogen distillation, 72—73
Hydrogenous shielding materials, 149—151
Hydrogen sulfide, 76—78

I

Immobilization, 176—177
Implosion forces, 198
Induced radioactivity, 115
Inelastic scattering, 146
Inertial confinement, 198—199
Infrared radiation, 144
Intermediate-level waste, 174—177
Ion beam focus, 199—201
Iron, 207
Irradiation, see Radiation
Isotope coating, 85
Isotope enrichment, 192
Isotope exchange, 76
Isotope separation, 70

K

Kinetic energy, 193

L

Laser fusion, 198
Lawson's criteria, 191
Leaching, 122, see also specific types
Lead, 118, 119
Letdown cooler, 108
Light water breeder reactor (LWBR), 53
Light water graphite-moderated reactor (LWGR), 107
Light water reactor (LWR), 53, 56, see also specific types
burnable poisons and, 130
containment in, 99
coolant systems for, 106—107
ordinary water and, 66, 67
pressure vessels in, 51
radiation and, 56
Liquid metal coolant, 118—123, see also specific types
Liquid metal fast breeder reactor (LMFBR), 50, 53, 54, 56
coolant system for, 101, 107, 117, 118
radiation and, 56
sodium and, 120
Liquid waste, 174—176, see also specific types
Lithium, 119, 203—205
Lithium blanket, 192
Lithium salt, 201
LMFBR, see Liquid metal fast breeder reactor

Logarithmic energy decrement, 62
Longitudinal fin type fuel elements, 110
Low-level waste, 174—177, see also specific types
LWBR, see Light water breeder reactor
LWGR, see Light water graphite-moderated reactor
LWR, see Light water reactor

M

Magnesiothermic beryllium, 92
Magnesium, 3—9, 33, 111
  corrosion and, 116
  extraction of, 4—5
  properties of, 5
Magnetherm process, 5
Magnetic field, 195
Magnetic mirrors, 196
Magnetic pinch confinement, 196—198
Magnets, 196, 206, 210, see also specific types
Magnox, 5—9, 53, 111
Makeup and purification system, 108
Martensitic stainless steel, 43, 46
Matter, see also specific types
  neutron interaction with, 146
  plasma state of, 190
  radiation interaction with, 143—146
MCN, see Mean collision number
MD, see Moderating density
Mean collision number (MCN), 62
Mean free path, 62
Metallic reactor vessel (MRV), 52
Metallurgy, see also specific types
  of boron, 133—136
  chemical, 133—136
  extractive, 34—35
  physical, 38
Metals, see also specific types
  hydrogen in, 26
  rare earth, 140—142
  refractory, 205, 207—208
MIBK, 33, 34
Migration length, 63
Moderately heavy elements, 149, see also specific types
Moderating density (MD), 62
Moderating power (MP), 62, 88
Moderating ratio (MR), 62, 88
Moderator materials, 203—205, see also specific types
Molten salt breeder reactor (MSBR), 54
Molybdenum, 206, 207
Monothermal exchange, 80
Mothballing, 166
Movable control rods, 125—129
MP, see Moderating power
MR, see Moderating ratio
MRV, see Metallic reactor vessel
MSBR, see Molten salt breeder reactor

N

Neutral beam injection, 194, 196
Neutron absorption, 128
Neutron multiplication, 205
Neutron poisons, 129—131, see also specific types
Neutron power plants, 156—157, see also specific types
Neutrons, 146, see also specific types
Nickel, 204, 206, 207, 209
Niobium, 34—42, 205, 207
  applications of, 36—38
  extractive metallurgy and, 34—35
  fabrication and, 38—39
  properties of, 35—36, 38—42
  structure of, 39—42
Niobium alloys, 40, see also specific types
Nitrates, 12, see also specific types
Norsk hydro process, 5
Nuclear chain reaction, 97
Nuclear power plants, see also specific types
  chemistry of, 156
  operation of, 154—158
  shutdown of, 156
  start-up of, 154—156

O

Ocean bed disposal of waste, 183
Ohmic heating, 194
Once-through coolant system, 98
Ordinary water, 65—69

P

Pair production effect, 144, 145
Particle emission, 146
PCI, see Pellet-clad interaction
PCR, see Prestressed concrete reactor
PCRV, see Prestressed concrete reactor pressure vessels
Pellet-clad interaction (PCI), 29—32, 34
Perovskite, 183
Phosphate control method, 109
Photoelectric effect, 144
PHWR, see Pressurized heavy water reactor
Physical metallurgy, 38
Pidgeon process, 5
Pinch-type systems, 196—197
Plasma burn, 196
Plasma confinement, 210
Plasma first-wall materials, 207—209
Plasma heating, 194—195
Plasma particle density, 191
Plasma state of matter, 190
Plateout, 133
Poisons, see also specific types
Potassium, 119
Power distribution transients, 157—158
Precipitation hardened stainless steel, 43, 47
Pressure vessels, 51—57, see also specific types
Pressurized heavy water reactor (PHWR), 53, 100—101
Pressurized water reactor (PWR), 20—22, 32, 53, 54, see also specific types

boiling temperature of water in, 99
burnable poisons and, 130, 131
coolant system for, 98—99
coolant systems for, 105—107
fuel elements in, 21, 29
ordinary water and, 67
power distribution transients and, 157
pressure vessels in, 51, 52
Prestressed concrete reactor (PCR), 54
Prestressed concrete reactor pressure vessels (PCRV), 52, 54, 56, 110
Primary coolant system, 98
Primary water coolant chemistry, 107—109
Proloidal magnetic field, 195
Properties, see also specific types
of aluminum, 3
of beryllium, 93, 94
of boron, 136, 137
of cadmium, 138
of control materials, 128, 134
of graphite, 85—88
of hafnium, 139
of inless steel, 43—45
of magnesium, 4
of niobium, 35—36, 38—42
of rare earth elements, 141
of zirconium, 17—21
Pumps, 120, see also specific types
Purification system, 108
PWR, see Pressurized water reactor
Pyrocarbon (pyrolytic carbon), 85
Pyrolytic carbon (pyrocarbon), 85

R

Radial field gradients, 195
Radiation, see also specific types
alpha particle, 143
beryllium and, 93
beta particle, 143—144
carbon dioxide and, 112
corrosion under, 112—115
effects of, 56—57, 210—211
gamma ray, 144—145, 206
graphite and, 89—90
heavy water and, 70
infrared, 144
matter interaction with, 143—146
swelling from, 207
tensile properties and, 42
ultraviolet, 144
Radiation growth, 32
Radiative capture, 146
Radioactive waste, see also specific types
classification of, 164
combustible, 176
deep underground burial of, 180—181
disposal of, 179—184, see also specific types
geological, 182—183
ocean bed, 183
gaseous, 177—179
high level, see High level waste
highly active, see Highly active waste
intermediate-level, 174—177
isolation of, 163
liquid, see Liquid waste
low-level, 174—177
management of, 161—187, see also specific types
isolation in, 163
objectives of, 162—163
principles of, 162—163
noncombustible, 174—176
shallow burial of, 180
solid, see Solid waste
sources of, 163—164
storage of, 168—170
treatment of, 166—179, see also specific types
drying in, 170
high level waste and, 167
highly active liquid waste and, 167—170
highly active solid waste and, 167
immobilization in, 176—177
intermediate-level, 174—177
low-level, 174—177
verification in, 170—174
volume reduction in, 174—176
Radioactivity, 115, see also Radiation
Radioactivity transport, 123
Radio-frequency waves, 194—195
Radiolysis, 68—69, 108, 110
Radiolytic corrosion, 114
Radiolytic decomposition of water, 118
Rare earth elements, 140—142, see also specific types
Reactivity, 125
Reactors, 53—54, see also specific types
Reduction, 5, 35, see also specific types
Reflector materials, 203—205, see also specific types
Refractory metals, 205, 207—208, see also specific types

S

Samarium, 157
SAP, see Sintered aluminum product
Scattering, 146, see also specific types
SCC, see Stress corrosion cracking
Secondary coolant system, 98, 109—110, see also specific types
Selective leaching, 122
SGHWR, see Steam-generating heavy water reactor
Shallow burial of waste, 180
Shielding, see also specific types
biological, 201, 206
materials for, 143—149, 205—206, see also specific types
boron-containing, 151
categorization of, 149—151
hydrogenous, 149—151
Shim, 131—133, see also specific types
Shock heating, 198
Silicon carbide, 209
Sintered aluminum product (SAP), 209

Sintering, 92
SNS, see Spallation neutron source
Sodium, 119—123
Solbor, 132
Solid waste, see also specific types
  highly active, 167
  noncombustible, 174—176
Solvent extraction, 12, 13
Spallation neutron source (SNS), 201
Spent fuel, 179
Spiral magnetic field, 195
Stainless steel, 42—51, 116
  applications of, 45—51
  composition of, 46—47
  properties of, 43—45
  types of, 43—45
Steam-generating heavy water reactor (SGHWR), 53
Steels, 116, 208—209, see also specific types
  austenitic, 43, 46, 47, 204, 208
  carbon, 204
  ferritic, 43, 46, 116, 204
  stainless, see Stainless steel
Stress corrosion, 116
Stress corrosion cracking (SCC), 30, 33, 34, 44
Stress fatigue, 207
Structural materials, 1, 206—209, see also specific types
Structure-property correlation, 39—42
Sulfates, 13, see also specific types
Superconducting magnets, 196, 206, 210
Superconducting materials, 211, see also specific types
Swimming pool reactor, 54
SYNROC, 183—184

T

TBP, 33, 34
Tensile properties, 42
Thermal corrosion, 112
Thermal gradient transfer, 122—123, 205
Thermal insulation, 206
Thermal worth, 127—128
Thermic reduction, 5
Thermonuclear fusion, 190, 201
Theta pinch concept, 198
Thiocyanates, 12, see also specific types
Threshold energies, 192
Titanium, 207
TNOA, see Tri-n-octyl amine
TOKAMAK, 195—197
Toroidal magnetic field, 195
Toxicity, 93, see also specific types
Training, research and isotope production reactor (TRIGA), 54
Transients, 156—158, see also specific types
Tributyl phosphate, 12
TRIGA, see Training, research and isotope production reactor
Tri-n-octyl amine (TNOA), 13, 33
Triple coolant system, 98
Tritium, 70, 191, 205, 207

U

Ultraviolet radiation, 144
Uranium mill tailings, 179

V

Vacuum distillation, 70
Vacuum hot-pressing, 93
Vanadium, 205, 207
Volumetric heat capacity, 104

W

Waste, see specific types
Water, see also specific types
  alkalinity of, 109
  heavy, see Heavy water
  hydrogen and, 78
  hydrogen sulfide and, 76—78
  ordinary, 65—69
  radiolytic decomposition of, 118
Water coolants, 106—110, see also specific types
Water distillation, 73—74
Water electrolysis, 72, 74—76
Wigner effect, 90
Wigner energy, 90

X

Xenon, 157
X-rays, 144, 206

Z

Zircaloy-2, 33
Zircaloy-4, 33
Zirconium, 10—34, 207
  applications of, 21—23
  corrosion and, 116
  extraction of, 10—17
  fabrication and, 23—29
  performance of, 29—32
  properties of, 17—21
Zirconium alloys, 39, see also specific types
Zirconium hydrides, 24, see also specific types
Zirconolite, 183
Z pinch concept, 198